Integrating Geographic Information Systems and Agent-Based Modeling Techniques for Simulating Social and Ecological Processes

Integrating Geographic Information Systems and Agent-Based Modeling Techniques for Simulating Social and Ecological Processes

Editor

H. Randy Gimblett
University of Arizona

Santa Fe Institute
Studies in the Sciences of Complexity

2002

OXFORD
UNIVERSITY PRESS

Oxford New York
Athens Auckland Bangkok Bogotá Buenos Aires Cape Town
Chennai Dar es Salaam Delhi Florence Hong Kong Istanbul Karachi
Kolkata Kuala Lumpur Madrid Melbourne Mexico City Mumbai Nairobi
Paris São Paulo Shanghai Singapore Taipei Tokyo Toronto Warsaw

and associated companies in
Berlin Ibadan

Published by Oxford University Press, Inc.
198 Madison Avenue, New York, New York 10016

Oxford is a registered trademark of Oxford University Press

Library of Congress Cataloging-in-Publication Data
Integrating geographic information systems and agent-based modeling techniques
for simulating social and ecological processes / editor, H. Randy Gimblett
p. cm. — (Santa Fe Institute studies in the sciences of complexity)
Includes bibliographical references and index.
ISBN 0-19-514336-1; ISBN 0-19-514337-X (pbk.)
1. Human geography—Mathematical models. 2. Human ecology—
Mathematical models. 3. Social ecology—Mathematical models. 4. Regional planning—
Mathematical models. 5. Geographic information systems. I. Gimblett, H. Randy.
II. Santa Fe Institute studies in the sciences of complexity
(Oxford University Press)
GF23.M35 I58 2001
304.2'01'5118—dc21 00-069204

This volume was typeset using TEXtures on a PowerMAC G3 computer.
Camera-ready output from a Hewlett-Packard Laser Jet 4M Printer.

9 8 7 6 5 4 3 2 1

Printed in the United States of America
on acid-free paper

Contributors List

Robert Allred, *Northern Arizona University, Flagstaff, AZ 86011-5717; e-mail: rpa@dana.ucc.nau.edu*

John Anderson, *Department of Computer Science, University of Manitoba, Winnipeg, Canada R3T 2N2; e-mail: andersj@cs.umanitoba.ca*

Joanne Bieri, *Northern Arizona University, Flagstaff, AZ 86011-5717; e-mail: jab34@dana.ucc.nau.edu*

Rian Bogle, *Northern Arizona University, Flagstaff, AZ 86011-5717; e-mail: Rian.Bogle@Colorado.edu*

Paul Box, *Utah State University, Geography and Earth Resources, UMC 5240, Logan, UT 84321; e-mail: paulbox@cc.usu.edu*

Eduardo Brondisio, *Indiana University, Center for Study of Institutions, Population and Environmental Change and Anthropological Center for Training and Research on Global Environmental Change, Bloomington, IN 47045; e-mail: ebrondiz@indiana.edu*

Susan Cherry, *University of Arizona, BSE Building, Room 325, Tucson, AZ 85718; e-mail: slcherry@u.arizona.edu*

Terry C. Daniel, *University of Arizona, BSE Building, Room 325, Tucson, AZ 85718; e-mail: Daniel@u.arizona.edu*

Peter J. Deadman, *University of Waterloo, Department of Geography, Faculty of Environmental Studies, Waterloo, Ontario, N2L 3G1 Canada; e-mail: pjdeadma@fes.uwaterloo.ca*

Scott M. Duke-Sylvester, *University of Tennessee at Knoxville, Ecology and Evolutionary Biology, 569 Dabney Hall, Knoxville, TN 37996–1610; e-mail: sylv@tiem.utk.edu*

Randy Gimblett, *University of Arizona, BSE Building, Room 325, Tucson, AZ 85718; e-mail: gimblett@ag.arizona.edu*

Louis J. Gross, *University of Tennessee at Knoxville, Ecology and Evolutionary Biology, 569 Dabney Hall, Knoxville, TN 37996–1610; e-mail: gross@tiem.utk.edu*

Steven J. Harper, *Department of Zoology, Miami University, Oxford, OH 45056; e-mail: harpersj@muohio.edu*

Robert M. Itami, *University of Melbourne, Department of Geomatics, Victoria, Australia 3010; e-mail: b.itami@eng.unimelb.edu.au*

Bin Jiang, *University College London, 1-19 Torrington Place, WC1E 6BT, London, UK; e-mail: b.jiang@ucl.ac.uk*

Dana Kilbourne, *University of Arizona, BSE Building, Room 325, Tucson, AZ 85718*

Kevin Lim, *University of Waterloo, Department of Geography, Faculty of Environmental Studies, Waterloo, Ontario, N2L 3G1 Canada; e-mail: klim@fes.waterloo.ca*

Stephen McCracken, *Indiana University, Center for Study of Institutions, Population and Environmental Change and Anthropological Center for Training and Research on Global Environmental Change, Bloomington, IN 47045; e-mail: stmccra@indiana.edu*

Michael J. Meitner, *Forest Sciences Centre, 2045-2424 Main Mall, University of British Columbia Vancouver, BC, Canada V6T 1Z4; e-mail: meitner@interchange.ubc.c*

Emilio Moran, *Indiana University, Center for Study of Institutions, Population and Environmental Change and Anthropological Center for Training and Research on Global Environmental Change, Bloomington, IN 47045; e-mail: moran@indiana.edu*

Michael Ratliff, *Northern Arizona University, Flagstaff, AZ 86011-5717; e-mail: Mike.Ratliff@nau.edu*

Merton T. Richards, *Northern Arizona University, PO Box 15018, Flagstaff, AZ 86011-5018; e-mail: mtr@nau.edu*

Catherine A. Roberts, *Northern Arizona University, Flagstaff, AZ 86011-5717; e-mail: Catherine.Roberts@NAU.EDU*

Edella Schlager, *School of Public Administration and Policy, 405 McClelland Hall, University of Arizona, Tucson, AZ 85721; e-mail: eschlager@bpa.arizona.edu*

Ann-Marie Shapiro, *US Army Corps of Engineers Construction Engineering Research Laboratory, PO Box 9005, Champaign, IL 61826–9055; e-mail: ann-marie.l.shapiro@erdc.usace.army.mil*

Doug Stallman, *Northern Arizona University, Flagstaff, AZ 86011-5717; e-mail: dws@dana.ucc.nau.edu*

James D. Westervelt, *Department of Agriculture and Consumer Economics, University of Illinois at Urbana/Champaign, MC-710 326 Mumford Hall, 1301 W Gregory Dr, Urbana, IL 61801; e-mail: westerve@uiuc.edu*

About the Santa Fe Institute

The *Santa Fe Institute* (SFI) is a private, independent, multidisciplinary research and education center, founded in 1984. Since its founding, SFI has devoted itself to creating a new kind of scientific research community, pursuing emerging science. Operating as a small, visiting institution, SFI seeks to catalyze new collaborative, multidisciplinary projects that break down the barriers between the traditional disciplines, to spread its ideas and methodologies to other individuals, and to encourage the practical applications of its results.

All titles from the *Santa Fe Institute Studies in the Sciences of Complexity* series will carry this imprint which is based on a Mimbres pottery design (circa A.D. 950–1150), drawn by Betsy Jones. The design was selected because the radiating feathers are evocative of the outreach of the Santa Fe Institute Program to many disciplines and institutions.

Santa Fe Institute
Studies in the Sciences of Complexity

Lecture Notes Volume

Author	*Title*
Eric Bonabeau, Marco Dorigo, and Guy Theraulaz	Swarm Intelligence

Proceedings Volumes

Editor	*Title*
James H. Brown and Geoffrey B. West	Scaling in Biology
Timothy A. Kohler and George J. Gumerman	Dynamics in Human and Primate Societies
Lee A. Segel and Irun Cohen	Design Principles for the Immune System and Other Distributed Autonomous Systems
H. Randy Gimblett	Integrating Geographic Information Systems and Agent-Based Modeling Techniques

Contents

Preface

In March of 1998, with sponsorship from Intel Corporation, the Santa Fe Institute (SFI) hosted a workshop entitled "Integrating Geographic Information Systems (GIS) and Multi-Agent Modeling Techniques" to which a small group of computer scientists, geographers, landscape architects, biologists, anthropologists, social scientists, and ecologists were invited. Presentations of invitees related to the workshop theme were undertaken to help form a set of coherent, cross-referenced perspectives on incorporating the spatial representation and analytical power provided by researchers employing the use of GIS with those developing agent-based technologies to dynamically simulate evolutionary and nonlinear phenomena.

The workshop was convened because researchers in these areas—who come from several disciplines—do not have a shared society or journal to unite them, and partly as a consequence, many recent advances in software algorithms for dynamically capturing, displaying, and manipulating geographic data as well as modeling social and ecological behaviors, and successes in their application, have not been adequately reported in the literature. My goal in conjunction with a variety of contributing authors is to produce a standard volume in this area for researchers, natural resource managers, and students on both sides of the Atlantic.

This volume is broadly focused on spatially explicit simulation modeling as a crucial component of decision making in natural resource management. In recent years, there has been a strong drive to improve the detail and realism of such models, due, in part, to the very real gap between the many models made for research purposes and the practical needs of decision makers. In addition, as the alternatives available in resolving resource management problems become increasingly expensive, and the resources themselves become increasing scarce and valuable, such models become vital tools not only in the direct management of resources, but in the control of expenses associated with resource management as well.

Two of the most significant recent developments in simulation modeling for these areas are the use of the individual-based perspective, and arising directly from this, the ability to include intelligent agents as entities within spatially explicit simulation models. The individual-based perspective does away with the traditional assumption of mean individuals and even distribution of events, and allows global phenomena to emerge as a result of interaction between individuals and other aspects of the environment, analogous to the physical world whose complexities are captured in a realistic manner. The inclusion of intelligent agents provides an added advantage of modeling the participation of humans in their environment, and thus the disturbances and changes induced by human activities that are central to many problems in natural resource management.

This volume is a collection of papers representing some of the participants from the workshop and others known to be developing the technologies and applying them to solve resource management or ecological phenomena.

Chapters in this volume will emphasize the use of intelligent agents as entities within spatially explicit simulation models. Chapters will be technically driven for the audience interested in algorithm and system development; application oriented for those readers exploring the use of intelligent agents for modeling both social and physical processes; and management implications resulting from applying the simulations.

H. Randy Gimblett
University of Arizona

Integrating Geographic Information Systems and Agent-Based Technologies for Modeling and Simulating Social and Ecological Phenomena

H. Randy Gimblett

1 INTRODUCTION

To acquire a more thorough understanding of the complexity of natural systems, researchers have sought the assistance of advanced computer-based technologies in the development of integrated modeling and simulation systems. Computer simulations have been utilized in a variety of natural resource management applications from modeling animal populations [53], to forest fires [9, 34, 56], to hydrologic systems [62]. Computer models may be developed to understand more about how a real system works, as when scientists develop models of ecological processes [59]. Such models may facilitate predictions of a real system's behavior under a variety of conditions, or a greater understanding of the structure of a real system. There are numerous advantages to developing and experimenting with models of real-system phenomena. Experimenting with the real system itself may be too costly and time consuming, or even impossible [59]. Simulations are completely repeatable and nondestructive. The data produced by simulations is often easier to interpret than data from a real system [59].

Geographic information systems (GIS) technology has led these developments providing powerful databases for storing and retrieving spatially referenced data [55]. Spatial information is stored in many different themes representing quantitative, qualitative, or logical information. These data can have different resolutions that range from detailed local information to small-scale satellite imagery. GIS operators provide the means for manipulating and analyzing layers of spatial information and for generating new layers. Since it allows distributed parametrization, a GIS is useful for ecological models that need to explicitly incorporate the spatial structure and the variability of system behavior [2, 3, 48]. A raster-based GIS represents spatial information as a grid of cells, and each cell corresponds to a uniform parcel of the landscape. Cells are spatially located by row and column and the cell size depends on the resolution required. GIS provides an excellent means of capturing real-world data in multiple layers (three dimensional) and resolutions (spatial scales) over time.

Due to the complexity of ecosystem dynamics, interest has increased in using GIS for simulation of spatial dynamic processes. According to Itami [38] "a great part of the challenge of modeling interactions between natural and social processes has to do with the fact that processes in these systems result in complex temporal-spatial behavior." Vegetation dynamics, surface and subsurface hydrology, soil erosion processes, wild fire behavior, population dynamics, disease epidemics, and urban growth are examples of processes that exhibit complex behavior through time. It can be a daunting task to model each component of a system, much less simulate the interactions between them. Yet with the rising awareness and concern for local effects on global climate and the need to understand how to maintain sustainable landscapes, it is more important than ever to develop theories and methodologies for understanding natural and human processes as complex dynamic systems."

Cellular automata (CA) theory has been successfully demonstrated for modeling landscape dynamics. Cellular automata modeling has assisted researchers in exploring natural phenomena in fields such as geography, mathematics, and ecology to mention a few. CA has been successfully used to implement spatial dynamic ecological models in cellular spaces [34, 37, 38]. They can be seen as discrete models of spatiotemporal dynamics obeying local laws. By employing more qualitative state representations in the CA methodology, it is easier to express the dynamics of interacting discrete units in space. From a conceptual point of view, all these approaches are used to study and simulate the behavior of complex spatial behavior based on the interaction of simpler elemental processes.

The majority of studies that use GIS data have focused on modeling biophysical interactions [33]. Human dimensions of ecosystems have been largely neglected. Specifically, very little has been done to:

- Incorporate social science data into spatially explicit models of human behavioral phenomena; and
- Incorporate intelligent systems research and object-oriented programming techniques for simulations.

The rapid development of accurate spatial databases using GIS has opened opportunities for resource scientists to extend the use of this data by integrating GIS with emerging technologies. Exploratory studies [5, 32, 30, 49] which have incorporated artificial intelligence techniques with GIS have shown potential for improving resource-management decision making. Emerging simulation technologies such as robotics and artificial life have been slow to embrace the advantages of using georeferenced spatial data [54].

While cellular automata (CA) have proven to be useful for dynamic modeling, studies that have attempted to include GIS have had varying success. Because GIS are inherently static processes, they are limited for use in dynamic modeling in both their asynchronous updating of cellular data and implicit cellular nature. They do not include procedures for handling time and they are designed to process entire arrays of data, and cannot easily address varying localized operations across the spatial grid [3, 48]. Maidment [44] and Itami [38] have both summarized the major deficiency of GIS as the need to incorporate some form of time-dependent data structure if it is to be capable of tracking the evolution of spatial phenomena. In order to model spatially explicit ecological phenomena, these restrictions need to be addressed.

2 OBJECT-ORIENTED APPROACHES FOR SOLVING ECOLOGICAL PROBLEMS

Previous research [11, 51] make clear the limitations of the CA framework. Its homogeneous cellular structure and synchronous time advancement are too rigid to easily accommodate the diversity of processes that interact at overall ecosystem scale. A generalization of the CA methodology, viz. discrete-event object-oriented simulation of hierarchical, modular models, has been shown to offer the flexibility to include the full spectrum of process description levels ranging from individuals to aggregated populations [4].

Simulation modeling of any landscape ecosystems requires the integration of the two technologies, GIS and simulation, into a new single environment with both database and simulation strengths. The task of simulation software is to provide the dynamic state projection necessary for understanding complex system behavior and ultimately providing decision support for examining ecosystem dynamics. (See ch. 9 for an example of such integration.)

Object-oriented modeling (OOM) has emerged as a serious modeling and knowledge representation paradigm for applied sciences. Object-oriented modeling is a combination of object-oriented analysis, design, and programming. As discussed by Saarenmaa et al. [49] "the power of OOM lies in representing the domain with concrete objects that have as much similarity with their real-world counterparts as possible. OOM combines the features of artificial intelligence (AI) and knowledge-based systems (KBS) to form a powerful tool for studying complex dynamic systems." Durnota [19], Bousquet et al. [6], Saarenmaa et al. [50], and others have all employed OOM to research ecosystem problems. Modularity, a reduction in programming complexity and software reusability, are commonly noted as advantages of OOM over other approaches.

Other attempts have been made to incorporate information about landscape directly through the use of object-oriented programming approaches. Models developed by Loehle, Saarenmaa et al., Stone [53], Folse et al. [22], and others were designed to describe the behavior of nonhuman agents (animal or insect predator-prey relationships) in natural habitats. Many of these approaches model individuals directly as autonomous entities using object-oriented methodologies. To simulate large populations, the individuals are replicated and simulated, including their interactions with one another and with the environment. While modeling individuals as populations of autonomous agents is fruitful, to be useful, the simulation requires that realism and attention must be focused on a thorough examination of the behavioral mechanisms of individuals and to more accurate spatial representations of the environment, as is made possible in GIS.

Research has demonstrated how a simulation engine and a GIS database can be interfaced to provide ecosystem simulations of fire spread and forest succession [56, 57]. Zeigler [59, 60] has developed a uniform model description language based on the discrete-event system specification (DEVS) formalism. This formalism has been implemented in a number of environments and shown to be a workable code-independent model description language. One of the current implementations of the DEVS formalism is seen in DEVS-scheme [61], an object-oriented (OO), LISP-based simulation environment for hierarchical, modular models. The strength of this approach has been demonstrated with models describing landscape succession and fire growth [56, 57]. Here the DEVS-scheme modeling and simulation environment is interfaced to the widely used geographic resource and analysis simulation system (GRASS). Although LISP/OO implementations appear the most attractive candidates from a pure capability standpoint, other considerations have an impact on implementation. Although many newer versions of LISP-based languages have made significant improvements, there is the question of space/time efficiency of LISP execution [61].

Work by Marti [45] and Catsimpoolas [8] using case-based reasoning in a DEVS-like environment simulates intelligent behavior of military units across GIS linked terrain data. Such work provides good evidence that object-oriented simulations can provide a strong foundation for complex systems design. Their work demonstrates that the discrete event, object simulations are "particularly suited to situations involving movement and interactions over terrain." The Rand integrated simulation environment (RISE) capitalizes on merging goal-directed behavior with geographic information into the planning mechanism

Object-oriented programming and modeling techniques have revolutionized the computer science world. These techniques provide added flexibility and make possible the development of novel solutions to environmental and ecological problems. The object paradigm is well suited to aid in developing solutions to many types of problems and, when merged with other techniques such as GIS, can prove to be extremely powerful in solving complex ecological and behavioral problems. Object classes and message passing provide excellent ways to communicate information back and forth as well as portability and reusability of code. Such systems can be used to model complex interactions.

3 SPATIALLY EXPLICIT, INDIVIDUAL-BASED MODELING APPROACHES FOR ECOLOGICAL MODELING

There is a growing interest in the research community for using GIS for modeling spatially explicit dynamic processes [7, 52]. The use of individual-based models (IBM) is one of the popular approaches to modeling spatially explicit ecological phenomena. Researchers (i.e., DeAngelis and Gross [14], DeAngelis et al. [15], and Huston et al. [36]) utilize IBM approaches in the study of a variety of ecological problems. In this work IBMs correspond to two fundamental principles as outlined in Slothower et al. [52]: "The first principle is that individual organisms are behaviorally and physiologically distinct because of genetic and environmental influences. The second principle is that interactions between individuals are inherently localized, i.e. organisms are influenced mostly by nearby organisms." When applying interaction rules to many individuals over time, they are capable of generating phenomenologically realistic and complex behavior [14, 36]. The advantages of IBMs according to DeAngelis et al. [15] include the following:

- A variety of types of differences among individuals in the population can be accommodated;
- Complex decision making by an individual can be simulated; and
- Local interactions in space and the effects of stochastic temporal and spatial variability are easily handled.

IBMs offer potential for studying complex behavior and human/landscape interactions within a spatial framework. Since spatial information about some known phenomena is stored on a georeferenced coordinate system, space within a grid is implicit and relative to the origin of the grid [52]. IBMs offer some basic advantages over current cellular automata and other dynamic spatial modeling approaches for examining spatially explicit phenomena. Since space is continuous and location is explicit in IBMs, individuals can be simulated, independent of the environment. This provides the modeler with the ability to define an individual's behavior, personality traits, and interaction rules when encountering other individuals. Computer modeling of most ecological phenomena evolves in simulated time. Since space is continuous and individuals are represented independently, temporal and spatial variability in IBMs can be handled asynchronously (individually updated) versus synchronously (global update) common to most raster-based GIS systems.

Artificial intelligent agents or what has been referred to as "agent-oriented programming" is being used to capture behavioral conditions and sets of intercommunication among and within agents that coexist in a environment. The recent explosion in the use of intelligent agent simulations has come to the forefront of AI research. Object-oriented agents are in many cases programmed with very simple rules of behavior and set adrift in an environment to perform some predetermined task. These agents may be roaming databases, searchers who wander through the world wide web (WWW), or programmed for more generic, search-finding missions [20, 24].

These search agents or active objects have been "challenged" by work in reactive agents. These agents interact and react with their environments—hence, they are active objects with some partially determined set of goals. This school of thought professes that agent behavior is an intensive interaction and emergent property. The environment is taken into account dynamically. Its characteristics are exploited to serve the functioning of the agent. A complex agent has complex goals, but behavior to meet the goals develops or emerges as the agent learns more about its surroundings. While these simulations have been successfully implemented in contrived environments, real biophysical environments have not yet been explored.

Included in this area of agent simulations is the study of complex adaptive systems. One important aspect of complex systems research has been the study the emergent behavior using developed tools and techniques [10, 17, 26]. There are many examples of models and tools developed and in use today. These include Gensim [1], Swarm [35, 43], and ECHO [23, 40]. Emergent behavior of interest is that which develops from the simulations as a result of simple rules imposed on individual agents. The essential characteristic of this kind of simulation is a collection of relatively autonomous entities interacting with each other in a dynamic environment. Each of the many individuals making up the "swarm" makes its own behavioral choices on the basis of information it receives from the external world, its own internal state, and through communication with other individuals.

Work by Anderson (ch. 2) has contributed extensively to this area of research in the development of an intelligent-agent-simulation test-bed referred to as Gensim. Anderson's chapter describes Gensim as a tool for developing spatially explicit simulation models incorporating intelligent agents that attempts to encompass much of this breadth. Gensim divides the problem of supporting intelligent agents into the two parts: providing support for the integration of agents and their specific needs within a spatially explicit simulation tool, and providing a means to model the computational abilities (decision making, perception) of intelligent agents. Both components attempt to cover as much breadth both in terms of the variety and sophistication of agents supported, the ease of their definition, and the range of their spectrum. While most intelligent agent development is still utilizing theoretical environments,

Anderson provides some initial steps to make direct linkages to real-world environments.

Using GIS to represent real-world environments is an important contribution of this book. While most multiagent systems have had limited capabilities to provide a two-dimensional lattice for agents to move across, few to date have taken advantage of the power of GIS. Those that have struggled to incorporate the functionality of GIS have had to build their own linkages between the simulation models and GIS. With the growing development of digital landscape information and the GIS to manage and manipulate this information, simulation modelers now have this information at their fingertips. With the diversity of GIS and agent development tools in use today and the worldwide interest in spatial data, it is imperative that standard spatial data objects be developed that link agent-based simulation models with a variety of GIS. These spatial data objects need to be flexible and interoperable, and can be directly linked to a variety of spatial data formats in existence.

This book contains several excellent examples of this integration. First, work by Box (ch. 3) outlines a general-purpose tool that was developed for conducting simulation-based experiments on landscape models built from GIS data layers. It was conceived originally as a method for representing GIS layers in a way that is accessible for agent-based simulations (where the agents roam around in the model landscape), but it was discovered early on in software development that representing the landscape itself as a set of agents allowed for simulation of dynamic landscape processes using CA, cellular stochastic, or spatial Monte Carlo rules. The GIS-CA toolkit was developed from the Swarm libraries, which were developed at the Santa Fe Institute [46]. The Swarm libraries are a reusable set of software tools designed for agent-based simulations. Their primary utility is for management of many of the more difficult programming issues that arise when one tries to implement an agent-based simulation. The most important issue addressed by the Swarm libraries is enforcement of an accepted concurrency model for the agents' actions: in essence, how to simulate truly independent behavior of a multitude of individual entities (agents) who are acting in parallel when the simulation environment is processed through the single sequential processor of a computer. In this chapter Box provides a detailed design and applications of such a GIS-based CA. The software design presented is a toolkit that has been implemented using the Swarm simulation software; while the examples will be necessarily implementation-specific, emphasis will be placed on the general aspects of object design that should allow for a user to reproduce this software design in any object-oriented environment.

Westervelt (ch. 4) clearly demonstrates how to efficiently utilize existing GIS data in the operation of entity-based landscape simulation models. The approach described in Westervelt was originally used in the integration of the GRASS GIS with the IMPORT/DOME programming language [58]. IMPORT/DOME is an object-oriented language designed to support agent-based simulation modeling. The integration was accomplished to easily provide agents with digital landscapes recorded through measurement and observation in the GRASS GIS. IMPORT/DOME language-based software encapsulated GRASS GIS map input and output functions that allowed IMPORT/DOME simulation models to access GRASS GIS data. The success of that effort was repeated in the integration of the GRASS GIS with the Swarm programming environment. IMPORT/DOME and Swarm are both agent-based simulation environments that provide common environments for managing the passage of simulation time while objects interact with one another. An example of the application of this programming environment can be seen in Harper et al. (ch. 5). The model was developed using GRASS (version 4.2) GIS coupled with Swarm (versions 1.0.4 and 1.4). This chapter reports on an applied simulation in which independent agents (i.e., individual cowbirds, cattle herds, and traps) interacted with one another in space and time. During a simulation each agent developed a unique history according to the rules assigned to its type of object. The Swarm model was written in Objective-C and consisted of an observer swarm (interface control, display-animation, and time schedule) and a model swarm (equations describing ecological and behavioral features of modeled agents). While not yet quantitatively validated, the map of cowbird visitations to feeding areas appeared to be relatively stable over a wide range of initial conditions (data not presented).

Duke-Sylvester (ch. 6) provides another exciting approach to integrating spatial data together with agents. The across-trophic-level system simulation (ATLSS) is characterized by the integration of several distinct agent-based models and spatially explicit data into a single modeling system. One of the goals of the ATLSS (pronounced like "atlas") project is to investigate the relative response of various interconnected trophic levels of the South Florida (SF) Everglades to different hydrologic scenarios over a thirty-year planning horizon. The ATLSS approach consists of several distinct component models each of which represents different biotic components of the Everglades system, linked together as a multi model. Each model within ATLSS is agent-based, but different levels of aggregation are used for different models. The level of aggregation is a function of many considerations, including computational efficiency, availability of empirical data to support a model, and the level of spatial and temporal resolution needed from the model. This chapter describes an approach for understanding larger spatial extent aggregate behavior of systems by modeling localized dynamics of component objects in the Florida Everglades.

4 HUMAN/LANDSCAPE INTERACTION MODELING

While many researchers and programmers have successfully used linked agent-based simulations with GIS to examine animal interactions, only recently have researchers seeking new ways to understand human/landscape interactions been exploring agent simulations as a tool for developing models of human behavior [10, 18, 21]. Computer simulation according to Conte et al. [10] "is an appropriate methodology whenever a social phenomenon is not directly accessible, either because it no longer exists (as in archaeological studies) or because its structure or the effects of its structure, i.e. its behavior, are so complex that the observer cannot directly attain a clear picture of what is going on (as in some studies of world politics)." He goes on to say that an "important objective of this work is to realize, observe, and experiment with 'artificial societies' in order to improve our knowledge and understanding, but through exploration, rather than just through description." While simulation provides excellent opportunities to explore and test models and their underlying theory, it also allows the researcher to observe and record behavior of the system. As Conte et al. [10] conclude "the simulation is based on a model constructed by the researcher that is more observable than the target phenomena itself."

Recent studies [10, 18, 21, 41] clearly demonstrate the potential for agent-based modeling techniques to examine human/landscape interactions. These studies utilize a general model of multiagent simulations based on computation agents that represent individual organisms (or groups of organisms) in a one-to-one correspondence. These studies seek to understand the process of evolution in the study of ecological and sociological systems. As Drogoul et al. [18] state, "we are interested in the simulation of evolution of complex systems where interactions between several individuals at the micro level are responsible for measurable general situations observed at the macro level. When the situation is too complex to be studied analytically, it is important to be able to recreate an artificial universe in which experiments can be done in a reduced and simulated laboratory where all parameters can be controlled precisely." This work and others examining emergent processes in societies is extremely exciting and is yielding interesting results that would have been hard to obtain without the use of such simulations.

A considerable body of work exists devoted to understanding the behavior of the institutions that people have developed and evolved to manage natural resources. Specifically, a large number of studies have been undertaken in an effort to understand how common pool resources (CPRs) have been managed in differing natural and institutional environments [47]. Numerous field studies and laboratory experiments using human subjects have supported the evolution of a theoretical foundation for the study of resource management institutions. But while field studies and experiments have been useful tools for exploring the management of natural resources, to date little effort has been devoted to exploring the potential role of modeling and computer-based simulation for understanding the behavior of resource management institutions.

The research work by Deadman and Schlager (ch. 7) is an exception to this rule. They provide an example of how to combine the theoretical foundations devoted to understanding the behavior of the institutions that humans have developed and evolved to manage natural resources with recent advances in human systems modeling and simulation. This chapter discusses models of individual decision making in the social sciences and in simulations. The discussion facilitates the construction of a theoretical framework for modeling individual action, drawing upon a series of simulations of intelligent agents interacting within the context of a common pool resource. This framework is built upon a number of theoretical tools for the study of institutions, including the grammar of institutions, and the institutional analysis and development framework developed by Elinor Ostrom and colleagues [47].

To explore the effects of different values of the characteristics of individuals outlined above on the choices they make and the outcomes groups collectively achieve, the authors performed a series of laboratory experiments and simulations of those experiments. These experiments were motivated by a glaring contradiction between theoretically predicted outcomes and actual outcomes in CPR settings. Interest in the study of CPRs is fueled in part by the desire to understand how the apparent conflict between individual rationality and group rationality, referred to as a commons dilemma, can be avoided. The multiagent simulation model was built using Swarm [46]. In this CPR model, individual agents are created, with their own set of unique characteristics, to represent individual participants in an institution and the natural resource itself. The initial simulation was developed to explore the baseline laboratory experiments. The base model for these simulations includes a collection of eight agents, who utilize a set of simple heuristics, referred to here as strategies, to make decisions about the investments they make in each of the two available markets. These agents display adaptive characteristics in the sense that they select, from amongst a collection of sixteen available strategies, the one that best achieves their goal. In this case their goal is to maximize their return given information about the returns they received in previous rounds, and the actions of the other members of the group. In the initial simulations, no communication occurred between agents. The model of bounded rationality used in the simulations performed better than the model of perfect rationality in the noncommunication experiments, at least in terms of simulating the behavior of laboratory subjects. Economic theory, based on the model of perfect rationality, predicted that individuals in the common-pool-resource setting would achieve the outcome described as a Nash equilibrium. The intelligent agents of the simulation consistently achieved such an outcome. However, perfectly rational individuals were predicted to individually pursue Nash equilibrium strategies. In other words, the group outcome would simply be a sum of the individuals' strategies. This did not occur among the intelligent agents of the simulations, who individually pursued a variety of strategies from which emerged a Nash equilibrium. The boundedly rational agents of the simulation behaved much more like the laboratory subjects.

The model of bounded rationality used in the simulations also performed better than the model of perfect rationality in the communication experiments. The behavior of perfectly rational individuals should not be affected by communication opportunities that do not affect payoffs. Decades of communication experiments have demonstrated the importance of communication for developing and sustaining cooperation. While the behavior of bounded rational intelligent agents changed when allowed to communicate, the intelligent agents, lacking a means of developing norms of reciprocity, failed to sustain cooperation, unlike the laboratory subjects. Comparing the behavior of laboratory subjects who bring norms of behavior with them into the laboratory experiment with the behavior of intelligent agents who do not possess norms of behavior, clearly demonstrate the importance of such norms for developing and sustaining cooperation. This work clearly demonstrates a method to use laboratory experiment data and, through agent-based simulation, closely mimic the decision-making process of individuals in a common pool resource.

As the field of agent-based social simulation continues to develop, modeling efforts expand in an attempt to tie models of individual human decision making to real-world settings and established theory. To date, few studies have harnessed the power of GIS for representing the real or spatial worlds which individuals or societies reside in and interactions with those worlds. Since human behavior is inherently spatial, GIS can provide the worlds that individuals could respond to and function within. Currently, there is no GIS system with IBM capabilities. Dibble [16] states that "individual-based models do currently exist (Santa Fe Institute, Swarm 1995) and in many ways these systems may offer far deeper insights into human geographic phenomena than any current GIS."

A few researchers [13, 16, 29, 30, 42] have sought to combine spatially explicit IBMs and GIS. In their work Kohler et al. [42] suggest that "traditional inferential approaches fail to capture the dynamics and coevolutionary nature of human settlement decisions as they respond to shifting resources and the presence of actions of other people." He goes on to say: "if only we had a laboratory to study the outcomes of various processes as they might play themselves out through hundreds of years on realistic landscapes." The incorporation of GIS with IBMs is imperative for this type of application.

Jiang and Gimblett (ch. 8) provide an example of modeling pedestrian movement in urban spaces. Virtual pedestrians (agents) explore the open space locally and learn themselves from what they have explored. The destination is provided to the agent and it is up to the agent to explore local space and work out its movement pattern. Vision is additionally provided to the agent as it wanders around in the two-dimensional set of streets and buildings. This work is extremely exciting as it begins to clearly examine the use of cognition in the agents to make sense of and function in their environment. The environment can be used as a design or test-bed to examine alternative configurations to optimize human movement patterns.

Decisionmakers such as resource managers, environmental planners, and designers, faced with realistic environmental problems, would substantially benefit from simulation techniques that enabled them to explore alternatives and test ideas or theories, before expensive management plans are implemented. What is crucial, however, is that if decisionmakers have confidence in the use and results of agent-based simulations, the design of humanlike agents must be bounded by what can be synthesized from actual behavior and grounded in this reality. Work to date has clearly illustrated the importance of calibrating artificial agents that will perform humanlike tasks in the real world. Most importantly, as in any simulation or modeling endeavor, results need to be reliable and reproducible if they are to be of any immediate value. In addition, if the researcher or decisionmaker using agent-based simulations for solving complex, real-world problems are to have a high degree of confidence in the results, the agents must perform like their humanlike counterparts and flexibly built so they can be applied in a variety of settings. In order to meet these needs, the design of humanlike agents must be calibrated with actual behavioral data and grounded in this reality.

Itami (ch. 9) and Gimblett, Itami, and Richards (ch. 10) discuss an agent-based simulation system and a recent application that merge social data (collected to build and calibrate agents), GIS, and autonomous agents to explore the complex interactions between humans and the environment [27, 28, 29, 30, 39]. The recreation behavior simulator (RBSim) is a computer program that simulates the behavior of human recreationists in high-use natural environments. Specifically RBSim uses concepts from spatially explicit agent-based simulation system in recreation research for exploring the interactions between different recreation user groups within geographic space. The agents are autonomous because, once they are programmed, they can move about the landscape like software robots. The agents can gather data from their environment, make decisions from this information, and change their behavior according to the situation they find themselves in. Each individual agent has its own physical mobility, sensory, and cognitive capabilities. This results in actions that echo the behavior of real animals (in this case, human) in the environment. Unique to this simulation system is the use of spatial analytic operators from standard GIS systems to provide the agents with a spatial exploratory and reasoning system. By continuing to program knowledge and rules into the agent and watching the behavior resulting from these rules and comparing it to what is known about actual behavior, a rich and complex set of behaviors emerge. What is compelling about this type of simulation is that it is impossible to predict the behavior of any single agent in the simulation and, by observing the interactions between agents, it is possible to draw conclusions that are impossible using any other analytical process. Chapter 10 discusses the recreation behavior data collected to build and calibrate agents in the simulation environment and provides an analysis of simulation alternatives.

RBSim is important because until now, there have been few tools for natural resource managers and researchers to systematically collect temporal and spatial recreation behavior data to explore different recreation management options. Much of the recreation research is based on interviews or surveys, but this information fails to inform the manager/researcher how different management options might affect the overall experience of the user. For example, if a new trail is introduced into the backcountry, one might expect that encounters might be reduced, but to what extent? If one goes to a system of scheduling use, what is the impact on the number and frequency of users? More importantly when you have different, conflicting recreation uses, how do different management options increase or decrease the potential conflicts? None of these questions can be answered using conventional tools. These questions all pivot around issues such as time and space, as well as more complex issues such as intervisibility between two locations. By combining human agent simulations with geographic information systems, it is possible to study all these issues simultaneously and with relative simplicity.

RBSim has been modified to allow park management to explore the consequences of change to any one or more variables so that the goal of accommodating increasing visitor use is achieved while maintaining the quality of visitor experience. RBSim II provides both a qualitative understanding of management scenarios by the use of map graphics from a GIS as well as a quantitative understanding of management consequences by generating statistics during the simulation. Managers are able to identify points of overcrowding, bottle necks in circulation systems, and conflicts between different user groups. RBSim II is designed to be easy to use by Parks Victoria staff. This is facilitated through a tight integration with MapInfo GIS which allows a practical solution for quickly building complex simulation models. Simulation techniques provide methods for evaluating details of management decisions as they impact visitors and the environment. Innovations include the use of network topology to represent road and trail systems, analytical hierarchy process (AHP) to rate the attractiveness of site features and generate recreational personality types, and object-oriented database management techniques to allow portability of RBSim II to any other park or recreation setting [39].

In Chapter 11, Gimblett and others apply the same type of methodology as described earlier to develop an agent-based model for simulating the movement patterns and social interactions among rafting trips down the Colorado River through the Grand Canyon National Park in the United States. The modeling system employs statistical analyses and mathematical models based upon and calibrated with existing river trip itinerary data, as well as new data collected from river trip reports that document spatial and temporal movement patterns along the river. The Grand Canyon River Trip Simulation (GCRTS) combines intelligent agent modeling with GIS in an interactive system that provides the manager with advanced visualization of individual trip progress, as well as interactions among trips during specified time periods. Locational information includes specified river reaches, camps and attraction sites, exchange points, and restricted areas. The goal of this model system is to provide the National Park Service managers (and other potential users) with an effective decision support tool for representing and evaluating the distribution and volume of use along the river. Unique in this work is the use of fuzzy logic in conjunction with intelligent agents to mimic day-to-day decisions that commercial and private boatmen utilize when traversing the river. Decisions related to daily attraction site visits, campsite stopovers, takeouts, etc., are critical to providing the visitor with a high-quality recreation experience. At the writing of this chapter, version 2 of GCRTS is being developed with exciting enhancements to the agents' decision-making framework. Advanced reasoning ability in the agents is being developed to provide more autonomy in overall trip and daily planning, attraction, and campsite selection.

Finally, in Chapter 12, Lim and others describe some initial efforts in the development of a spatially referenced, intelligent agent-based simulation of land-use change in the Amazon (LUCITA). The simulation described here is focused on a case study in Altamira, Pará, Brazil in which demographic data on individual households and the land-use decisions that they make regarding deforestation have been documented. A simulation based on data collected in the study area were created using Swarm. Agents were created to represent the actions of individual households over time. These agents consider the status of their natural and institutional environment, as well as their own resources using genetic algorithms, when making decisions about how to utilize the land on their individual properties. This chapter describes the initial results of these simulations, while making observations on the utility of spatially referenced, agent-based simulations for modeling human-environment interactions.

5 SUMMARY

It is without doubt that individual-based modeling techniques coupled with GIS for examining human/wildlife/landscape interactions is a powerful tool for decisionmakers. Aside from representing individual behaviors and their associated environments, exploring both the social and ecological impacts in a dynamic system has tremendous potential for assisting decisionmakers in understanding and protecting extraordinary landscapes. From all the chapters outlined in this book, there are several points that can be summarized from this work.

Using agents to represent individuals or parties that are provided with communication and negotiation abilities (for example, to decide to stop and camp or to alter their plans to avoid encounters with other trips) can provide exciting opportunities to model individual behaviors like never before. Agents who are programmed with strategies, goals, and intentions (where they want to hike and how they want to achieve that goal, how long they plan to spend and how difficult a route they wish to choose) can provide a decisionmaker with a dynamic tool to examine and measure changes in both social and ecological systems. While agents can be programmed to perform a multitude of tasks, the ability to learn from their environment and derive adaptive strategies to adjust to their surroundings and others that they encounter (coping behavior) provide many exciting opportunities for researchers and decisionmakers to construct and implement many types of planning and management strategies.

With the inclusion of GIS to represent a spatially, georeferenced environment, dynamic behaviors and subsequent impacts of agents can be linked to spatial location critical in natural resource planning and management. To view agent interactions over time, in a spatially referenced environment and provides a powerful tool for policymakers, planners, managers, and the public to understand the outcomes of the simulation. Since landscapes are dynamic, GIS can be used to update the scenarios developed in the simulation with changing environment conditions (that is, as new information is captured in relationship to land-use changes). Spatial analytic GIS classes (neighboring statistical summaries, intervisibility analysis, distance measure, travel time, travel cost, slope, solar aspect, slope direction) provide a "tool box" of analytical capabilities that can be programmed into the agents. This decreases computation time and provides the agent with an enormous pool of spatial reasoning abilities.

Finally, using agent technology as a decision-making tool allows one to develop "what if" scenarios and provide options that will guide decisions in resolving conflicts of use. Using a simulation environment composed of agents derived from data synthesized from the real world provides a solid foundation for calibration and refinement to ensure they are humanlike. This is essential if any decisionmaker is to have confidence in applying agents for planning purposes. Finally, agent-based simulations provide new and exciting ways for managers and researchers to explore and compare alternative scenarios before they are implemented and evaluate them in terms consequences of policy actions and social, ecological, and economic impacts.

REFERENCES

[1] Anderson, J., and M. Evans. "A Generic Simulation System for Intelligent Agent Designs." *Appl. Art. Intel.* **9(5)** (1995): 527–562.

[2] Band, L. E., and E. F. Wood. "Strategies for Large-Scale Distributed Hydrologic Simulation." *Appl. Math. Comp.* **27(1)** (1988): 23–38.

[3] Band, L. E., D. L. Peterson, S. W. Running, J. Coughlan, R. Lammers, J. Dungan, and R. Nemani. "Forest Ecosystem Processes at the Watershed Scale: Basis for Distributed Simulation." *Ecol. Model.* **56** (1991): 171–196.

[4] Baveco, J. M., and R. Lingeman. "An Object-Oriented Tool for Individual-Oriented Simulation: Host-Parasitoid System Application." *Ecol. Model.* **61** (1992): 267–286.

[5] Berry, J. S., G. Belovsky, A. Joern, W. P. Kemp, and J. Onsager. "Object-Oriented Simulation Model of Rangeland Grasshopper Population Dynamics." In *Proceedings of Fourth Annual Conference on AI, Simulation, and Planning in High Autonomy Systems*, held September 20–22, 1993, in Tucson, AZ, edited by B. Zeigler and J. Rozenblit, 102–108. Los Alamitos, CA: IEEE Computer Society Press, 1993.

[6] Bousquet, F., C. Cambier and P. Morand. "Distributed Artificial Intelligence and Object-Oriented Modeling of a Fishery." *Math. & Comp. Model.* **20(8)** (1994): 97–107.

[7] Briggs, D., J. Westervelt, S. Levi, and S. Harper. "A Desert Tortise Spatially Explicit Population Model." In *Proceedings of the Third International Conference/Workshop on Integrating GIS and Environmental Modeling*, January 21–25, 1996, Santa Barbara, CA: National Center for Geographic Information and Analysis. ⟨http://www.ncgia.ucsb.edu/conf/SANTA_FE_CD-ROM/main.html⟩.

[8] Catsimpoolas, N., and J. Marti. "Scripting Highly Autonomous Simulation Behaviour using Case-Based Reasoning." *IEEE* (1992).

[9] Cohen, P. R., M. L Greenberg, D. M. Hart, and A. E. Howe. "Trial by Fire: Understanding the Design Requirements for Agents in Complex Environments." *AI Mag.* **10(3)** (1989): 32–48.

[10] Conte, R., and N. Gilbert, eds. "Computer Simulation for Social Theory." In *Artificial Societies: The Computer Simulation of Social Life*, 1–15. London: UCL Press, 1995.

[11] Costanza, R., F. H. Sklar, and M. L. White. "Modeling Coastal Landscape Dynamics." *BioScience* **40(2)** (1990): 91–107.

[12] Csikszentmihalyi, M., and I. S. Csikszentmihalyi. *Optimal Experience: Psychological Studies of Flow in Consciousness.* New York: Cambridge University Press, 1988.

[13] Deadman, P., and H. R. Gimblett. "A Role for Goal-Oriented Autonomous Agents in Modeling People-Environment Interactions in Forest Recreation." *Math. & Comp. Model.* **20(8)** (1994): 121–133.

[14] DeAngelis, D. L., and L. J. Gross, eds. *Individual-Based Models and Approaches in Ecology.* New York: Chapman and Hall, 1992.

[15] DeAngelis, D. L., D. M. Fleming, L. J. Gross, and W. F. Wolff. "Individual-Based Modeling in Ecology: An Overview." In *Proceedings of the Third International Conference/Workshop on Integrating GIS and Environmental Modeling*, January 21–25, 1996, Santa Barbara, CA: National Center for Geographic Information and Analysis. ⟨http://www.ncgia.ucsb.edu/conf/SANTA_FE_CD-ROM/main.html⟩.

[16] Dibble, C. "Representing Individuals and Societies in GIS." In *Proceedings of the Third International Conference/Workshop on Integrating GIS and Environmental Modeling*, January 21–25, 1996, Santa Barbara, CA: National Center for Geographic Information and Analysis. ⟨http://www.ncgia.ucsb.edu/conf/SANTA_FE_CD-ROM/main.html⟩.

[17] Doran, J., and N. Gilbert, eds. "Simulating Societies: An Introduction." In *Simulating Societies: The Computer Simulation of Social Phenomena*, 1–23. London: UCL Press, 1994.

[18] Drogoul, A., and J. Ferber. "Multi-agent Simulation as a Tool for Studying Emergent Processes in Societies." In *Simulating Societies: The Computer Simulation of Social Phenomena*, edited by. N. Gilbert and J. Doran, 127–142. London: UCL Press, 1994.

[19] Durnota, B. "Defining Relationships in Ecology using Object-Oriented Formal Specifications." *Math. & Comp. Model.* **20(8)** (1994): 83–96.
[20] Etzioni, O., and D. Weld. "A Softbot-Based Interface to the Internet." *Comm. ACM* **37(7)** (1994): 72–76.
[21] Findler, N. V., and R. M. Malyankar. "Emergent Behaviour in Societies of Heterogeneous, Interacting Agents; Alliances and Norms." In *Artificial Societies: The Computer Simulation of Social Life*, edited by N. Gilbert and J. Conte, 212–236. London: UCL Press, 1995.
[22] Folse, L. J., J. M. Packard, and W. E. Grant. "AI Modelling of Animal Movements in Heterogeneous Habitat." *Ecol. Model.* **46** (1989): 57–72.
[23] Forrest, S., and T. Jones. "Modeling Complex Adaptive Systems with Echo." In *Complex Systems. Mechanism of Adaptation*, edited by R. J. Stonier and X. HuoYu, 3–20. The Netherlands: IOS Press, 1994.
[24] Genesereth, M. R., and S. P. Ketchpel. "Software Agents." *Comm. ACM* **37(7)** (1994): 48–53.
[25] Georgeff, M. P., and F. F. Ingrand. "Real-Time Reasoning: The Monitoring and Control of Spacecraft Systems." In *Proceedings of the Sixth IEEE Conference on Artificial Intelligence*, held in Santa Barbara, California, 1990.
[26] Gilbert, N., and K. G. Troitzsch. *Simulation for the Social Scientist.* Buckingham, UK: Open University Press, 1999.
[27] Gimblett, H. R. "Simulating Recreation Behavior in Complex Wilderness Landscapes using Spatially Explicit Autonomous Agents." Ph.D. diss., University of Melbourne, Parkville, Victoria 3052, Australia.
[28] Gimblett, H. R., and R. M. Itami. "Modeling the Spatial Dynamics and Social Interaction of Human Recreators using GIS and Intelligent Agents." In *Proceedings of MODSIM 97—International Congress on Modeling and Simulation*, held December 8–11, 1997, in Hobart, Tasmania.
[29] Gimblett, H. R., B. Durnota, and R. M. Itami. "Conflicts in Recreation Use in Natural Areas: A Complex Adaptive Systems Approach." In *Proceedings of the Third Australian Complex Systems Conference*, held July 14–18, 1996, at Charles Stuart University, in Albury, Australia.
[30] Gimblett, H. R., R. M. Itami, and D. Durnota. "Some Practical Issues in Designing and Calibrating Artificial Human Agents in GIS-Based Simulation Worlds." *Complex. Intl. J.* **3** (1996).
[31] Gimblett, H. R., and G. L. Ball. "An Exploration and Application of Neural Network Architectures for Monitoring and Simulating Changes in Forest Resource Management." *AI Appl., Natl. Res., Agric. & Env. Sci.* **9(1)** (1995): 23–51.
[32] Gimblett, H. R., G. L. Ball, and A. W. Guisse. "Autonomous Rule Generation and Assessment for Complex, Spatial Modeling." *Landscape & Urban Plan. J.* **30** (1994): 13–26.
[33] Goodchild, M., B. O. Parks, and L. T. Steyaert, eds. *Environmental Modeling with GIS.* Oxford: Oxford University Press, 1993.

[34] Green, D. G. "A Generic Approach to Landscape Modelling." In *Proceedings from Eighth Biennial Conference of Simulation Society of Australia, Canberra.*

[35] Hiebeler, D. "The Swarm Simulation System and Individual-Based Modeling." In *Decision Support 2001. 17th Annual Geographic Information Seminar and the Resource Technology '94 Symposium*, edited by J. M. Power, M. Strome, and T. C. Daniel, vol. 1, 474–494. Toronto, Canada: American Society for Photogrammetry and Remote Sensing, 1994.

[36] Huston, M., D. L. DeAngelis, and W. M. Post. "New Computer Models Unify Ecological Theory." *BioScience* **38** (1988): 682–691.

[37] Itami, R. M. "Cellular Worlds: Models for Dynamic Conceptions of Landscapes." *Landscape Arch.* **78(5)** (1988): 52–57.

[38] Itami, R. M. "Simulating Spatial Dynamics: Cellular Automata Theory." *Landscape & Urban Plan. J.* **30** (1994): 27–47.

[39] Itami, R. M., H. R. Gimblett, R. Raulings, D. Zanon, G. MacLaren, K. Hirst, and B. Durnota. "RBSim: Using GIS-Agent Simulations of Recreation Behavior to Evaluate Management Scenarios." Aurisa 9—The Spatial Information Associations, held November 22–26, 1999, at the Fairmont Resort, Blue Mountains, NSW, Australia.

[40] Jones, T., and S. Forrest. "An Introduction to SFI Echo." Working Paper 93-12-074, Santa Fe Institute, Santa Fe, NM, 1993.

[41] Kohler, T. A., and George J. Gumerman, eds. *Dynamics in Human and Primate Societies.* New York: Oxford University Press, 2000.

[42] Kohler, T. A., C. R. Van West, E. P Carr, and C. G. Langton. "Agent-Based Modeling of Prehistoric Settlement Systems in the Northern American Southwest." In *Proceedings of the Third International Conference/Workshop on Integrating GIS and Environmental Modeling*, January 21–25, 1996, Santa Barbara, CA: National Center for Geographic Information and Analysis. ⟨http://www.ncgia.ucsb.edu/conf/SANTA_FE_CD-ROM/main.html⟩.

[43] Langton, C., N. Minar, and R. Burkhart. "The Swarm Simulation System: A Tool for Studying Complex Systems." 1995. Santa Fe Institute, Santa Fe, New Mexico, USA. ⟨http://www.santafe.edu/projects/swarm/swarmdoc/swarmdoc.html⟩.

[44] Maidment, D. R. "GIS and Hydrologic Modelling." In *Proceedings of the First International Conference/Workshop on Integrating Geographic Information Systems and Environmental Modelling.* held September 15–19, 1992, in Boulder, CO.

[45] Marti, J. "Cooperative Autonomous Behaviour of Aggregate Units over Large-Scale Terrain." In *Proceedings of AI, Simulation, and Planning in High Autonomy Systems*, edited by B. Zeigler and J. Rozenblit. Los Alamitos, CA: IEEE Computer Society Press, 1990.

[46] Minar, N., R. Burkhard, C. Langton, and M. Askenazi. "The Swarm Simulation System: A Toolkit for Building Multi-agent Simulations." 1996.

Working paper 96-06-042, Santa Fe Institute, Santa Fe, NM. Available at ⟨http://www.santafe.edu/projects/swarm/overview.ps⟩.

[47] Ostrom E., Gardner R., and J. Walker. *Rules, Games, and Common Pool Resources.* Ann Arbor: The University of Michigan Press, 1994.

[48] Running, S. W., R. R. Nemani, D. L. Peterson, L. E. Band, D. F. Potts, L. L. Pierce, and M. A. Spanner. "Mapping Regional Forest Evapotranspiration and Photosynthesis by Coupling Satellite Data with Ecosystem Simulation." *Ecology* **70(4)** (1989): 1090–1101.

[49] Saarenmaa, H., J. Perttunen, J. Vakeva, and A. Nikula. "Object-Oriented Modeling of the Tasks and Agent in Integrated Forest Health Management." *AI Appl. Natur. Res. Mgmt.* **8(1)** (1994): 43–59.

[50] Saarenmaa, H., J. Perttunen, J. Vakeva, and M. J. Power. "Multi-agent Problem Solving and Object-Oriented Decision Support Systems for Natural Resource Management." *AI Appl.* **9(1)** (1995).

[51] Sklar, F. H., R. Constanza, and J. W. Day. "Dynamic Spatial Simulation Modeling of Coastal Wetland Habitat Succession." *Ecol. Model.* **29** (1985): 261–281.

[52] Slothower, R. L., P. A. Schwarz, and K. M. Johnson. "Some Guidelines for Implementing Spatially Explicit, Individual-Based Ecological Models within Location-Based Raster GIS." In *Proceedings of the Third International Conference/Workshop on Integrating GIS and Environmental Modeling*, January 21–25, 1996, Santa Barbara, CA: National Center for Geographic Information and Analysis. ⟨http://www.ncgia.ucsb.edu/conf/SANTA_FE_CD-ROM/main.html⟩.

[53] Stone, N. D. "An Object-Oriented Approach to Modeling Arthropod Predator-Prey Interactions." In *Proceedings of Resource Technology '90*, held November, 1990, in Washington, D.C.

[54] Tobler, W. R. "Cellular Geography." In *Philosophy in Geography*, edited by S. Gale and G. Olsson, 379–386. Boston, MA: D. Reidel, 1979.

[55] Tomlin, D. C. *Geographic Information Systems and Cartographic Modeling.* Englewood Cliffs, NJ: Prentice Hall, 1990.

[56] Vasconcelos, M. J., and B. P. Zeigler. "Simulation of Forest Landscape Response to Fire Disturbances." *Ecol. Model.* **65** (1993): 177–198.

[57] Vasconcelos, M. J., B. P. Zeigler, and I. A. Graham. "Modeling Multi-scale Spatial Ecological Processes under the Discrete Systems Paradigm." *Landscape Ecol.* **65** (1993): 199–213.

[58] Westervelt, J., and L. D. Hopkins. "Facilitating Mobile Objects within the Context of Simulated Landscape Processes." In *Proceedings of the Third International Conference/Workshop on Integrating GIS and Environmental Modeling*, January 21–25, 1996, Santa Barbara, CA: National Center for Geographic Information and Analysis. ⟨http://www.ncgia.ucsb.edu/conf/SANTA_FE_CD-ROM/main.html⟩.

[59] Zeigler, B. P. *Theory of Modeling and Simulation.* New York: John Wiley, 1976.

[60] Zeigler, B. P. *Multifaceted Modeling and Discrete Event Simulation.* London: Academic Press, 1984.

[61] Zeigler, B. P. *Object-Oriented Simulation with Hierarchical, Modular Models: Intelligent Agents and Endormorphic Systems.* London: Academic Press, 1990.

[62] Zhang, X., V. Lopes, and G. Ball. "Integrating Hydrology Models and GIS: A Preliminary System Design." SRNR Report 94-1, School Renewable Natural Resources, University of Arizona, 1994.

Providing a Broad Spectrum of Agents in Spatially Explicit Simulation Models: The Gensim Approach

John Anderson

1 INTELLIGENT AGENTS AND INDIVIDUAL-BASED MODELING

Research in natural resource management may be characterized as a search for an understanding of patterns and processes relating to a particular resource. Modeling is a crucial tool to these efforts: resource scientists use such models to help them conceptualize, understand, test, predict, or assess various aspects of the resource being studied [52]. One central function, however, underlies all of these uses: a model simulates the way in which a real system would behave under conditions of interest to the user, and illustrates changes over time. Such a model may be used to determine the consequences of particular situations, leaving judgment of the attractiveness of those consequences to the user [25]. Particularly in the case of complex ecosystems, such a model may also serve to clarify interactions and contribute to a deeper understanding of ecological phenomena [26].

In recent years, computer-based models have become the most significant tool of resource managers, for two reasons. First, any model must accurately portray the real system it represents if research based on the model is to have any reliability. The use of computer technology has greatly increased the extent and the detail to which ecosystems can be modeled, and thus the accuracy of these models. The other reason for the extensive use of computer models is the flexibility that the computer as a tool brings to the modeling process. Many ecosystems are poorly understood, and complex models for such poorly understood systems are almost never completed. Rather, modeling such a system is an iterative process, with a partial understanding generating new hypotheses, which in turn generate changes to the model based on further research [42]. Computer technology brings flexibility and ease of modification to the modeling process, naturally supporting this iterative development. In addition, as the alternatives available in resolving resource management problems become increasingly expensive, and the resources themselves become increasingly scarce and valuable, such models become vital tools not only in the direct management of resources, but in the control of expenses associated with resource management as well.

Despite the widespread use of computer technology in modeling and the added complexity and realism that this technology brings, there is still a large gap between the models made by researchers and those used by natural resource managers. Many of the models created for specific research projects are not practically useful to natural resource managers, often because they deal only with simplified aspects of particular problems. In recent years there has been a growing realization that resource management science needs to tackle more significant, directly applicable problems, and ensure that the models used in such research capture the complexity of real problems in a manner which is useful to natural resource managers.

The desire to fill this gap, and to, in general, provide more powerful computer modeling tools capable of significant real-world simulations, has led to important developments in modern computer modeling. Two of the most significant developments form the foundation of this chapter: individual-based modeling and the inclusion of intelligent agents as inhabitants of simulation models.

Individual-based modeling is a computational approach to modeling a system through the interaction of atomic models of each individual inhabiting the system. Scientifically, individual-based models provide many advances over more traditional ecosystem models. They do away with the assumption that there is some mean individual that can adequately represent each and every individual in a large population, as well as the assumption that significant interactions take place evenly across populations [17, 36, 37]. Moreover, most individual-based models are spatially explicit, allowing interaction between individuals to occur over a wide range of space (and analogously, doing away with the assumption that significant interactions take place evenly across space). As opposed to a simple mean, individual-based models have the ability to represent the biological, physiological, and behavioral distinctions seen in individuals in the real world. Moreover, because the individual is the atomic unit, the simulation naturally takes localized interactions into account. A model of higher-level entities (communities, populations) thus emerges from the dynamics of individual interactions, in the same manner as the high-level phenomena we observe (and in which we also participate) in the real world.

There are many approaches to individual-based modeling, from creating individuals as cellular automata with simple transition and interaction rules [49], to modeling individuals with broader perceptual/effectory abilities (e.g., Gecko [13] and the earlier Swarm system [33] on which it is based). However, computing resources and associated restraints on the complexity of individual models are still the largest limiting factors in individual-based models, along with a lack of suitable tools for developing such models. Many individual-based ecosystem models manage large numbers of individuals (e.g., Wilson et al. [54] and Schmitz and Booth [46]) while maintaining only a very small amount of information on each individual or supporting a fairly limited variation on individual physiology and behavior. The need to maintain more than a few parameters of interest for each individual is also recognized as one of the most significant factors in deciding whether individual-based modeling is feasible [36].

The need for intelligent agents in natural resource management simulations arises in part directly out of the individual-based approach. Since the behavior of an individual-based model emerges from the individual behaviors and interactions of its inhabitants, the accuracy and ultimate utility of the model itself can be no better than the accuracy of its components. A more comprehensive, flexible, and ultimately more accurate model is obtained by modeling the intelligence inherent in individual inhabitants via their implementation as intelligent computational agents.

While this motivation for including intelligent agents as part of an ecosystem model is readily apparent, accuracy in modeling is really a goal underlying a host of additional concerns. Foremost among these is the growing recognition that the major source of change in many ecosystems (and indeed the major reason for studying many ecosystems at all) is the direct and indirect impact of human activity within the system. In order to reflect this, an ecosystem model must have the ability to directly include changes and disturbances induced by humans as a part of the model itself. While this statement has been recognized historically (Spafford [50], for example, argues that this is absolutely essential for any ecosystem model to be useful in a management context), there has been little hope of including models of human behavior of any realistic variety until the relatively recent development of intelligent agent technology. To date, there has still been very little work done in this context (but see Deadman and Gimblett [18] and Gimblett et al. [27]).

Similar arguments can also be made for nonhuman agents. Animals in general respond to their environment with some degree of intelligence: the representations and processes that form these creatures' decision-making capabilities are simply not as extensively evolved as our own. The need to model intelligent information processes in nonhumans will, of course, vary widely from model to model. In an attempt to model the entire population of mountain gorillas in an individual-based manner [47], for example, it will be much more significant than, say, an individual-based model of a fish population. While it has been demonstrated that the ongoing behavior of extremely simple animals (generally well below the latter case) can be modeled using extensive stimulus-response systems (e.g., Beer [11]), it is also generally accepted that these have bounds, and that more sophisticated methodologies are necessary to model more advanced behaviors. Even in the latter case, techniques used in simple intelligent agents are applicable: rule-based systems, for example, have been used to model the grazing habits of foxes in a very limited manner [15], and similar examples of using low-level artificial intelligent (AI) techniques to construct simple behavioral models can be readily found (e.g., Folse et al. [23]).

As we have seen in this section, spatially explicit individual-based models incorporating intelligent agents do indeed represent a potentially powerful solution to many problems. This application represents a significant area of overlap between the fields of artificial intelligence and ecosystem management; much more in fact than simply an application of the former in the domain of the latter. The general development process involved in the creation of intelligent agents for most any domain generally relies heavily on software-based testing [21]. Software environments are created to examine and test agent principles and components, and indeed, the completed agent may itself function entirely within a software development. As such, the development of intelligent agents for most other problem-solving environments essentially incorporates the development of individual-based models on a scale that is small from the perspective of the number of agents involved. Compared to most ecologically related individual-based models however, the environment itself is highly sophisticated, as the focus of the model is generally upon the problem-solving efforts of the agent itself (and the associated low-level interactions with the domain) as opposed to emerging characteristics of the environment as a whole. What this means is that AI can benefit greatly from the many other modeling advances and general rigor of ecological modeling, and ecological modeling incorporating intelligent agents can benefit from many analogous problems that AI has already had to deal with.

To begin reaping these benefits, however, we must move from the realm of theory to that of physical embodiment. Thus, we must turn to examining the characteristics of these models from the point of view of a tool for building them, with the ultimate goal being the construction of such a tool. Section 2 outlines these needs, and the remainder of this chapter describes Gensim, a tool that attempts to meet many of these needs though an integrated facility for defining and working with a broad range of intelligent agents.

Before moving on however, two important clarifications regarding the nature of agent technology should be made. First, the near-ubiquitous nature of the term itself. Agents are to contemporary software technology what objects were until comparatively recently. The term "agent technology" has become a "buzz" word that is associated with many expensive software packages, and touted as the future solution to many problems, in much the same way that the term "object technology" used to be employed. As with any contemporary darling of the computer software industry, the use of the term should be viewed with some suspicion, and rightly so. We hear of programs associated with web sites, for example, that are relatively simple search programs one day and "agents" the next, with no changes to the underlying software whatsoever. Similarly, many "agent-based" help facilities are simply heavily anthropomorphized keyword indices.

That is not at all to say that all agents are such vapor ware: indeed, it is, to some degree, in the nature of the term for such misuse to occur. At its simplest, the idea is that an agent is a systems metaphor—a perspective from which to view and study a particular system. An agent is simply a program that perceives its environment and acts upon it [45]. While the agent approach has been an incredibly useful tool for bringing areas together and encouraging a wider viewpoint of particular problems, the generality of the term, like that of object technology before it, has left it open to abuse. Russell and Norvig [45] point out that under this definition, the daylight-savings-time adjustment routine that waits in the background on personal computers and alters the clock on the correct days of the year is an agent. Indeed, so are many other simple machines. A true intelligent agent, however, is exactly that: one that adopts the agent perspective in its design (which is a natural one for anything inhabiting an artificial world) and displays intelligence to some degree in its behavior. The nature of the term "intelligence" is also a vague one, but several helpful definitions, from avoiding unnecessary search [40] to adopting a rational stance [45], can easily be employed to ensure some substance exists where the term is implied.

It should also be noted that the use of intelligent agents within a natural resource management setting is by no means limited to intelligent inhabitants in individual-based models as discussed here. The interpretation of such models, for example, is also an important part of overall resource management. The need has long been recognized for the inclusion of not only the concise, measurable, decision-theoretic aspects of traditional computer simulations, but also the informal judgment and intuition of the resource manager in the interpretation of such models [38, 52]. Agent-based techniques can be used in conjunction with more standard expert systems techniques to achieve these goals. Other diverse applications such as data mining and information filtering are also good applications of this technology.

2 MODELING TOOLS AND BREADTH

In order to allow functioning intelligent agents within an ecosystem model, two things are required. First, a model of the behavior of the desired agent (that is, an intelligent agent design) must be constructed, and some underlying facility for the implementation of that agent's decision-making abilities must be available. As discussed earlier, these abilities will differ greatly with the domain and the expectations of the sophistication of the agent. This translates into a great deal of variation in the underlying implementation of the agent model. From the perspective of a physical implementation, such models may range from architecturally simple, purely reactive approaches for extremely primitive agents (e.g., Agre and Chapman [2]) to sophisticated systems integrating reactive and deliberative reasoning (e.g., Anderson [4]).

Secondly, and more significantly from the point of view of general natural resource management, the modeling tool itself must provide for the inclusion of intelligent agents. This requires great sophistication in comparison to traditional modeling tools and demands many features of a modeling tool that are not generally associated with ecosystem modeling. Not only must an environment be modeled in a spatially explicit manner, for example, but a realistic perspective of this model must be given to each agent in order that it may make decisions in a manner similar to that of its physical counterpart. That is, the simulated world functions as both an environment unto itself, and a virtual reality to the agents that inhabit it. Likewise, the competing effects of many different agents must be carefully managed, all under the constraints of reasonable timing and reasonable consumption of computational resources. Such tools must also bridge the practicality gap described in the previous section: they must be useful for realistic problem-solving situations, as well for simple research models.

While it is certainly possible to construct completely specialized environments from scratch for each new project (though this occurs in both ecosystem management and AI with unfortunately high regularity), we would ideally like a tool that is flexible enough to support a wide range of agents and environments. The reason for this is not only the desire to avoid reinventing the wheel. Control and verifiability aspects, for example, are involved as well: a tool allows us to ensure that implementations under comparison have identical constraints placed upon them, and helps to ensure scientific validity in the results obtained. More importantly, however, a flexible tool is required because of the complexity of the agents and environments themselves. Complex software systems can neither be completely understood nor completely designed in advance. This will sometimes necessitate changes in design approach in midstream, and if our original tool does not support such a change, the tool must be abandoned and a new one chosen. In the worst case (designing specialized environments from scratch for each new project), we have to completely rebuild the entire environment and agent *de novo* with each small design change.

This is an extremely common problem in AI, which is arguably the branch of computer science whose systems are by necessity the most poorly understood at the outset of the design process. As such, many specialized design techniques have been developed in this field to alleviate the complexity associated with the applications being constructed, most notably prototyping techniques for commercial expert systems. Like the problems associated with a changing ecosystem model or agent design, changes occur in the model of how an expert performs in his or her domain. These, in turn, necessitate either small changes, if a tool is flexible enough, or a complete abandonment of a given tool and reimplementation in a tool more suitable to the revised model. The latter is termed a *paradigm shift* [32, 53], and is analogous to what would be necessary in an ecosystem model if we do not have a tool that can bend and flex with the changing needs of the modeler as a more accurate picture of the domain emerges. In ecosystem modeling (and in intelligent agent research apart from expert systems as well) paradigm shifts are exacerbated by the experimental nature of the domain. In intelligent agent research, we are often experimenting with different categories of agents in the same domain, or the effects of varying domain characteristics on a particular agent design, putting even more emphasis on the flexibility required of a modeling tool. Similar ongoing alterations in agent and environmental characteristics will also affect ecological models.

Out of this general desire for flexibility, we can construct several more specific requirements of a modeling tools:

- Interaction between agents must be supported, to as sophisticated a degree as possible.
- The nature or implementation of intelligent decision making in individuals should not be assumed. That is, the simulator should support an interface between agents and their environment, and a mechanism for agents to make decisions in time sync with the rest of the environment, rather than putting constraints on specific agent designs or methodologies.
- The nature of the domain should also be as assumption free as possible. Many simulators for intelligent agents, for example, provide only for grid-based domains (e.g., Pollack and Ringuette [43] and later work). The environment must be able to include a wide range of objects, including agents of limited intelligence as discussed previously, and also the inclusion of nonliving objects that can directly affect the environment and the other organisms inhabiting it.
- A simple but realistic model of perception must be provided: agents cannot simply react to other species by virtue of the simulator's knowledge that they occupy the same location in an ecosystem model, for example. Agents must be able to scan around themselves, and gather information as their perceptual abilities would normally allow them. Similarly, the effectory capabilities of agents should be realistic and individually limited in scope. Agents in an individual-based model should not be forced to implement the same actions as other agents in precisely the same ways.
- Last but certainly not least, the simulator must as much as possible provide means to lessen the computational load required to support large numbers of agents.

These points are by no means independent: for example, supporting varying effectory abilities for each intelligent agent is the most realistic approach, but demands that more information be maintained for each particular agent, increasing the computational load.

There is a general theme of *breadth* among these points: one that makes significant demands on the underlying structure of our simulation tool. The most obvious sense of this breadth is in the broad agents that such a tool must be expected to support. In AI, the term "broad agents" is most often used to refer to agents that can be employed for several purposes or that combine several aspects of intelligence, such as reasoning and communication abilities [10]. That is, the term is used to refer to the breadth of the capabilities of the agent itself. Here, on the other hand, we refer to the huge range of possible agents which a single tool may be asked to support. This includes not only between simulation applications, where it might be feasible to switch simulator modules geared toward specific types of agents, but also within a simulation, where we may have human-level intelligent agents[1] as well as a range of animal agents with varying degrees of intelligence, all inhabiting and interacting with each other in the same environment cotemporally. This requires that the approach used to support the agents involved in a simulation be capable of dealing an enormous breadth of possible agent models, and have the computational ability to feasibly support many such models (and many instances of each) simultaneously during a simulation. Moreover, the support of individual agents within this spectrum must be done in such a manner as to balance realism and computational tractability.

The fact that agents at one end of sophistication require very different facilities than those at the other, in tandem with an inability to assume the needs of any one particular simulation application (required population of different agent types, granularity of the environment, sophistication of agent interactions, sophistication of timing, etc.) poses a significant challenge for a modeling tool to say the least. Consider just one example: providing a visual interface to an intelligent agent. In an AI simulation we might be interested in the formation-following abilities of agents in a crowded area. In such a simulation, a great richness of vision (such as the ability to pick out color differences between every single tiny component of every object in the environment) would not only be relatively useless from the point of view of the design of the simulation, but would slow the simulation down tremendously. In a second instance more closely involving ecological modeling, we might allow intelligent agents to roam an area, and be interested in the visual contacts they have with one another, or with aspects of the environment that might be less than pleasant. The latter would require, from the same tool, a significant visual acuity and the representation of most concepts in the environment from a visual perspective.

Also significant from the perspective of real-world problems is the trade-off between the diametrically opposed metrics of realism and computational tractability themselves. On one hand, the sophistication of resource manage-

[1]Or really, as close to human-level performance as technology currently permits.

ment problems requires detailed models; on the other hand, we model in part to explicitly avoid representing everything: We place controls on the complexity of the environment and examine (hopefully) those parts that affect the context of interest. Using the same example of a visual interface, we would likely want to skip above the pixel-level of vision, and describe the situation in a more object-level manner.

The many levels of breadth described here act in concert to define the ultimate desirable breadth of our modeling tool itself: what the tool will and will not support, and much more importantly, what kind of effort will be required on the part of the developers to construct models. Tools in general are extensions of our own abilities, designed to enhance those abilities by way of speed, accuracy, or extent. As such, a tool is designed for a specific purpose, and while improvised use of such a tool may also allow it to be used in a variety of scenarios for which it was not initially intended, the effectiveness of that tool wanes as its use diverges from its original purpose. As the adage[2] states, "for every tool, there is a task it fits," and this is certainly as true with software tools as it is for hardware tools. When developing an expert system, for example, we can choose between many ready-made system-building tools (see Carnegie-Mellon University's AI Software archive for an extensive list of freely available tools alone), offering very powerful techniques used for specific applications. Each of these tools differs in features and, more importantly, in the degree to which the tool adequately matches a developer's task. Similarly, we can use powerful predefined modeling tools to construct an individual-based model (though the range of tools available at this point in time is comparatively much smaller). However, we will spend considerable time fighting the aspects of the tool that do not closely fit the representational and/or computational needs of the desired model.

Alternatively, a programming language can be considered to be a system-building tool, albeit at a very low level. It is certainly more flexible than the tools described above: rather than attempting to fit square pegs into round holes, we can build any computational structure we want. The price to be paid for this is great—this is essentially the case we started out with when justifying the need for flexibility in a modeling tool, and brings us all the difficulties already discussed above. However, these two basic alternatives are a common tradeoff, not only in individual-based modeling and expert systems, but over and over in AI in general. The tradeoff between weak, universal methods and strong, specific methods is well known and discussion on this topic can be found in any basic AI textbook (e.g., Rich and Knight [44]).

The important thing to recognize is that these two alternatives form two ends of a very wide spectrum, as shown in figure 1. While tools with completely predefined components leave little flexibility and thus are difficult to fit to domains that do not perfectly match the tools' original intent, tools that are completely programmable can be fit to any domain but lack the rapid

---

[2] Davis' Law.

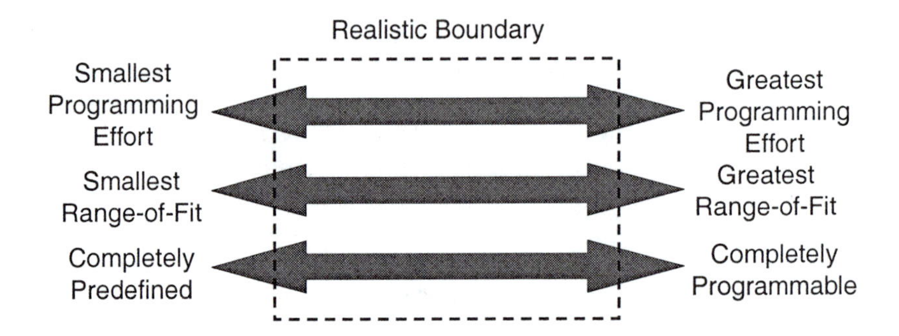

FIGURE 1 Spectrum of choices in a modeling tool.

alteration capabilities required for flexible modeling. While neither of these on its own is a good choice, the optimum lies somewhere between these two poles. Put another way, we would like a tool that has many predefined components, so that many different types of agents and environments can be built relatively quickly, with programmable aspects so that the tool can better fit our needs. Thus, there is a range between a domain-specific tool and a programming language that we would like our overall simulation tool to cover. Translating back to AI terminology, there is a desire to have the advantages of both weak and strong methods.

Our place on the spectrum of programmability directly translates to the amount of effort it will take to construct a model directly using the tool, as well as the likelihood we will be able to successfully construct the model using the tool. By "directly," we mean the amount of effort required that is directed completely toward constructing the model, as opposed to tweaking elements of the tool to try to get it do what we want. To most model builders, who are not primarily computer scientists, programming effort is also important in and of itself. Despite the desire for programmability to achieve the greatest range-of-fit, it is also desirable to minimize the overall programming effort—or at least make the necessary programming as straightforward (intuitive, and more important, standardized in methodologies across tasks) as possible. Thus, on these perspectives there is also a range we would like our tool to cover.

So the question becomes, why not develop a tool that will support the entire spectrum across all these interconnected facets? The answer lies in what current technology can hope to achieve. While having a single tool which may be useful across the entire spectrum depicted in figure 1 is certainly the ultimate goal, being able to develop such a tool would require solving most of the open problems in knowledge representation and reasoning in artificial intelligence, as well many in programming languages and, indeed, in simulation modeling itself. A much more reasonable short-term goal is to develop tools that will cover as wide a range of this breadth as possible, in a practical manner.

The remainder of this chapter describes Gensim: a tool for spatially explicit simulation models incorporating intelligent agents that attempts to encompass much of this breadth. Gensim divides the problem of supporting intelligent agents into the two parts discussed at the beginning of this section: providing support for the integration of agents and their specific needs within a spatially explicit simulation tool, and providing a means to model the computational abilities (decision making, perception) of intelligent agents. Both components attempt to cover as much breadth both in terms of the variety and sophistication of agents supported, the ease of their definition, and the range of the spectrum depicted in figure 1. The two major components of the Gensim approach and the connections between them are described in the following sections.

3 GENSIM AND LOW-LEVEL SUPPORT FOR INTELLIGENT AGENTS

Gensim is a LISP-based object-oriented simulation system designed explicitly around the features necessary to support intelligent agents in spatially explicit simulation environments. The system operates in a uniprocessor environment (though one of our future goals is to use the system as a basis for larger distributed simulations) and supports multiple intelligent agents, each of which may consist of multiple time-shared processes. The system has a concise interface between agents and the rest of the environment, clearly defines all restrictions and assumptions regarding the agent-simulator relationship (e.g., action timing, perception) and provides the ability to control many aspects of the simulation itself. Because the system was originally intended for developing and working with intelligent agent research [4], breadth in agent support was one of the foremost considerations of the system itself. Gensim thus incorporates many modular design aspects, making it relatively straightforward both to modify the features provided with the system and to add particular features required to support a given agent or domain structure.

Rather than simply providing a flexible domain or set of domains, as is done in many AI-based simulators for intelligent agents (e.g., Engleson and Bertani [20]), Gensim provides flexibility in the simulation process itself. That is, rather than providing extensive domain parameters, Gensim provides the features necessary for users to define their own domain and their interfaces with intelligent agents. The system also incorporates facilities for constructing intelligent agents, which will be described in the next section.

A high-level overview of Gensim is illustrated in figure 2. The simulator itself manages the environment, represented as a collection of objects. This environment is spatially explicit and *objective*: it is the universal set from which all intelligent agents' perspectives are defined. The simulator also pos-

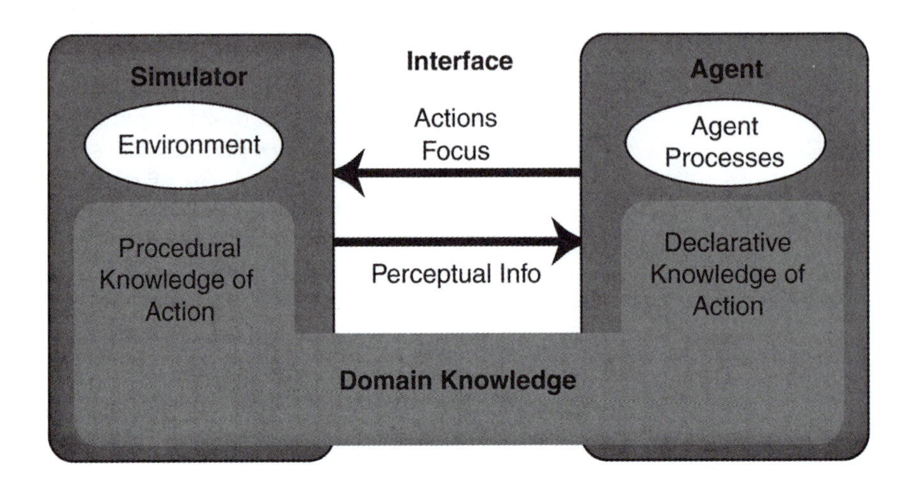

FIGURE 2 Conceptual view of Gensim.

sesses a collection of procedural knowledge describing the actions that agents can perform on these objects, as well as the physical events that can occur to these objects outside of the influence of any agent.

A collection of agents is also defined, each with its own view of the environment based on its sensing ability and memory. The reasoning abilities of an agent are implemented as a set of time-shared processes, which collectively allow the agent to perceive the environment around itself and act on the basis of those perceptions. Agents gather information passively and actively from the environment around them (i.e., from the simulator), and are given limited time periods in which to process that information and commit to actions. After each agent is given a time slice, a representation for the activities of each agent over that shared time slice is obtained and the simulator then manifests the effects of each agents' actions, both short term and long term, on the simulated world. These manifestations include both the effects of (from the agents' points of view) intended and unintended interactions between agent activities.

Agents are viewed as "black boxes" by the simulator, in that Gensim neither knows nor cares about their internals nor how the agents arrive at their decisions for action. As indicated in figure 2, some commonality in domain knowledge is required in order for agents to interact with the simulator. For example, the actions that the agent can select from (represented largely in declarative form) must be identifiable by the simulator, which then uses its own (largely procedural) knowledge of action to physically modify the environment appropriately. Agents also inform the simulator of their interests in objects in the environment (their *focus*) in order to provide appropriate sensory feedback. However, emphasis is on making agents' knowledge distinct from that of the simulator, and the illustration in figure 2 should not be interpreted as implying that agents somehow must physically share the simulator's knowledge.

The basic design goal of Gensim was to support as wide a variety of agent and domain designs as possible. To this end, Gensim was designed to keep agent and simulator knowledge as separate as possible, and thus limit the knowledge each must have of the other. A complete environment is defined for the simulator, and agents are expected to possess their own knowledge of the objects in the domain and how they operate. This is more complex than simply allowing an agent to share the same internal objects manipulated by the simulator, but allows much more flexibility in implementing complex domains. In particular, it allows agents' perspectives to differ, and limits agents' knowledge to that information the designer of the domain wishes them to possess. It also forms an experimental control on the developer: as one has to explicitly cause agents to share global knowledge, one is less likely to take this route with the naïve assumption it will make no difference to the end result of the simulation. On the other hand, there are indeed times when one does legitimately want agents to explicitly share information, and the ability to do so is provided within the simulator. This makes it possible to have agents with varying degrees of autonomy, allowing them to share knowledge of action, knowledge of the world, and even perceptual abilities [9]. It also allows us to declare information and abilities common to classes of agents separately from individuals within those classes. From an individual-based perspective, this in turn allows us to support large numbers of individuals without the work of redefining all the knowledge and abilities of each redundantly.

The simulator manages change through a global queue of events (representing future changes the simulator must manifest) to which all agents contribute. An event in Gensim is a packet of change for a specific period of time (which may be initiated by an intelligent agent, by some other object in the environment, or randomly in the environment itself). The system is clocked, and the length of time represented by an event will always tie directly to the clock-cycle length. Since the clock-cycle length can change, allowing for various levels of granularity desired in a simulation, some occurrence in the domain will generally consist of a series of events as defined here, with the number in that series varying by the length of the clock cycle. Actions of intelligent agents and other sources of change are defined as event generators: any sequence of events over time is represented by an event generator generating an event, which in turn invokes the same or another event generator to continue the series. Thus a movement event across distance involves the initiation of movement by an event generator, with the actual processing of that event by the simulator resulting in the invocation of another event generator to continue the movement across the next time cycle. This brings about two very important results from the point of view of efficiency and realism. First, the length of the simulator's event queue is shorter, in that an ongoing event broken into an event series still has only one element of that series in the queue at once: each element generates the next when it is processed. The latter point also allows realistic interaction of events in a computationally feasible manner: each event can have dynamic effects based on those in the queue at the same time, and once again, the shorter queue means that such interactions can be more efficiently managed. It also means that interactions can be handled efficiently at a finer-grained level than is possible if events were represented using more temporally extensive structures.

This approach also allows great flexibility in timing, in that if the time-cycle length in a simulator is shortened, each event simply represents a correspondingly temporally smaller packet of change. For example, in the movement event above, if the time cycle were shortened, each event simply represents a smaller distance traveled in that time. It does require that the definition event generators take time length into account, but in most events this is a fairly natural thing to do (and indeed, what one would desire for a realistic simulation regardless). The ability to support a variable-length system time cycle in turn brings another important aspect of flexibility to the process of individual-based modeling. The basic cycle length is irrelevant from the point of view of time within the simulation, in that shorter time cycles will simply result in a slower execution of the simulator itself. However, large numbers of sophisticated agents combined with short time cycles may slow the simulator down enough that the overall time of a simulation run would become impractically long. Allowing variance in time cycles supported directly in the representation for change in the simulator itself allows us to tune this level for the number of agents we have: discretizing time within the domain differently, but speeding the overall simulation. For a more fine-grained level of detail, short time cycles can still be used (to the degree one is tolerant of a slower simulation) without changing the definition of the environment itself.

Further details and analysis of this approach to timing may be found in Anderson and Evans [7]. This model of timing is extremely flexible and more than adequate for most ecosystem modeling domains (which tend not to operate on the low level of temporal granularity that AI simulations require). It certainly improves upon the many simple simulators that manage time at the level of whole actions rather than a resource unto itself (e.g., Agre and Horswill [3]). It can be criticized, however, on one significant point: Since the processes of an agent (and indeed, the agents themselves) are serially time-shared, they cannot simulate some of the interactions possible in truly parallel processes. For example, an agent might consist of two processes: one to recognize objects in which it has an interest, and another to decide what to do based on the objects it has seen. When these run in parallel, the acting process would process input from the recognition process as it became available. This is a much more complex interaction than that possible under time-sharing in a uniprocessor system, where the recognition process must recognize all the objects it can in a given interval and then pass the entire collection to the acting process at once. This is a legitimate criticism, but like the accuracy of a sensory interval, where a short enough interval becomes equivalent to the agent to continuous data flow, the effects of this depend entirely on the interval size. Like other applications of time-sharing, the smaller the time slices involved, the more transparent the time-sharing will appear.

The timing model which Gensim is designed around is, at a higher level, representative of the modeling approach adopted by the system in general. Referring back to figure 1, this approach attempts to take in most of what would fall within the realistic boundaries of expectations of a simulation tool. The model is flexible, straightforward to modify, and encompasses most of what would be needed for the majority of modeling scenarios. Situations such as the one described above can be identified that do not fit the model well. However, including them in Gensim's model of timing would result in a system that is only marginally more capable than it would be otherwise, at a cost of a significant decrease in ease of use and short-term flexibility in general.

A Gensim environment is managed in an object-oriented fashion. When an event is processed, for example, messages are sent to the objects involved, which in turn evoke changes to the environment, and possibly new events. Say, for example, a particular agent performs a throw action on a particular object that is part of the environment. That decision would be based on the agent's declarative knowledge of the object that it's throwing, what its throw action does, and the other overall knowledge the agent has that motivates it to perform this action. That fact is communicated to the simulator by the agent during its decision-making time slice. When the simulator process runs, the initial changes are made by sending a throw message to the object being thrown (the simulator's version of this object), which contains knowledge dictating the objects' behavior (e.g., how fast it would travel, whether the trajectory could change spontaneously). A series of traveling events for the object would ensure over time, possibly interacting with other events in the queue at the same time. For example, if two objects were traveling and entered the same physical space, a hit event could be generated for each. Once again, the simulator could send this message to each object, resulting in the manifestation of change in the environment based on the physics defined. A more detailed description of this overall process may be found in Anderson and Evans [7].

In addition to performing actions, agents also interact with the simulated world through perception. Perception is the most difficult design aspect of any simulator for intelligent agents: on one hand, it is a crucial aspect of virtually any domain while, on the other, simplification of perception is in AI research one of the primary motivators for using a simulator, and the greater the simplifying assumptions, the smaller the range of agents we can accurately support.

The perceptual abilities of intelligent agents are much more active processes than might first be assumed. We do not simply see whatever our eyes happen to be directed toward, for example: we pick and choose from the image, discerning what interests us. What we have already seen also alters further perception, helping to define a focus for perceptual efforts. In more computational terms, a set of *anticipatory schema* (domain-specific knowledge indicating the sensory information in which the agent has an interest) acts as a filter for the vast amount of knowledge available from the world. This knowledge directs the agent to explore the world, looking for given pieces of information. This exploration directs the agent to specific information (objects) in the environment, which then modifies the anticipations that the agent has for future sensations [39]. Within this cycle, one of the basic problems of implementation is that an agent must be allowed to specify its interest in given aspects of the environment, but must also be given access to information independent of those interests. Perception must exist to confirm the agent's expectations of the world, but also must provide the agent with new information (not necessarily what it expects to see or is directed toward) in order to form the expectations of the world that guide its sensory abilities.

Gensim attempts to follow this model as closely as is practical. However, some aspects of perception are simplified, in order to simplify the simulator itself and to conform to one of the basic design goals of Gensim: a simple interface between agents and the simulated environment. One of the major simplifications concerns the internal processing within this cycle. Rather than beginning with low-level vision, Gensim operates at the object level. That is, an agent senses a combination of complete objects and specific sensory aspects of those objects (e.g., a ball, or the fact that it is red or round), rather than examining the individual edges and features that make up an image of the object itself. Agents perceive objects through a symbolic description of the visible aspects of the object. They make sensory requests by describing the direction which they want to look in, or attributes of the objects which they wish to look at. A bandwidth can be set for sensory information in each agent, limiting the number of objects that the agent can focus on and the amount of information that can be received at once. To go lower than the object-attribute level would be outside the realm of most ecological modeling work, as well as most work in AI from the agent perspective, and would immensely overcomplicate the process of defining and working with a domain from the modeler's point of view.

The ability of an agent to express a focus for its sensory abilities is provided in Gensim through a sensory request mechanism. This mechanism allows an agent to express interest in certain aspects of its environment, and the methods by which Gensim fulfils sensory requests allows the agent to receive limited sensory information not only about its focus of interest, but also about objects outside that focus. An agent makes a sensory request for some particular set of information, and the simulator records this request. After all agent processes have been run on a particular cycle, the simulator updates the environment according to the actions and events that have occurred during the interval, and then prepares sensory information based on each agent's requests. Gensim has defined sensory actions that allow an agent to look in a given direction or angle in the environment, and to look for objects matching particular templates.

Like physical actions, sensory requests form the interface between the agent and the simulator: they are the agreed-upon vocabulary that the agent will use to interact with its environment. As such, like actions, part of the definition must lie within the agent (what it thinks the action will do), and part within the environment (the actual effects of the agent performing that action). Thus, when defining the domain, the physical results of a sensory request must be defined for an agent. That is, what kind of objects or attributes an agent will be biased to see over others: allowing larger objects to be seen first, for example, or decreasing the accuracy of sensing (the likelihood that particular attributes are perceived) as distance increases. This sounds time consuming, but is in fact quite simple compared to many other aspects of domain definition. All perception is a filtration process, here selecting a bandwidth-bound group of object attributes from the simulator's knowledge base. A simple set of rules can easily bias an agent toward one type of information, and many of these rules are similar from one perceptual action to another and can simply be duplicated (the angle representing an agent's perceptual boundaries, for example). Moreover, once again in keeping with the idea of supporting a broad range of simulations, this level of detail can simply be avoided entirely by increasing an agent's bandwidth and not biasing the agent away from receiving any particular attributes of an object over others. That is, limited, biased perspective can be easily avoided for the many simulation environments in the spectrum of figure 1 in which it is deemed unnecessary.

Defining a domain in Gensim involves constructing the objective environment that the agents inhabit, the knowledge base for each agent, and the behavioral mechanisms (the agent design) of each agent. Once again, Gensim attempts to take a middle-of-the-road approach to the spectrum of inflexible predefined elements vs. complete programmability as shown in figure 1. Many elements of an environment are defined using simple macros, making construction of a simulation environment fairly straightforward. For example, a defaction macro defines an action, names it, and defines a procedure describing how the action manipulates the object it uses. Similar macros are used to define potential events, as well as to define agents inhabiting the environment. One still must program the physical effects of the action; however, most aspects of this are common to many actions (transferring possession, creating an object, destroying an object, etc.) and have library elements that can be used to eliminate at least some of the low-level programming involved.

While macros define the objects that exist in any particular domain, other aspects of a domain are set using parameters. For example, a single parameter specifies the number of dimensions in a domain. The user can then define a function that maps the domain out in a regular mathematical pattern such as a grid (most common mappings for simple structures such as grid-based domains are provided), or as a collection of symbolically named locations. The latter allows domains where areas cannot be mapped into a regular grid structure: significant locations can be grouped and designated with a symbol, allowing an irregularly structured area as a single location. This is similar in functionality to the use of geographic information system (GIS) information by Folse et al. [23]: related areas (e.g., a single location with similar vegetation or soil structure) are grouped into a single area that is treated as a unit for simulation purposes.

As has been stated throughout this section, the major features of Gensim are all designed toward making the simulator available relatively easily to a very broad range of environments. While lines have been drawn in terms of facilities not available to intelligent agents associated with the simulator (e.g., parallelism, as mentioned above), in general the reasonably clean, extendable interface between simulator and agent allows for an extremely broad range of potential agents. While not directly interfaced with any commercial GIS tool, the general spatial organization underlying a Gensim environment is not unacceptably far removed from the approach used by such tools.

Gensim also directly supports the management of computational resources in individual-based models in several ways, the most significant of which arise from the system's object-oriented nature. Each intelligent agent in a Gensim environment belongs to a class that describes not only shared attributes, but shared abilities. It is thus possible for many different agents to share perceptual abilities, effectory abilities, and knowledge of the world. Classes are organized hierarchically, and so at any point lower-level classes or individuals can redefine knowledge or abilities assumed by higher-level classes. Moreover, the same code may also be used to make decisions for many agents at once. The system thus directly supports a wide variation of individual abilities and knowledge with minimal redundancy and greater conservation of computational resources. Gensim can also vary the time each agent is allowed to make its decisions, giving agents (or classes of agents) intellectual abilities relatively higher than others. Finally, these mechanisms are generic enough to easily support a broad range of agent abilities within a single model.

When outlining the basic requirements for incorporation of intelligent agents in a spatially explicit simulation (sec. 2), we began with a rough division into two major component groups: support for the design of agents themselves, and support for their realistic interaction in a simulated environment. Throughout this section, we have shown the applicability of many of the features of Gensim to the latter problem. However, considering the breadth of agents desired and the desired range of breadth necessary for the tool itself, the former remains a significant problem.

The problem is certainly not that intelligent agents cannot be employed using only the features of Gensim described thus far. Within any Gensim simulation, an agent can be any collection of LISP functions (simply because the simulator itself is currently implemented in LISP, a holdover from Gensim's origins as a testbed for intelligent agent research). This means that if one so desires, any agent design that can operate by describing actions to the simulator in the manner that Gensim expects, requesting and accepting perceptual information in the manner described above, and adhering to the Gensim timing model can be programmed for the system.

While several general classes of agents have been developed for use with the Gensim system, forcing modelers to work with the simulator in this fashion is contrary to the breadth desired of the simulator itself. By programming agents individually from scratch, we are back to the potential paradigm shifts described in section 2 at another level: a slight change in an agent's design or purpose requires potentially extensive reimplementation. What is required to deal with this is a framework for agents themselves that provides benefits at the agent implementation level analogous to those the features of Gensim bring to the agent-support level. That is, a framework that supports within a single basic design the ability to define a broad range of agents capturing a variety of degrees of intelligence; the ability to modify those agents quickly; and the ability to alleviate some of the difficulties of declaring and working with large numbers of agents. Such a framework is described in the following section.

4 IMPROVISATION AND INTELLIGENT AGENTS

The approach to agent design used to provide the breadth of agents in Gensim arose out of a study [4] attempting to explain and duplicate some of the breadth of ability displayed in everyday human behavior. This work created a model that, while certainly not claiming to implement and display the full spectrum of human behavior, implements techniques that permit a wide range of adaptable behavior, in a manner that can be employed to deal with other forms of breadth as well. To see the utility of the approach in supporting a broad range of agent types in an individual-based simulation, we must begin with the origins of the approach in extending traditional approaches used to implement agents.

Consider the range of agents required in a sophisticated individual-based model designed to examine human impact on an ecosystem of interest. We need to model the abilities of humans: the activities consistent with their motivations for the use of the ecosystem, and the mental processes underlying their decisions (including, for example, each individual's collective previous experiences in this environment). We might also have to represent a range of lesser creatures, with tendencies to particular behaviors, patterns of interaction with resources within the ecosystem, and to some degree the ability to make decisions on their actions. This ability can range from advanced agents that can employ some of the same intellectual resources as humans (again, such as previous experience within the environment), down to those that can operate only at a purely mechanical, stimulus-response level. Once again, this is what causes the difficulty in supporting intelligent agents within individual-based simulations: there is an enormous variance in features required for these various implementations.

The interesting thing about this breadth is that it is also observable within the course of the activity of a human agent alone. While activities such as working on an assembly line (where each potential problem has been experienced many times and there is a limited number of possible interactions) fit a purely reactive model nicely, as does any activity with which we are completely familiar, and which has fairly distinct boundaries around it (in that it interacts in only a limited fashion with other activities). This is one end of the spectrum of human activity. In virtually any human activity, there will be aspects or components of the activity with enough structure (experience on the part of the agent and physical structure in the environment [29, 41]) to support a compiled collection of responses. During the course of an activity such as preparing a meal, for example, much is routine in a general sense despite the complexity of the environment: we can immediately recall a routine that has been put together over the course of many previous episodes of behavior (compiled plan or routine knowledge). Portions of this routine with which we are extremely familiar may be compiled to the point of a complete collection of reactions. Conversely, there will also be a large component of any complex activity that cannot hope to have this kind of structure: anywhere where every possible contingency cannot be anticipated. This includes performing an activity with which we are not *completely* familiar, performing the activity in conjunction with others in the short or long term, or performing an activity in a different situation than usual [4].

In order to apply a routine effectively and flexibly in the face of greater variability than can be completely anticipated, we possess a vast collection of more general knowledge that allows us to integrate alternatives seamlessly with our routine. We divert from our routine when it makes sense to do so, and return to it without anything like the kind of effort known to be required (e.g., to alter a stored symbolic plan). We can also use our routine as a weaker guide in conjunction with background knowledge to cope in a satisfying[3] manner with even greater degrees of variation. For example, one can shop reasonably successfully even when in a hurry and in a strange supermarket; can prepare a meal easily in a friend's kitchen; and can sharpen a pencil without a great deal of intellectual work even if no sharpener is available. We commonly call the methodology behind such efforts *improvisation.*

Improvisation as creating minor to extensive variations on a routine in a satisfying manner in real time occurs in the vast majority of human behavior, from theatre and music to cooking, driving, and architecture [35]. In the realm of intelligent agents, the term has been used previously, most notably Agre's [1] definition of *continually redeciding what to do.* This use, however, leaves out

[3]Simon [48, p. 259] coined this common AI term to describe the behavior of humans and other organisms; as opposed to the ideal of optimizing behavioral strategies, organisms adapt well enough to *satisfy*, or obtain a "good enough" reward to suit their efforts.

much of what improvisation, as described above, embodies—most notably a basis upon which to improvise.[4]

This basis is the collection of routines through which we normally accomplish our activities and which are constructed through many episodes of similar activity. During the course of improvisation, this compiled knowledge represents a resource that we rely upon to reduce the intellectual effort that would normally be associated with complex activities, in order to perform in a timely manner. We can and do rely strongly on this resource in cases where the current situation follows our previous experience. In situations where our previous experience differs, we can use our routine as a weaker resource. That is, we can follow it as closely as possible (for economic reasons), improvising to obtain alternative actions where the routine is not appropriate by examining the associated background knowledge to the degree the situation warrants. The more novelties there are in a given situation, the less directly the routine can be used, and the more search is required.

This process of improvising on a routine can also occur when one wants to reason beyond the routine aspects that normally constrain our reasoning: when one wants to do better than one's normal routine, for example, or when one wants to come up with creative new solutions using the resources at hand as opposed to those commonly associated with the activity. Improvisation is a naturally satisfying process, allowing the agent to follow its routine and obtain immediate possibilities for action, or to devote as much time as is available or as the agent deems necessary for the task at hand to reason more deeply about alternatives for action.

The range of performance when employing improvisation correlates directly to the wide variety of agents we require in individual-based models. At its most sophisticated, we can have extensive routines for activity, including rich and diverse background knowledge, allowing the implementation of agents representing human-level adaptability in an environment. Yet in the same way that a human might be given less adequate routines or less of an efficient ability to make use of them (e.g., a very limited response time in decision making, a wide range of tasks requiring attention to be divided, etc.), defining less intelligent agents involves applying simpler routines in a simpler manner. This can be extended directly down to routines consisting only of low-level stimulus-response information for the most primitive agents. The end result is a *single* framework that can be used to define all agents irrespective of their abilities. This is one more example of the "middle of the road" approach to modeling within Gensim: by providing a single framework, we ease the burden of implementation required on the part of the modeler while still maintaining the breadth of models suitable to the tool.

Computationally, improvisation has several requirements: control over the extent to which background information is explored, both with regard

[4]Others, most notably Agre and Horswill [3] and Hayes-Roth et al. [31] (and more recent Stanford KSL work), have previously used the term *improvisation* in a more general (and differing) sense than that defined here. See Anderson [4] for further discussion.

to the time spent on any one decision for action and the extent to which such exploration is deemed more valuable than the agent's routine response; dealing with limitations on the amount of background information that can be considered at once, and organizing this information such that the most valuable information can be examined first; integration of multiple goals; and use of limited perception to recognize new opportunities in light of the agent's intended activities, to name a few.

Our approach to improvisation relies on representing and reasoning explicitly about these aspects of control using constraints, a form of knowledge representation mechanism based on restrictions. A constraint can represent what is and is not permitted, and to what degree, in particular situations. This naturally encompasses most of the knowledge used by an agent in decision making. Individual pieces of knowledge about the domain (what one has done before, specific warning signs, and so on) drive the agent toward or away from specific choices for action. These kinds of information are also hierarchical: we have "Don't Do X" constraints, which we can employ *prima facie*, but which themselves are based internally on the many more detailed constraints that originally led us to decide (or someone else to teach us) that action X was not a very good idea. These background constraints exist but remain largely dormant unless we need to enquire why action X is not appropriate. This naturally fits into the improvisational model: we would be performing such inquiries (outside of idle curiosity) when we need to reason more about action X than we would normally care to. That is, when the situation is more important, where action X has potential interactions that do not usually occur, or when our choice between X and other potential actions is not clear.

Similarly, the means of control of improvisation itself is also naturally constraint directed. How far to explore, when an answer is good enough, perceptual focus, and so on, can all be viewed as explicit constraints that describe the state of the agent's decision-making process and evolve as activity progresses. The process of arriving at a choice for action thus becomes a constraint-guided search process [24], where we consider the constraints on an agent's activity in a satisfying manner, controlled ultimately by other constraints on how far the process can go.

Our approach to improvisation, *constraint-directed improvising agents*, relies on a representation based on constraints combined with specialized internal mechanisms for processing constraints in this manner. The sophistication involved in this may sound like somewhat of an overload for something like a simple stimulus-response agent: however, stimulus-response bounds are also constraints at an atomic level, with very little above them. That is, such agents represent the absolute endpoint of simplicity within this model, and simply do not use much of the power (or require as much of the effort to define) as the model is nearly capable of. The agent model as a whole supports both these simplistic level intelligences, and can also be used to describe the vast majority of human behavior.

A constraint-directed improvising agent's compiled routine knowledge and background knowledge are incorporated into distributed constraint-based knowledge structures known as *intentions*,[5] essentially representing the agent's resources for performing some activity in both routine and nonroutine manners. A conceptual view of an intention is depicted in figure 3. In general, an intention consists partly of the agent's routine knowledge of a particular activity; that is, a general description of how a normative instance of the activity should proceed. This description is made up of the constraints that the activity places on the agent's behavior (for example, preferences for actions or further intentions, or requirements for the activity). An intention also contains links to the background knowledge out of which the agent's compiled routine has evolved. This knowledge also consists largely of constraints, and represents knowledge behind the preferences and other constraints that make up the agent's routine. This division allows the agent's routine to make suggestions for activity immediately where appropriate, and background information to do the same using a search process whose length will vary depending on how closely associated particular pieces of background knowledge are to the agent's routine. This in turn allows the agent access to immediate responses that are useful in the typical case of the activity, and the ability to search and deliberate as time and the significance of the situation permit.

Considered in more detail, the compiled knowledge representing the agent's normative routine in figure 3 is divided into ordered logical units which can conceptually be thought of as steps in the routine or plan. These may simply be connections to more specific intentions (just as "travel to location X" might be one step in an overall food-gathering routine, yet a whole intention with many lower-level activities in its own right) with constraints supplied to tie the intention to the current context, or direct recommendations for immediate actions. The background concepts are organized in a hierarchy of relationships with one another (taxonomic, geographic, component, etc.) and include information about the setting for this behavior and how it affects the routine, the objects required for its performance and how they contribute, and so on. Literally, this information is the collection out of which the routine itself would have been built as part of a learning process (and thus these background concepts have direct connections to the various elements of the compiled portion of the intention). These background concepts are also shared between a great many other intentions which an agent might have.

Constraints from both the routine itself and the background knowledge associated with it (to the degree the agent chooses to explore the latter) direct the agent toward individual actions over time. Actions may include adopting further intentions, thus changing the milieu of constraints that form a basis for the agent's moment-by-moment choices for action and the overall context

[5]The use of the term *intention* here arises from Boden's [12] use of the term, to indicate representation of both an agent's goals and an overall means for achieving activity, as opposed to a more basic plan structure. The term is also used in other systems, most notably that of Bratman et al. [14].

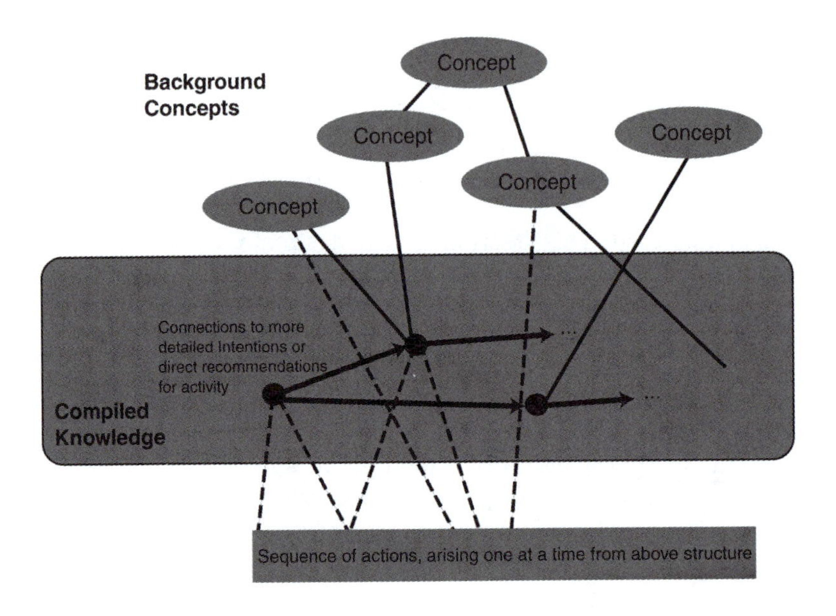

FIGURE 3 Simple example of an intention.

in which any adopted intention is considered. A series of actions thus emerges over time as a result of the initial adoption of an intention in conjunction with others adopted at the same time and independent events that occur in the environment as the intention unfolds. The nature of this process will be explained in much more detail shortly.

As a specific example of an intention, consider a sophisticated human activity such as preparing a common food or drink, such as coffee. A human would have a routine for doing this particular activity, assuming it was indeed a routine activity to that person, and that activity would have arisen out of (and be connected to) many common concepts: kitchens, coffee, coffeemakers, spoons, cups, water, and so on. If the individual in question was in his or her own home with familiar utensils, the steps in the compiled routine could be used more or less directly (barring unforeseen circumstances arising in the environment, such as dropping utensils and so on). If the agent was performing this activity in a different locale, however, or with tools of a different nature, the background concepts would become much more important (in proportion to the degree the current situation differed from the agent's routine). Some direct steps would still be applicable at certain times, but new ones would have to be recommended (based on, for example, differences in how coffee-making equipment operated in the new locale).

Intentions such as this particular example are, in general, much more complex than would usually be required for an ecosystem model. The intentions of human agents are not likely to be this detailed in an ecosystem model (nor, in fact, will the model itself likely be at this level of detail), and those of lesser agents are likely to be concerned with much more general activities: food gathering, following paths to shelter, hiding behavior, etc. [8]. However, intentions of the complexity of the above example have indeed been implemented [4, 5, 6] and applied in real-time environments. In the case of applying this mechanism to ecosystem modeling, it is simply easier to define such structures for simpler situations. The complex example above is illustrated to more adequately illustrate the nature of intentions than a simpler example might, and to show the sophistication of which this representation structure is capable.

Constraints in intentions represent regularities in the world around the agent and the influence of those regularities on agent behavior. Despite the wide range of knowledge being represented, we have found as have others (e.g., Fox [24] and Evans et al. [22]) that only a reasonably small number of distinct types of constraints are needed. Constraints are used in this approach to represent a broad range of concepts, from physical restrictions (*physical* constraints) and relationships between entities (*expectation* constraints, *requirement* constraints, and *temporal* constraints such as the ordering between routine steps depicted in figure 3) to abstract policies (*behavioral goals* and *preference* constraints), to representing normative responses to particular situations and control of internal agent components (*normative* and *focus* constraints) [4]. Each of these types has similar components, including a specified active lifetime, activation requirements, and an attachment to some larger knowledge structure. Constraints are organized in a loose hierarchy, abstracting both knowledge and control, and perform different functions depending on the level at which they are defined.

The use of constraints as the primary knowledge representation mechanism directly supports the ability of a improvising agent to perform in real time and to perform in a satisfying manner with the knowledge at its disposal. In an intention such as the one shown in figure 3, for example, the compiled routine may contain among other things a constraint indicating that the agent should prefer working with a standard drip coffeemaker as opposed to some other tool (an old-style percolator, or the microwave and instant coffee). This constraint expresses a preference that is normally applied in the course of the activity with no exploration as to the reasons behind the preference. When this tool is unavailable or the agent wishes to reason beyond the routine (due to error, knowledge of potential error, or to high-level constraints that affect how an agent performs an activity), the agent can make use of further constraints behind that preference (background knowledge) that describe the role and function of the coffeemaker in the overall routine. The agent can then use those constraints as a basis for reasoning about alternative ways of performing the activity, to the degree the agent wishes to devote intellectual effort to this. For example, constraints about making will lead the agent to a set of objects with characteristics suitable for this purpose (the tools mentioned previously or further improvised versions of these based on the functionality of these alternatives). Similarly an agent can prefer one path to another, or one form of prey to another, with more detailed knowledge representing those preferences still available to consult if the situation warrants.

Because constraints are modular, constraints external to an intention (e.g., from another, or more global constraints associated with the state in which the agent finds itself) can also have immediate effects on it. The presence of a constraint such as hurrying (brought on by a combination of intentions or some external event) may affect it in certain predictable ways that are part of the routine itself. That is, the presence of a *hurry* constraint from outside the intention may allow certain routine components to become active that would be otherwise ignored. Such a constraint will also affect the agent itself: how much information the agent considers, strategies for deliberation, automatic preference choices, etc.

In any significant domain, there will clearly be a large number of potentially relevant constraints available for an agent's consideration at any point in time.[6] However, the agent's cognitive effort in this approach is for the most part not spent on looking for constraint violations, as in most constraint directed reasoning systems. While we are concerned about violations in some cases (e.g., expectations), here most constraints act positively: their presence compels the agent toward or away from specific courses of reasoning or activity, just as the landscape influences the direction of one's travel. The key to real-time performance is the selective processing of constraints in order to make satisfying decisions in the time available. This is done through the multilevel organization of constraints in intentions, in tandem with facilities provided in the agent architecture for limiting the number of constraints considered.

The architecture of a constraint-directed improvising agent is shown in figure 4. The agent itself is divided into several computational processes (which would be time-shared by Gensim) and two major stores of knowledge. The agent's *long-term memory* contains all the possible intention and conceptual knowledge possessed by the agent. The central role in this architecture, however, is played by the agent's *working memory*, which directly supports the selective recall and processing of constraint knowledge over time. Working memory is of limited size, and represents the amount of information that the agent can process in parallel (in our implementation a time-shared simulator is used, and the size of working memory represents the amount of information the agent can process in a time slice). Any constraint in working memory is thus viewed as having immediate effects on reasoning or behavior, and relations among concepts in working memory are assumed to be immediately

[6]Once again, given much simpler intentions than that depicted in figure 3 this may not be significant, but this point must still be dealt with because of the potential for building very sophisticated agents.

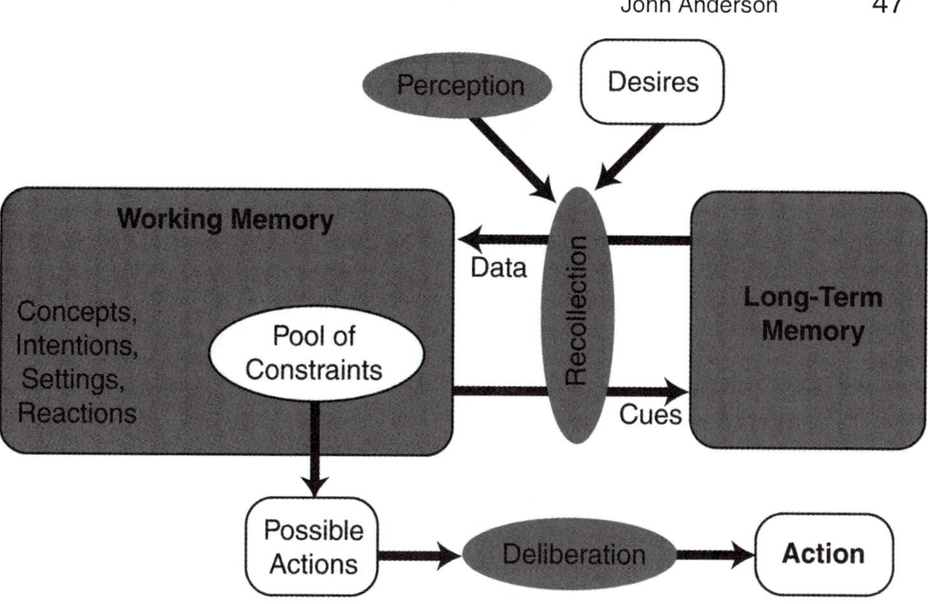

FIGURE 4 Computational aspects of constraint-directed improvisation.

realizable (via direct memory connections). These are not unrealistic assumptions, since the number of concepts or intentions physically allowed in working memory at the same time can always be set to a small enough limit to make this so.

Items migrate from long-term to working memory through the use of memory triggers. Each object, concept, or intention may have simple trigger conditions associated with it that, when satisfied, allow this item to be brought into working memory. Trigger conditions may be satisfied through perception, through new desires, or through objects already present in working memory. Trigger conditions are only considered when present in working memory, and are brought into working memory when adjacent or connected concepts are present in working memory [5]. This directly supports the ability to gradually move to more detailed background knowledge over time.

At any point, the collection of constraints in the agent's working memory will give rise to a particular set of alternatives for action, which the agent can deliberate over. The more constraints available to working memory, the fewer alternatives need be deliberated upon, and hence deliberation is secondary to working memory in this approach. However, because working memory is very limited, the agent will often have only an incomplete picture of its overall choices and have to decide if further exploration is warranted, and so deliberation is still significant. Deliberation in this architecture is dealt with using constraints on the amount of cognitive effort that can be directed toward a particular end. This illustrates the most significant aspect of this approach:

not only do constraints represent the structure of an agent's activities, they also control the architecture itself.

Individual constraints in working memory influence the agent toward (or away from) some alternative for action through a quantitative or qualitative measure of utility. This measure is composed of an innate estimate of the constraint's importance, modified by the importance of the chain of intentions through which the constraint has come into working memory and by specific conditions the constraint is concerned with. At any time, a given number of constraints (those that have passed through working memory) will have contributed their utility to particular alternatives, and a given number (those attached to relevant concepts not in working memory) will not have contributed. In order to make decisions using incomplete information such as this, the agent has in its working memory a constraint on utility, representing the minimum utility necessary for an alternative to be acted upon. This constraint may be global or local to a particular activity the agent is performing, and serves to limit deliberation by allowing the agent to select the most important alternatives. It is not static, and can be affected by intentions and concepts in working memory as well as higher-level knowledge. For example, when conflicting intentions are present, high-level knowledge may alter the agent's utility constraint to be higher so that the agent has greater confidence that it is selecting the right action in these situations. If no action can be performed, the agent may wait for more information to be processed through working memory. This affords the possibility of a higher utility for one or more alternatives, or for constraints in working memory to alter the agent's utility constraint, allowing less highly ranked alternatives to be selected. It also presents the possibility that the agent could miss some time-constrained opportunity. Utility may also be used in conjunction with activity-specific measures, as described in Anderson [4, 5]. Detailed examples of the use of this control scheme may also be found in these works.

Constraints are also used to control many other aspects of this architecture, such as memory management. The most appropriate memory management policy varies with the situation (e.g., if the agent is in a time-constrained situation with regard to some task, working memory should be biased toward concepts contributing to this task). Constraints associated with intentions may suggest an appropriate policy, or more likely, higher-level constraints (associated with the setting or more general background knowledge) will recognize when specific policies are required. Constraints also define a perceptual focus for the agent, and can also focus the agent toward retrieving particular concepts from long-term memory and thus follow a particular line of reasoning. Further details on the constraint-directed reasoning mechanisms internal to these agents may be found in Anderson [4, 5].

At a surface level, this approach is most immediately compared to that of Bratman et al. [14] and more recent derivatives of its use of intentions and the ability to derive new options and deliberate about them. There are several significant differences. First, intentions in this approach are much more sophisticated than those used by Bratman et al. A constraint-directed improvising agent's intentions serve as active guides: they contribute alternatives for action and connections to background knowledge that can indirectly supply alternatives, as opposed to influencing activity through a vague sense of commitment to the intention itself. This architecture also provides specific methods for generating and dealing with options as a core of the architecture itself, rather than leaving these and other components as implementation-dependent details. Finally, relying heavily on constraints as a knowledge representation mechanism allows an agent to reason as deeply or shallowly about a particular situation as time allows. Elements of this approach can also be compared to other work in real-time planning, including work in partial universal planning [19], case-based activity [30], and others. The interested reader is directed to Anderson [4] for much more detailed comparisons and analysis.

5 DISCUSSION

While it is possible (though certainly not easy) to interface a running program with a simulation tool and thereby support agent decision making in an individual-based model, Gensim is one of very few simulators that attempts to provide sophisticated agent-definition facilities within the tool itself. More important than making the development process user-friendly however, the previous section has shown the constraint-directed improvisation approach to agent design incorporated within Gensim to be an integral part of supporting a breadth of agents using a single simulation tool. While it is possible to develop very elaborate reactive/deliberative agents for inclusion in individual-based models using this approach (real-time agents have been developed for simulated cooking and other everyday human domains [4, 5, 6]), it is equally viable to use the same approach to quickly produce agents with scaled-down intelligences for simpler environments or to represent individuals of lesser intellect.[7] Moreover, the same approach can be used to define a broad range of agents within the same simulation.

The constraint-directed improvisation approach in tandem with the basic organization of Gensim itself also bring to bear several important methods for working effectively under the computational restraints associated with complex individual-based models. Because of the distributed nature of the intentions that form a basis for improvising agents, and the facilities in Gensim itself for sharing knowledge between agents, it is possible for agents to have shared common concepts in intentions, analogous to the vast shared cultural knowledge that humans possess. This not only allows agents to share some common social aspects, but from the point of view of implementation also allows us to define shared knowledge once only, thus greatly reducing the overhead associated with large numbers of agents. Because of the object-oriented

[7] Anderson and Evans [8] detailed some simple experiments on improvising agents vs. strict planning and reactive agents in an ecosystem model.

nature of agents in Gensim, a hierarchy of agents can be defined with varying amounts of shared knowledge. This in no way compromises the realism of the simulation: within the improvisational model, agents can be constructed to improvise on the same routines in very different manners simply by changing constraints associated with individual agents. That is, even though a large portion of intentions are shared, because intentions have distributed aspects it is still possible to define differences that alter individual behavior radically. Given some complex, knowledge-intensive behavior, the behavior itself can be represented once at the class or superclass level, and a few simple constraint changes can bias one agent to some aspects of the particular behavior, another agent to others.

The real power to support agent breadth is derived directly from the constraint-directed approach employed in the agent design. Beyond what is defined in intentions, constraints also completely control each agent's internal mechanisms. This means that a few simple constraint variations between one agent and another can provide enormous individual variations in both intellectual ability and behavior. Two agents sharing routines, for example, can have different sets of perceptual capacities, different abilities or inclinations to explore background knowledge, and so on. This translates into broad variations in individuals with less effort on the part of the developer and less demand on computational resources.

Having expounded on the abilities of this system to support a broad range of agents and a broad range of simulation scenarios in general, it is important to emphasize that despite these developments, Gensim still cannot directly support *every* agent or *every* simulation that could be imagined. While Gensim strives to cover the breadth of simulations and agents depicted in figure 1, it cannot support every simulation while remaining inside the boundaries of realistic expectations on users depicted in that same figure. To be accurate, there is one exception to this point: Gensim could actually be pushed directly to the extreme of programmability, beyond what we have examined in this chapter. Given that Gensim is written in LISP, a traditionally interpreted language, its source code is not separated from domains being designed for the simulator as compiled code would be, and a user could indeed reprogram any aspect of the simulator to handle anything that is not already handled. Whether it would be practical to do so is another matter—this would be moving directly to the right-hand side of rest of the spectrum of figure 2, and building a simulation from scratch (albeit beginning with the bulk of the source code that would be required). While possible, as discussed in section 2, this is not desirable from the points of view of control of experimentation and ongoing changes in agent design.

With regard to restrictions on what Gensim can adequately support, there are several assumptions made regarding the nature of agents in Gensim. As described in the previous section, a Gensim agent must still assume the same timing model as the simulator, and for simulations to operate at various levels of granularity (another important provision for computational feasibility in large simulation models), actions must be defined in a time-independent way. However, as has been previously discussed, these features are necessary and are implemented in such a way as to maximize breadth despite their necessity.

Beyond this, there is one more significant assumption regarding the nature of agents in Gensim. That is, agents as inhabitants of a simulated environment must describe their actions (and their perceptual focus) in a manner that the simulator can comprehend in order to manifest change and/or supply the appropriate perceptual focus. For example, if an agent picks up an object, it must somehow inform the simulator which object the action has been performed with; if the agent wants more information about "that thing over there," it must describe the object as best it can, in order for the simulator to differentiate it from other objects which the agent could be referring to and thus respond with the correct information.

This requirement limits agents to an understood grammar and vocabulary of communication with the simulator, and thus places a design assumption on the agents themselves. However, this is no more a problem in Gensim than it would be in any other simulation system, because it is a direct consequence of the simulation problem itself. Communication is one aspect of the relationship between an agent and the real world that can never be completely approximated by any simulator. The reason for this is that no communication takes place when an agent interacts with the real world (at least in the sense that communication is normally thought of). When an agent wants to pick up an object, for example, it just does so: it doesn't have to communicate this fact to the world. This is in part because the world does not exist for the benefit of an agent, the way a simulator does: any agent is simply an object in the world, like any other physical object. It is also due to the fact that the real world keeps track of itself: an agent's actions physically alter objects, rather than indirectly manipulating some virtual representation of those objects. A simulator, on the other hand, keeps a detailed representation of every object in the world, and changes in the world are manifested by alterations in these abstract entities. Because it no longer occurs naturally, change intended by agents must be communicated to the simulator in some way. A stronger tie between the agent and the rest of the world is thus necessary.

However, in most simulators, this tie, and the constraints on agent design it represents, is far too strong. Many simulators do not even use separate representations of objects for agent and simulator: The identical physical chunk of knowledge that describes an object to the simulator also describes it to any agent. When compared to the real world, this is like taking each object in a physical environment, labeling it with a unique agreed-upon symbol, and having everyone reference everything in that fashion.

Like its other facilities, Gensim attempts to take an approach to this problem that creates a compromise between breadth and ease of use. The system provides both an indexical-functional or deictic [1] means of describing objects, and also an objective one. The system's deictic representation allows an agent to construct a description from its own perspective in a communication language (something might become, in translation, the-animal-in-front-of-me, or the large-object-to-my-right, for example) which is then parsed by the simulator to determine the object desired. This is realistic from the point of view of human representation [1], but once again is more sophisticated than will often be needed. More importantly, if real-time aspects of the domain are of interest, the time required to construct such descriptions may significantly affect the agent. The objective method, where an agent simply specifies its internal designation for the object (some random symbol name), and the simulator then examines the agent's internal representation to match the object to one in the environment, suffices in these cases. From the point of view of accurate simulation, the latter is still suitable: in no case does an agent assume knowledge of the objective world over and above its own perspective, it simply saves time by constructing an artificial object description purely for the simulator. Details of the communication language used may be found in Anderson and Evans [7] and Anderson [4].

Within the realm of current and future development, there are further aspects of the agent-simulator relationship that can be made more realistic. As stated previously, agent vision, for example, is both an active and passive process, meaning that some elements are directly affected by the agent itself (what its looking for) and others are directly affected by the environment (what can be seen). Currently vision is defined within the agent itself, simply to enforce the separation of agent and simulator knowledge, one of the major design goals of the system. However, like agent-simulator communication, this is somewhat artificial. If an environment is affected in such a way that vision is impaired (e.g., fog), it is partly an environmental consequence (the moisture in the atmosphere interfering with light transfer) and an agent consequence (its vision can not adapt to these effects). We are working on ways of extending the breadth of agents supportable by defining another level between agent and simulator to cover situations such as this more realistically in situations that require it. This will be handled by an agent-simulator interface defined for each agent, where shared aspects such as visual acuity under environmental conditions can be defined. Once again, this can be shared between agents and classes of agents in order to facilitate the definition and support of significant numbers of agents. Also once again, it is a method to extend the breadth of support the simulator offers, and can be avoided in simple scenarios where it is not required.

Another issue of importance in agent and simulation breadth is support for social-level simulations. There are several aspects of this: agents can be formed into groups and manipulated in groups by the simulator; agents can communicate with one another; and explicit social-level interactions can be defined as well as being emergent from individual actions. The first two of these can already be handled to varying degrees by Gensim. As agents already communicate with the simulator, precisely the same communication methods can afford interagent communication as well. In addition, should other forms of communication be desired, communicative actions can be defined as easily as any other form of action within the system.

With regard to the aggregation and disaggregation of agents, and multiple levels of granularity in general, these aspects are also supported to a significant degree within the existing system. Multiple levels of granularity are supported in part by the object-oriented approach to the simulator itself, in that actions can be defined to affect groups of agents, and that groups of agents can be manipulated as a whole. Granularity in other simulation aspects is also supported by the system. As described in section 2, for example, time can also be implemented to the degree of granularity desired. One general desire in future development is to make the system move between levels of granularity as easily in terms of space as it currently can in terms of objects and time. Given that space is currently either defined as a grid-based mechanism or through specific user-defined locales, it is currently up to the user to perform any necessary spatial aggregation and disaggregation. Full support of this along with ease of use will likely necessitate the interfacing of the simulator and its objects with a more traditional GIS system to provide spatial representation and manipulation. Current work on this facet of the system involves converting Gensim's internal spatial representation to a more Gecko-like [13] representation (effectively a real-based grid structure as opposed to a discrete interval), and removing its dependence on a user-defined set of qualitative locations.

In order to facilitate explicit representation of social interaction, we are extending the improvisational paradigm described in section 3 to include collective shared intentions, with multiple agents as participants. This is currently supportable within this existing model, but requires explicit user definition of specific communication actions required, as well as the definition of internal agent concepts that go along with social planning: expectations, commitments, responsibilities, etc. Note that these also are easily viewed as constraints and can fit into individual intentions within the member agents contributing to social interactions. It is simply up to the modeler to currently define them, putting this aspect uncomfortably beyond the reasonable boundaries depicted in figure 1. Extending Gensim to better support this involves separating individual and social knowledge in intentions, defining explicit coordination mechanisms (e.g., a wider array of built-in communication actions) and also defining some of the many implicit coordination abilities social animals possess (e.g., interpreting one's own perception of other agents in light of shared intentions). By extending Gensim and its internal support for improvising agents to encompass the definition of these facilities, the breadth of supportable simulations within acceptable ease-of-use boundaries will further be extended.

In combination, the current version of the Gensim simulator and the constraint-based improvisation approach to agent design outlined in this chapter provide a unique tool for developing individual-based simulation models incorporating intelligent agents. We are still, however, using this tool exclusively in our own agent-based research for several reasons. While completely functional, its user interface is still primitive, as we have been concentrating limited resources on developing realistic agent-environment interfaces and on the flexibility of the agents themselves. The system's time-sharing implementation is both Macintosh- and Common LISP-specific, as are other portions of its system code, and so a major goal in the immediate future is to port the system to a more widely used platform (Java) before going on to develop such facilities.

While attempting to support breadth in intelligent agents is the obvious contribution of this work, it can be argued that the most significant contribution of this work is similar to that described in the other chapters of this book: to point out the commonalities between research in intelligent agency and that of simulation modeling in general and ecosystem modeling in particular, and to reap the benefits of the synergy that results. This chapter has expounded on the virtues of intelligent agents and the benefits a broad range of agents can bring to simulation models. These benefits are very promising: the inclusion of intelligent agents can offer improved accuracy of ecosystem models, as well as the ability to include in ecosystem simulations intelligent agents designed to give advice directly to resource managers. However, it is even more readily apparent to an AI researcher that studying how intelligent agents can assist in simulation models will also be of great benefit to the field of artificial intelligence as a whole. Gensim was originally constructed because existing simulation tools did not provide the support needed to construct flexible, complex domains for the purpose of testing intelligent agent designs. The detailed domains that will emerge from intelligent ecosystem modeling will be able to serve as such testbeds, providing a wide variety of complex and dynamic domains for intelligent agent research. The ability to support human-level intelligent agents specifically in simulation using such tools has the potential for great impact on many areas outside of ecosystem management (e.g., Tambe et al. [51]). Moreover, the possibility of using such systems for training humans to behave optimally in problem-solving situations has also been stressed as an important future goal for the field of artificial intelligence as a whole [28]. In addition, some of the issues described in this chapter, such as aggregation and disaggregation in models, are also very important issues in particular simulation applications (e.g., Hillestad and Juncosa [34]) that work such as this can also contribute toward.

The rigorous standards of ecological modeling and the control over experimentation that results will also serve AI well, ensuring that the models used to test intelligent agents are realistic. AI has much to learn about the formal methods of experimentation that fields such as ecosystem modeling hold as central, and it is only comparatively recently that any works have emerged advocating rigor and control in experimentation in AI (e.g., Cohen [16]). Overall, the addition of intelligent agents to complex ecosystem models has the potential to greatly advance both fields. Tools such as Gensim and the others described in this book are a significant step in this direction.

REFERENCES

[1] Agre, P. E. *The Dynamic Structure of Everyday Life.* Ph.D. diss., Department of Electrical Engineering and Computer Science, Massachusetts Institute of Technology, 1988.

[2] Agre, P. E., and D. Chapman. "Pengi: An Implementation of a Theory of Activity." In *Proceedings of the Sixth National Conference on Artificial Intelligence*, 196–201. Cambridge, MA: MIT Press, 1987.

[3] Agre, P. E., and I. D. Horswill. "Cultural Support for Improvisation." In *Proceedings of the Tenth National Conference on Artificial Intelligence*, 363–368. Cambridge, MA: MIT Press, 1992.

[4] Anderson, J. "Constraint-Directed Improvisation for Everyday Activities." Ph.D. diss., Department of Computer Science, University of Manitoba, 1995.

[5] Anderson, J., and M. Evans. "Constraint-Directed Improvisation." In *Proceedings of the Eleventh Biennial Canadian Society for the Computational Studies of Intelligence Conference (AI-96)*, edited by G. McCalla, 1–13. Lecture Notes in Artificial Intelligence, vol. 1081. Berlin: Springer-Verlag, 1996.

[6] Anderson, J., and M. Evans. "Real-Time Satisficing Agents for Complex Domains." In *Proceedings, Ninth Florida AI Symposium*, edited by J. H. Stewmann, 96–100. 1996.

[7] Anderson, J., and M. Evans. "A Generic Simulation System for Intelligent Agent Designs." *Appl. Art. Intel.* **9(5)** (1995): 527–562.

[8] Anderson, J., and M. Evans. "Intelligent Agent Modelling for Natural Resource Management." *Math. & Comp. Model.* **20(8)** (1994): 109–119.

[9] Anderson, J., and M. Evans. "Supporting Flexible Autonomy in a Simulation Environment for Intelligent Agent Designs." In *Proceedings of the Fourth Annual Conference on AI, Simulation, and Planning in High Autonomy Systems*, edited by J. Rosenblit, 60–66. Tucson, AZ. 1993.

[10] Bates, J., A. B. Loyall, and W. S. Reilly. "Broad Agents." *SIGART Bull.* **2(4)** (1991): 38–40.

[11] Beer, R. *Intelligence as Adaptive Behaviour.* Boston, MA: Academic Press, 1990.

[12] Boden, M. "The Structure of Intentions." *J. Study Soc. Behav.* **3(1)** (1973): 23–46.

[13] Booth, G. "Swarm Gecko: A 2-D Floating World for Ecological Modelling." Technical Report, Center for Computational Ecology, Yale Institute for Biosphere Studies, New Haven, CT, 1996.

[14] Bratman, M. E., D. J. Israel, and M. E. Pollack. "Plans and Resource-Bounded Practical Reasoning." *Comp. Intel.* **4** (1988): 349–355.

[15] Carter, J. "Moab: A Spatially Explicit, Individual Based, Expert System for Creating Animal Foraging Models." 1997. Poster presentation at Technology 2007, Boston, MA, 1997. ⟨http://www.abpi.net/T2007/posters/moab.html⟩.

[16] Cohen, P. R. *Empirical Methods for Aritificial Intelligence*. Cambridge, MA: MIT Press, 1995.

[17] DeAngelis, D. L., D. M. Fleming, L. J. Gross, and W. F. Wolff. "Individual-Based Models in Ecology: An Overview." In *Proceedings, Third International Conference on Integrating GIS and Environmental Modeling*, held in Santa Fe, NM on January 21–26, 1996. Santa Barbara, CA: National Center for Geographic Information and Analysis, 1996. ⟨http://www.ncgia.ucsb.edu/conf/SANTA_FE_CD-ROM/main.html⟩.

[18] Deadman, P., and R. H. Gimblett. "A Role for Goal-Oriented Autonomous Agents in Modeling People-Environment Interactions in Forest Recreation." *Math. & Comp. Model.* **20(8)** (1994): 121–131.

[19] Dean, T., L. Kaelbling, J. Kirman, and A. Nicholson. "Planning with Deadlines in Stochastic Domains." In *Proceedings of the Eleventh National Conference on Artificial Intelligence*, 574–579. Cambridge, MA: MIT Press, 1993.

[20] Engelson, S. P., and N. Bertani. "Ars Magna: The Abstract Robot Simulator Manual." Technical Report, Department of Computer Science, Yale University, New Haven, CT, 1992.

[21] Etzioni, O. "Intelligence without Robots: A Reply to Brooks." *AI Mag.* **14(4)** (1993): 7–13.

[22] Evans, M., J. Anderson, and G. Crysdale. "Achieving Flexible Autonomy in Multi-Agent Systems using Constraints." *Appl. Art. Intel.* **6(1)** (1992): 103–126.

[23] Folse, L., H. Mueller, and A. Whittaker. "Object-Oriented Simulation and Geographic Information Systems." *AI Appl. Natl. Res. Mgmt.* **4(2)** (1990): 41–47.

[24] Fox, M. S. "Constraint-Directed Search: A Case Study of Job Shop Scheduling." Ph.D. diss., School of Computer Science, Carnegie-Mellon University, New York, 1983.

[25] Friedland, E. "Values and Environmental Modelling." In *Ecosystem Modelling in Theory and Practice*, edited by C. Hall and J. Day, 115–132. New York: John Wiley and Sons, 1977.

[26] Gigon, A. "A Hierarchic Approach in Causal Ecosystem Analysis." In *Potentials and Limitations of Ecosystem Analysis*, edited by E. Schulze and H. Zwölfer, 228–244. Berlin: Springer-Verlag, 1987.

[27] Gimblett, R. H., R. M. Itami, and D. Durnota. "Some Practical Issues in Designing and Calibrating Artificial Human-Recreator Agents in GIS-Based Simulated Worlds." *Complexity Intl.* **3** (1996).

[28] Grosz, B., and R. Davis, eds. "A Report to ARPA on Twenty-First Century Intelligent Systems." American Association for Artificial Intelligence. 1998. ⟨http://www.aaai.org/Policy/Papers/arpa-report.html⟩.

[29] Hammond, K., and T. Converse. "Stabilizing Environments to Facilitate Planning and Activity." In *Proceedings of the Ninth National Conference on Artificial Intelligence*, 787–793. Cambridge, MA: MIT Press, 1991.

[30] Hammond, K., T. Converse, and C. Martin. "Integrating Planning and Acting in a Case-Based Framework." In *Proceedings of the Eighth National Conference on Artificial Intelligence*, 292–297. Cambridge, MA: MIT Press, 1990.
[31] Hayes-Roth, B., E. Sincoff, L. Brownston, R. Huard, and B. Lent. "Directed Improvisation." Report KSL-94-61, Stanford Knowledge Systems Laboratory, Stanford, CA, 1994.
[32] Hayes-Roth, F., D. Waterman, and D. Lenat. *Building Expert Systems.* Reading, MA: Addison Wesley, 1983.
[33] Hiebler, D. "The Swarm Simulation System and Individual-Based Modelling." In *Proceedings, Decision Support 2001*, edited by J. M. Power and M. Strome, 474–494. Bethesda, MD: American Society for Photogrammetry and Remote Sensing, 1994.
[34] Hillestad, R., and M. Juncosa. "Cutting Some Trees to See the Forest: On Aggregation and Disaggregation in Combat Models." Technical Report MR-189-ARPA, Rand Corporation, Santa Monica, CA, 1993.
[35] Jencks, C., and N. Silver. *Adhocism: The Case for Improvisation.* New York: Doubleday, 1972.
[36] Judson, O. P. "The Rise of the Individual-Based Model in Ecology." *Trends in Ecol. & Evol.* **9** (1994): 9–14.
[37] Kester, K. "Individual-based Simulation Modelling: Approaches and Application in Insect Behaviour and Ecology." Paper presented at a symposia at 1996 Annual Meeting of the Entomological Society of America, Louisville. 1996. ⟨http://www.inhs.uiuc.edu/cbd/ESA-Annual/ESA_WWW_design.html⟩.
[38] McArthur, D. "Decision Scientists, Decision Makers, and the Gap." *Interfaces* **10(1)** (1980): 110–113.
[39] Neisser, U. *Cognition and Reality.* San Francisco, CA: W. H. Freeman & Co., 1976.
[40] Newell, A., and H. Simon. *Human Problem Solving.* Englewood Cliffs, NJ: Prentice Hall, 1972.
[41] Norman, D. A. *The Psychology of Everyday Things.* New York: Basic Books, 1988.
[42] O'Neill, R. "Management of Large-Scale Environmental Modelling Projects." In *Ecological Modeling in a Resource Management Framework*, edited by C. Russell, 251–282. Washington: Johns Hopkins University Press, 1975.
[43] Pollack, M., and M. Ringuette. "Introducing the Tileworld: Experimentally Evaluating Agent Architectures." In *Proceedings, Eighth National Conference on Artificial Intelligence*, 183–189. Cambridge, MA: MIT Press, 1990.
[44] Rich, E., and K. Knight. *Artificial Intelligence*, 2d ed. New York: McGraw-Hill, 1991.
[45] Russell, S., and P. Norvig, eds. *Artificial Intelligence: A Modern Approach*, 932. Upper Saddle River, NJ : Prentice Hall, 1995.

[46] Schmitz, O. J., and G. Booth. "Modelling Food Web Complexity: The Consequence of Individual-Based Spatially Explicit Behavioural Ecology on Trophic Interactions." *Evol. Ecol.* **11(4)** (1997): 379–398.

[47] Scahill, M. "Individual-Based Simulation of Mountain Gorillas." 1996. ⟨http://larch.ukc.ac.uk:2001/gorillas/simulation/⟩.

[48] Simon, H. A. "Rational Choice and the Structure of the Environment." In *Models of Bounded Rationality*, edited by H. A. Simon, vol. 2, 259–268. Cambridge, MA: MIT Press, 1982.

[49] Slothower, R. L., P. A. Schwartz, and K. M. Johnson. "Some Guidelines for Implementing Spatially Explicit, Individual-Based Ecological Models within Location-Based Raster GIS." In *Proceedings, Third International Conference on Integrating GIS and Environmental Modeling.* Santa Barbara, CA: National Center for Geographic Information and Analysis, 1996. ⟨http://www.ncgia.ucsb.edu/conf/SANTA_FE_CD-ROM/main.html.⟩.

[50] Spofford, W. Jr. "Ecological Modeling in a Resource Management Framework: An Introduction." In *Ecological Modeling in a Resource Management Framework*, edited by C. Russell, 13–48. Washington: Johns Hopkins University Press, 1975.

[51] Tambe, M., W. L. Johnson, R. M. Jones, F. Koss, J. E. Laird, P. S. Rosenbloom, and K Schwamb. "Intelligent Agents for Interactive Simulation Environments." *AI Mag.* **16(1)** (1995): 15–40.

[52] Walker, H., and W. Cuff. "Scientists, Models, and Resource Managers." *Memoirs Entomol. Soc. Canada* **143** (1988): 11–17.

[53] Waterman, D, ed. *A Guide to Expert Systems*, 419. Reading, MA: Addison-Wesley, 1986.

[54] Wilson, W. G., A. M. de Roos, and E. McCauley. "Spatial Instabilities within the Diffusive Lotka-Volterra System: Individual-Based Simulation Results." *Theor. Pop. Biol.* **43** (1993): 91–127.

Spatial Units as Agents: Making the Landscape an Equal Player in Agent-Based Simulations

Paul Box

1 INTRODUCTION

Agent-based modeling has generated considerable interest in recent years as a tool for exploring many of the processes that can be modeled as bottom-up processes. This has accelerated with the availability of software packages, such as Swarm [20] and StarLogo [28], that allow for relatively complex simulations to be constructed by researchers with limited computer-programming backgrounds. A typical use of agent-based models is to simulate scenarios where large numbers of individuals are inhabiting a landscape, interacting with their landscape and each other by relatively simple rules, and observing the emergent behavior of the system (population) over time. It has been a natural extension in this sort of a study to create a landscape from a "real world" example, typically imported through a geographic information system (GIS). In most cases, the landscape is represented either as a static object, or a "stage" upon which the agents act (see Briggs et al. [6], Gimblett et al. [15], and Remm [27]). In some cases, an approximation of a dynamic landscape has been added to the simulation in a way that is completely exogenous to the population being simulated; the dynamic conditions are read from historical records, in effect "playing a tape" of conditions, to which the population reacts through time (such as Dean et al. [10] and Kohler et al. [18]).

There has also been many simulations where dynamic landscape processes have been modeled through "bottom up" processes, where localized processes in landscapes are simulated, and the global emergent processes are observed. Topmodel [3] is a Fortran-based implementation of this concept for hydrologic processes; and PCRaster has used similar software constructs to simulate a variety of landscape processes, with sophisticated visualization and data-gathering tools [31]. In both of these examples, the landscape is represented as a regular lattice or cell structure. There are also many examples of "home grown" tools (simulations created for a specific project), applying cellular automata (CA) rules to landscapes to simulate urban growth [8], wildfire [7], lava flows [21], and groundwater flow [19]. There are also examples of how agent-based modeling tools were employed to model dynamic landscape processes such as forest dynamics, i.e., Arborgames [29]. In these models the landscape was the object of the simulation, and free-roaming agents were not considered as part of the model.

It is much more difficult to find examples of simulations combining both dynamic population interactions and a dynamic landscape or environment. Part of this has been a necessary limitation on the scope of systems being modeled: If your objective is to study group interactions of populations who are not significantly altering their environment, addition of dynamic environmental components would simply make the simulation unnecessarily complex. In a similar vein, CA-like models of landscape processes, such as the ones cited in the previous paragraph, have provided concise representations of their respective systems, whose validity has passed at least the test of peer review in their respective journals. Parsimony is required for effective explanation.

However, there are a number of systems where population dynamics and environmental interaction are so fundamentally interrelated that a modeler cannot satisfactorally represent one without the other. The concept of populations and environment being so fundamentally linked is certainly not new, as virtually any introductory textbook on ecology will tell you. Recent discussions of the need to consider the dynamic, nonlinear nature of the relationship between populations and their environments have come out in discussions of systems such as grazing animals and the plant communities that they graze [24, 25]. Books such as the one by Pahl-Wostl [23] have described similar situations for ecosystems in general. Similar statements have been applied to human systems, especially in recreational contexts; Hammon et al. [17] described the complex nature of recreational boat users in North Carolina and pronounced such systems as practically impossible to simulate given the state of the art of computer technology of the day, and Drogin [11] repeated these concerns decades later. More recent uses of individual-based models for simulating human systems have already been cited in this introduction.

While there has been the desire in past years to integrate GIS-based CAs with agent-based models, the lack of examples in the literature of actual implementations may be due to the difficulty in merging existing software designed for either of the two modeling paradigms. Software that is optimized for simulating CA-like functions with GIS-generated data may not be compatible with code that is optimized for agent-based simulation, and vise versa. A considerable number of "tricks" or *hacks* are often required to create an application of appreciable complexity, and hacks that work for one purpose (CAs or agent-based simulations) often make the code incompatible with other designs.

2 INTEGRATION OF CELLULAR AUTOMATA AND AGENT-BASED MODELS

One way of accommodating various modeling paradigms into a single simulation or application is to express one modeling framework in terms of another. If an agent-based model (ABM) can be expressed in terms of CA rules, that would in principle make for easy integration of agents into the CA. Likewise, if a cellular automaton can be constructed from an agent-based modeling framework, then a similar software base can be used for the agents *and* the CA landscape that they inhabit. There is ease of integration, at least in principle, since both CAs and ABMs are conceived from the principle of an entity (a cell or an agent) acting "autonomously" in a world inhabited with like entities. There are some examples of such integrations available in existing simulation software demos, at least for "theoretical" landscapes.

The integration of GIS data into this framework is a little less straightforward. This is due in part to a basic difference in GISs and CAs: A CA is built from the fundamental unit of the cell, and a GIS is built from the fundamental unit of the *data layer* or *theme*. There is little problem in reconciling the difference between these two representations of the world when implementing a CA where cells have simple rules that refer to only a few GIS data layers. The problem becomes more unwieldy as the modeler tries to add more complex rules to the behavior of to the cell, or to increase the number of geographic themes that must be referenced at each time step as the cell in the CA makes its operational decisions. As complexity increases, the code is likely to become more and more specific to a particular scenario being simulated, making modification, portability, and documentation more difficult.

This problem can be significantly reduced by taking advantage of the modular structure afforded by object-oriented design. Agent-based models have benefited greatly from object-oriented programming, allowing modelers to "hide" code that is not immediately relevant to many of the programming tasks that a modeler may be working on in a simulation, and to segregate functions and state variables that are germane only to the internal workings of certain components of their simulation. For a well-designed agent-based simulation, the world is represented by a bunch of individuals (*entities* or *agents*) who have any level of internal complexity, but whose notion of other agents is limited strictly to a *message interface* (messages that they pass to one another). This allows for each agent to be represented as a unique creation in the course of the simulation; different agents can have any variety of internal individual details, but are able to interact in the simulation as long as the messaging interface is consistent.

The same advantages that object-orientation provides to agent-based design can be extended to a CA structure if one constructs the CA surface where all the cells are represented as discrete *objects*. Each cell object can contain any variety of internal state variables, which are equivalent to the layers of a GIS within that cell. In this construct, the cell itself represents a particular geographic location (or entity), and its internal state variables represent qualities of that location. As there is no need in principle to have cells contain the same list of state variables (or even have the same internal structure), then each cell need only contain those qualities (variables) that are relevant to that location. In an object-oriented design, as long as the messaging interface between objects is capable of handling the same messages, the objects themselves may have significant variability. This can be very advantageous if one is constructing a CA landscape where some areas have great internal complexity and others are relatively featureless (imagine constructing a landscape that is largely "featureless" desert with a few complex, lively, feature-rich oases interspersed).

While this discussion is being presented in terms of a cellular model, it should be noted that a cell is just one of many possible abstractions of spatial units or entities. With careful encapsulation of functions within a cell, there is no need for a cell itself to be aware of a specific geometry between its neighbors, as long as the topological relationships are specified. When the rules of cellular automata are implemented, the cells themselves do not care if they are a regular lattice, a hexagonal lattice, irregular polygons, or links on a network. This chapter focuses on examples of cellular models, but it should be realized that the most of the issues discussed here are applicable to a wide range of geographic objects or spatial entities.

Many of the advantages of representing the world in this way will become apparent as applications to particular examples are discussed. In this chapter, a detailed design of such a GIS-based CA will be presented, and applications of the design will be discussed in the following section. The software design presented will be a toolkit that the author has implemented using the Swarm simulation software; while the examples will be necessarily implementation-specific, emphasis will be placed on the general aspects of object design that should allow for a user to reproduce this software design in any object-oriented environment.

3 THE GEOGRAPHIC INFORMATION SYSTEM-CELLULAR AUTOMATA TOOLKIT

The GIS-CA toolkit is a software library that was developed at Utah State University's Remote Sensing and GIS Laboratory. It was developed as a general-purpose tool for conducting simulation-based experiments on landscape models built from GIS data layers. It was conceived originally as a method for representing GIS layers in a way that is accessible for agent-based simulations (where the agents roam around in the model landscape), but it was discovered early on in software development that representing the landscape itself as a set of agents allowed for simulation of dynamic landscape processes using CA, cellular stochastic, or spatial Monte Carlo rules.

The GIS-CA toolkit was developed from the Swarm libraries, which were developed under the direction of Chris Langton at the Santa Fe Institute [20]. The Swarm libraries are a reusable set of software tools designed for agent-based simulations. Their primary utility is for management of many of the more difficult programming issues that arise when one tries to implement an agent-based simulation. The most important issue addressed by the Swarm libraries is enforcement of an accepted concurrency model for the agents' actions: in essence, how to simulate truly independent behavior of a multitude of individual entities (agents) who are acting in parallel when the simulation environment is processed through the single sequential processor of a computer. The principles of concurrency have been addressed in the computer science literature [1], and the issues of implementing them are far from trivial. By using a standardized and public library such as Swarm, the researchers and those evaluating the research can concentrate on higher-level details of agent behavior rather than the minutiae of how sequences of agents' behaviors are handled. Swarm also handles many of the details of list management, memory allocation, and scheduling that must be addressed when constructing an agent-based simulation, plus a very useful set of methods for probing (examining) the internal details of objects while the simulation is running.

The Swarm environment provides a generalized infrastructure of lists, schedules, and probes (collectively called a *swarm*) within which a model can be built. Within this model, a generalized *SwarmObject* is made available, which is a software object that is capable of responding to messages to create and destroy itself and do the things that are required for being a member of a *swarm*. A modeler takes a *SwarmObject* as a base object and, through inheritance, adds on all the functionality and attributes (methods and variables) that are required to turn the object into an agent for his/her simulation. The examples used for discussion in this section are all derived from fundamental *SwarmObjects*, and rely on Swarm's environment for management of the objects.

Both the GIS-CA toolkit and the underlying Swarm libraries are licensed under the GNU general protection license; they may be used and distributed free of charge as long as all of the source code is made freely available.

The remainder of this chapter will describe how a dynamic landscape object can be created from Swarm's building blocks, how free-roaming agents can be placed into the fray, and the implications that this has for the modeling environment.

4 AN AGENT-BASED CA MODEL FOR THE LANDSCAPE

4.1 BASIC DESIGN

The landscape object described in this chapter was conceived from a single fundamental idea: *Any given portion of the Earth's surface is a unique location, which reacts to localized processes with no notion of the global behavior of the landscape.* For a given portion of land, the processes that govern such things as whether vegetation will grow on it, soil will erode, or material will burn is largely a function of factors that have to do with localized conditions on that portion of land and its immediate neighborhood. It is safe to say with some generality that vegetation growth in a particular piece of land is more a function of the local conditions of soil type and moisture than the fact that the majority of the landscape is populated with similar vegetation. Groundwater flow through a (small) unit area of land is ultimately a function of the permeability and saturation of that parcel of land, and the hydrolic head of the parcels of land immediately adjoining it. While some global factors (say, windspeed, air pressure, or rainfall) may matter, much of the behavior of a unit area of land is going to be decided at the most local level.

Given this basic idea, the idea was to construct a landscape whose basic design would be like a raster image. In a raster image, the entire "world" or picture is described by the coordinates of a particular cell (or *pixel*) and the value (color) of that cell. The cell represents a unit of surface area, and for the purposes of the model, the properties of the landscape are homogeneous within the boundary defined by that cell. Landscape variability is represented by having a sufficiently large number of cells, much the same way as subtleties of color in an image can be captured with a large numbers of pixels.

In the landscape object described here, each raster cell would be replaced by an object that contains a number of state variables that describe the qualities of that cell that one wishes to model, as well as whatever functions are required for the cell to act upon those variables. In short, each cell would become an independent "model" of the processes that one wishes to study that exact location (fig. 1). The model within the cell itself may be aspatial in nature, and described as a flow chart much as the Fortran-based Forrester models; as with the raster image, spatial variability would be captured in the large number of cells running concurrent models over the landscape. In this conceptualization, each cell is a virtual processor capable of executing a model relevant to its (potentially unique) situation; the landscape then is a collection of thousands of cells acting in parallel, each one accepting conditions of its neighbors as inputs into its own execution, but executing its functions according to its internal rules. If the conditions and behavior of a cell is a function of its neighbors' states, then the actions of the landscape can be characterized by CA rules.

In practice, implementation of this idea can be constructed from a combination of a few basic building blocks:

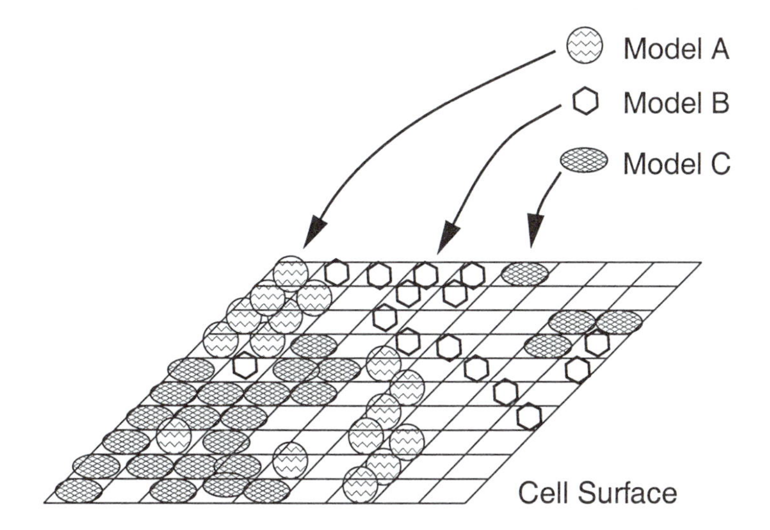

FIGURE 1 Cell-surface object with models in cells.

- a basic cell object which performs operations,
- a list for keeping track of the objects,
- a scheduling mechanism for calling the operations in the objects, and
- a data file manager for file input and output.

In the case of the CA surface, these generalized objects became:

1. the cell,
2. the landscape,
3. the scheduler, and
4. the data file manager.

The purpose of each of these objects should be explained in greater detail.

5 PRINCIPAL OBJECTS

5.1 CELL

The cell is the fundamental building block of this model. It contains all of the state variables that describe that particular piece of the Earth's surface, as well as methods for interpreting those variables. A cell also has methods for setting and altering these variables, and for sending requests to other cells that they provide information or execute methods.

It is possible for a cell to possess other objects, including objects that encapsulate complete models. These objects may also respond to requests from other objects that they run, change parameters, etc. There is also a facility to have each cell contain variables for its future state; execution of each cell's actions would finish with the cell transferring its future state to its present state. By having this functionality unique to each cell, it's possible to have cells (or the models or components encapsulated in the cells) run and update themselves at their own rates, allowing a cell to operate at a temporal scale completely different from its neighbors, or to perform simultaneous updates in complete synchronicity with the rest of the cells in the landscape at the discretion of the modeler.

In essence, all the "interesting" (application specific) behavior of the model occurs at this level. The cell becomes the place where landscape models are handled. The importance of this role should become obvious as specific applications are discussed.

5.2 CELL SURFACE

As the cell is a platform for executing specific landscape processes in a location, the cell surface can be envisioned as a platform for managing the cells. In agent-based simulations, it is common practice to manage the multitude of independent agents collectively by a list. In this case, the cell surface can be thought of as a *two-dimensional list*, responsible for knowing the location of the cell, knowing which cells are neighbors of that cell, and facilitating interaction between cells.

5.3 SCHEDULER

The scheduler, which manages the order in which the cells and their models will execute at each time step. This scheduler is taken directly from the Swarm libraries with no fundamental modification. The scheduler allows for dynamic scheduling; it is in the scheduler where the modeler specifies whether cells will update simultaneously, asynchronously, or in a random order.

5.4 DATA FILE MANAGER

Since this object is designed to work with geographic datasets, there needs to be some capability to import data from GIS data sets, and to export the output to the same GIS format. In this implementation, file input/output is handled by a separate datafile object, which is responsible for knowing the format of the GIS data files that it is importing, and converting them to and from the format that the GIS-CA object expects.

One of the most important aspects to mention of the datafile object is what it does *not* do: it will *not* convert data from one spatial resolution or data format to another. It is designed to simply take a raster data set, read the data set grid cell by grid cell, and load those values into cell objects of the *same resolution*. It will not handle raster files of differing resolutions. This is a very deliberate omission: combination of data sources of varying scales or degrees of resolution, while a common function in most commercially available GISs, introduces possible errors, which can be difficult to trace after the fact. It is desired in this case that if a modeler wishes to incorporate data of varying resolutions that s/he explicity conduct the necessary resampling of the varying themes *before* they are introduced into a spatially explicit model such as this one.

As of this writing, the GIS-CA datafile object reads and writes GIS files that are in geographic resource and analysis simulation system (GRASS) or Arc/INFO GRID export format. The input/output code is segregated from the rest of the code in such a way that a user wishing to write his/her own import/export filter will be able to do so with a minimum of coding.

To better illustrate the utility of modeling landscape processes with this framework, an example of a landscape simulation using the GIS-CA toolkit will be presented here. This is a work in progress, and presented here for illustrative purposes only.

6 SAMPLE APPLICATIONS: A CA-BASED WILDFIRE MODEL

One example of a landscape process that is commonly represented by cellular modeling is wildfire propagation. Wildfire is commonly modeled as a spatial stochastic process [26], some models emphasize the use of coupled differential equations [14], while others either embed known fire models within a cellular context [4], and still others utilize CA rules [7]. All of the models cited here share a common idea: the likelihood of ignition of a section of vegetation is partially a function of how and how much of the surrounding area is burning. For data sets of high resolution and small unit areas (a few square meters), it is common to include explicit terms and actions representing physical processes in fire burning (such as heat flux, dessication, etc.) to represent transmission of heat from one cell to another. With coarser spatial data (tens of meters), the direct effects of fire from one cell to another are confounded with other processes, weakening the predictive power of many of the physical process models. In these cases, it is more common to approach transmission of flame from one cell to another through a cellular stochastic processes.

In this section the design of an elementary wildfire propagation model will be presented. This model design here is part of a pilot simulation for a fire ecology project at Utah State University. It has not yet been formally published, but some information is available from the project web page [5]. The details presented here are to illustrate the applicability of the described modeling technique to wildfire, as they are considered particularly illustrative of the modeling concepts described earlier.

An under-riding concern for creation of this model was to have every cell in the landscape act with relative independence, accepting inputs from neighbors as required but otherwise an entity unto itself. A cellular model is a logical way to implement this, with many precedents in the fire modeling literature. However, an additional concern was to accommodate any level of heterogeneity within the landscape, allowing each cell to act by its own rules with little concern for the internal details of its neighbors. In the landscape, a stand of trees, a meadow, a building, and an agricultural field will have distinctly different characteristics of how each reacts to a burn, and each will follow a different set of rules as to its burn characteristics. The desired effect was to have each cell have not just a customized set of parameters, but rather each cell embed its own *model* of how it burns, which can be radically different for various landcover types.

Some aspects of this pilot model were determined by data availability. The simulation was created with vegetation and landcover inputs from the Utah GAP Analysis [12], and topographic information from the U.S.G.S. digital elevation models (DEMs) [30]. Since both of these data sources are in latticed (grid) format at 30 m resolution, this was selected as the base grid cell size for the simulation. At this resolution, it was not considered practical to attempt a model of propagation based on first-order physics, as many of the crucial processes such as heat flux are acting at a much more localized scale. Additionally, other factors, such as ignition by flying embers (known as *spotting*), can be modeled entirely as physical processes, but for practical reasons are modeled according to probabilistic or stochastic processes. For these reasons, it was decided to model wildfire as a cellular stochastic process.

6.1 MODEL DESCRIPTION

The pilot model was constructed using the GIS-CA toolkit, with several levels of detail working in its hierarchy (fig. 2). It is fundamentally a cellular model, with each grid cell running copies of various embedded fire models, and the cell surface managing message passing between the models in the cells. To best understand the structure, it is best to start at the most fundamental level, which is the fuel object (the thing that burns).

6.2 THE FUEL COMPONENT

A fuel component is considered to be a collection of material that reacts in a relatively uniform fashion to fire. In its most rudimentary form, it consists of a quantity of fuel (called the *fuel load*), and has moisture, density, and other inherent "burnability" characteristics which are *attributes* of the fuel component.

Each fuel component has the ability to receive and evaluate a request to ignite or burn; the component's decision to burn will be a function of how much burnable material does it have available, and how "burnable" is the material (how moist is it, how densely packed, whether it's already burning, etc.). The fuel component will compare the weight (urgency) of the request to

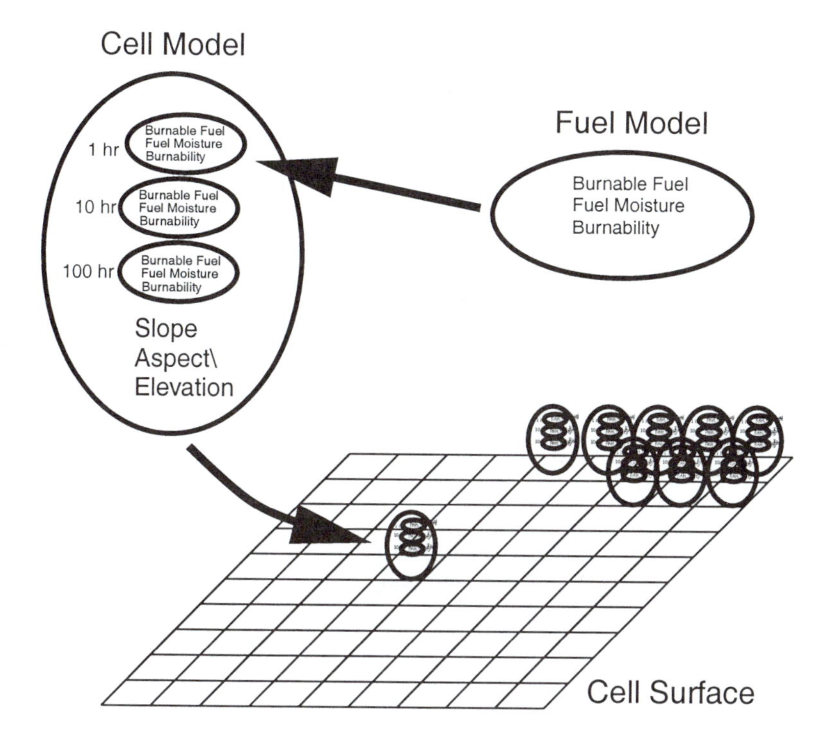

FIGURE 2 Relationship of fuel model objects, cell objects, and cell surface.

its own burnability, and "decide" (via a random number) whether to ignite or not. If the component is dry grass, a very light request could result in a burn; if the component is a moist log, only the most "urgent" of requests would result in it igniting.

Once the component is ignited, it "burns" at each time step by decreasing its fuel load, and possibly changing its burnability characteristics (dessication, heating, etc.) as a function of the intensity of the burn. Additionally, it will start sending out messages that it is burning, which get translated at higher levels in the simulation as "requests" to other fuel components to burn.

In this preliminary model, the fuel components are modeled roughly after the fuel models in the BEHAVE fire simulation [2], which is one of the more trusted fire simulation packages used by public land managers in the United States. The fuel classes used are *1-hour, 10-hour*, and *100-hour* fuels. These classes are named according to the estimated time that it takes for them to dry to equilibrium with the atmosphere in idealized conditions; for this discussion it is safe to think of them simply as "grasses," "twigs," and "branches" or "wood," respectively. Estimates of loads for each fuel type was taken from simple estimates based on vegetation type from GAP landcover types.

Each component is encapsulated into an object, with its own (possibly unique) burn models and characteristics contained within. While each component is capable of responding to the same requests, the actual way that it responds becomes an internal detail of the object. Thus, an object representing grass can have a burn model that is appropriate for grass, while twigs and branches in the same immediate area can burn according to their own rules.

This model presently is only implementing the BEHAVE fuel types. However, there is no limit on the number of fuel components that a cell may contain, nor are there any constraints on the internal details of the components as long as the "interface" (protocols for accepting messages) is consistent. Thus, other kinds of fuels (ground litter, slash from a clearcut, structures, and pickup trucks) can respond to the same requests to ignite, though their internal details of *how* they burn can be quite variable.

6.3 THE CELL

The cell is the object that fixes the instances of fuel component models to a geographic location. The grid cell knows its surface area, geographic coordinates, landcover type, slope, aspect, and elevation. It also contains any number of fuel component objects that represent its ability to burn. Since, in this simulation, a grid cell is 30 m resolution, or 900 m^2 area, the 1-hour fuel component that it contains is a model of all of the grassy material in that 900 m^2 area. The 10-hour object represents all of the twigs, and so on.

While the actual burning of material in each component is conducted within the object, the spatial dynamics of the burn is dependent on messages that are sent between objects. The cell is responsible for coordinating messages between its various fuel components, forwarding messages from its burning fuel components to other cells, and delivering messages from other cells to its fuel components. The important communication is between fuel components; since the cell bundles all messages from components together to send them to other cells, it is convenient to refer to this message passing as communication between cells. This abstraction will be used for the remainder of this discussion; the reader should be aware that references to "messages between cells" are in fact "messages between fuels components in cells."

In this simulation, all of the messages are requests to burn. In a flat, homogeneous landscape, the urgency U of the request from one cell i to another cell j to burn would simply be a function of the intensity of the burn in cell $i(I_i)$, and inversely proportional to the distance d_{ij} between the cell centroids.

$$U_{ij} \sim \frac{I_i}{D_{ij}} . \tag{1}$$

If a cell is burning intensely, a neighboring cell should receive a correspondingly urgent request to burn. The neighboring cell's decision to act on that request will be a function of its internal state; if it's burnable, it will likely honor the request to burn. If it's just not flammable, or has no fuel left to

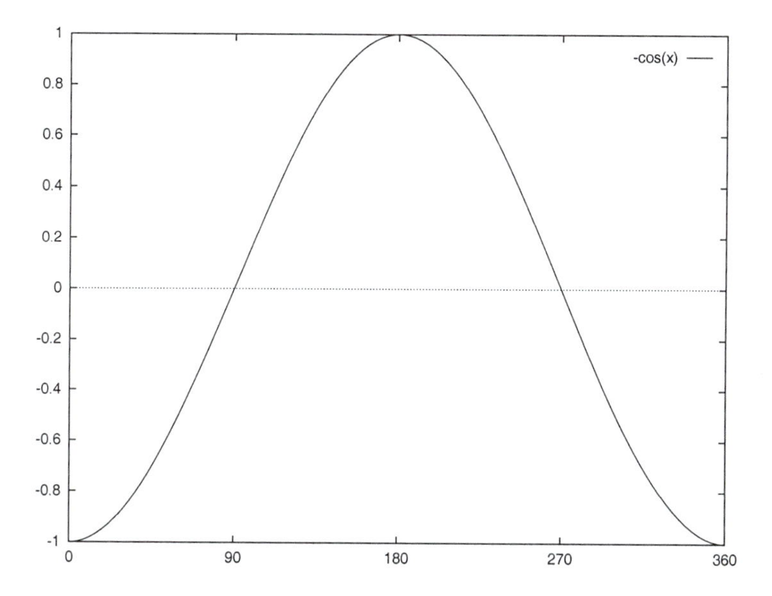

FIGURE 3 The inverse cosine function.

burn, it will stay dormant regardless of the urgency of the requests it receives. In an alternate formulation, the distance parameter may be replaced with a simple connectivity function c_{ij}, reflecting whether cells are adjoining or not.

Fire spread is highly influenced by wind and topography. In this simulation, these factors are implemented by adjusting the urgency of requests from a burning cell to its neighbor according to windspeed, wind direction, slope, and aspect. It was considered important, in keeping with the principles of cellular modeling, that the cell sending the request be responsible for adjusting the urgency of the request; the receiving cell should have no notion of the request other than it being "urgent" or "not urgent."

Scaling of urgency according to wind and topography was accomplished by taking the inverse cosine of the difference in direction to cell j (azimuth A_{ij}) and to wind direction W and slope aspect a, respectively.

$$-\cos(A_{ij} - W) \qquad\qquad -\cos(A_{ij} - a)\,. \tag{2}$$

The inverse cosine gives a value of -1 "head on" (0° or 360°), 1 in the opposite direction (180°), and 0 at right angles to the wind or slope (fig. 3). This inverse cosine of wind and slope direction gives the effect of deflating the level of urgency upwind (down slope) and correspondingly inflating the urgency downwind (up slope). The degree (amplitude) to which the urgency of the request gets deflated or inflated is a function of windspeed V_w and steepness of slope S, respectively. The function that the cell uses to adjust the urgency of the request from cell i to $j(U_{ij})$ then becomes

$$U_{ij} = \frac{\alpha I_i - \beta V_{ij}\cos(A_{ij} - W) - \gamma S_i \cos(A_{ij} - a_i)}{d_{ij}^{\delta}} \tag{3}$$

where I is the intensity of the burn in cell i, V_w is the wind velocity, A_{ij} is the azimuth to cell j, W, and a is the wind direction and slope aspect, S is slope percent, d_{ij} is the distance between cells i and j, and α, β, γ, and δ are scaling parameters. Determination of the appropriate values of α, β, γ, and δ is an ongoing problem beyond the scope of this discussion.

In the absence of wind or slope (when V_w, and $S \to 0$), this equation reduces to

$$U_{ij} = \frac{aI_i}{d_{ij}^{\delta}} \tag{4}$$

which is a simply parametrized version of eq. (1).

This urgency is calculated for every burning cell at every time step. If any parameter were to change from one time step to the next, the cell would simply respond by changing the urgency of its request to its neighbor at the next time step. This way, if there were to be a sudden change in windspeed or direction, every cell would simply react accordingly.

In summary, the cell acts as a broker for the fuel components, taking the burn messages, adjusting the magnitude of the message for localized conditions, and transmitting the messages to the appropriate neighbors.

6.4 CELL SURFACE

At the next highest level in the hierarchy is the cell surface, or the "world" that the cells inhabit. As was mentioned in section 2.2, the cell can be envisioned as the actual surface of the ground, or as a two-dimensional list of cell agents. The cell surface is responsible for knowing the geographic locations of every cell, as well as the topological relationships between them. In a cellular lattice, this means that for any cell, the cell surface can identify the neighbor who is at the north, northeast, south, etc. of the cell. This function is a direct extension of the role of *linked lists* for keeping track of agents in multiagent simulations—knowing the location, and being able to traverse the list or retrieve the ith member of the list is a fundamental tool for managing multiple agents.

Another important function of lists is managing execution of functions of the agents in a way that ensures an acceptable model of concurrency. In this simulation, when a cell burns, it sends messages to neighboring cells. Since its neighboring cells are likely to be burning as well, any cell will also need to be able to receive and process messages from several neighbors as it sends its own messages out.

This was addressed in this simulation by splitting up the *burn actions* and the *sending and processing of messages*. When a cell is getting ready to "do its thing" (burn a bit), it needs to be sure that it has processed all the

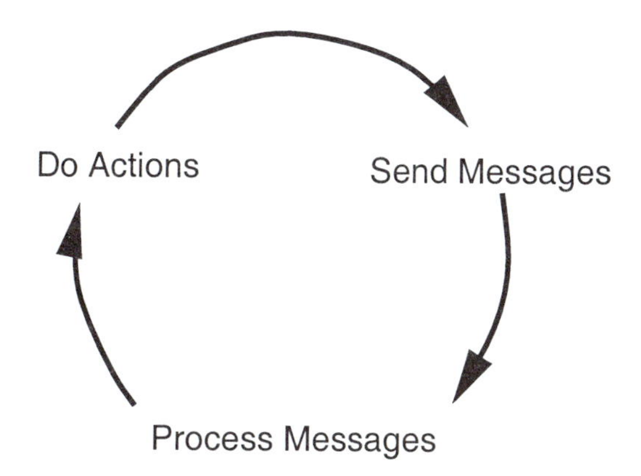

FIGURE 4 The burn cycle of the cell object.

inputs that it is to receive from its neigbhors. When it burns, it needs to make sure that its messages have been delivered to the relevant neighbors. To accommodate this, each cell is provided with a "message cache" (think of it as a mailbox). When the cell burns, it sends messages to some of its neighbors. Simultaneously, it will be receiving messages from its neighbors. Before its next time step, the cell will need to evaluate these messages and convert them to inputs for its own burn model.

By segregating the burn function of the cell into a *doActions* and a *processMessages* method, the cell surface can give a reasonable approximation of concurrency. In the *doActions* phase, the cell surface first traverses the list of all its member cells, asking them to do their burn thing. As each cell burns, it sends messages to neighbors, which get stored in the message cache of each cell for future reference. During the *processMessages* phase, each cell is traversed again and the message caches are evaluated. Each cell evaluates its messages and decides on its inputs for its fire model(s) at the next burn step. Cells that have "decided" to ignite at this time step merely flag themselves as ignited, and wait for the next *doActions* phase to actually start burning. Only after this second traversal is complete, and every cell has processed its inputs, is the action phase permitted to start again. The cycle is illustrated in figures 4 and 5.

6.5 THE DYNAMIC MODEL

The cell surface is the object that receives the instructions to do a "time step." However, at the time step, the cell surface sends requests to its multitude of cells, and the cells in turn send requests to their fuel components. The fuel components do the actual burning, but the cellular structure coordinates the burn activity and produces an overall picture. The fuel components work by

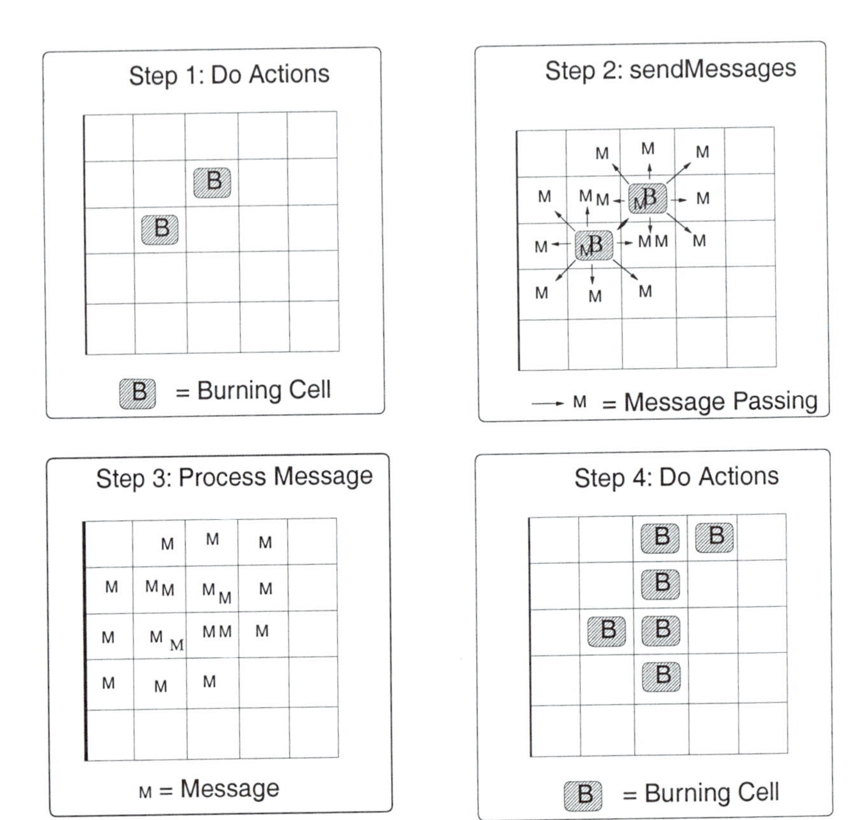

FIGURE 5 The cell surface as the burn cycle progresses. Step 1: cells are burning. Step 2: burning cells send messages to neighbors. Step 3: cells evaluate messages, decide whether to burn. Step 4: some of the cells have accepted the request and are burning.

their internal rules, while a global "fire" behavior emerges at the cell surface (landscape) level.

7 ADDING FREE-ROAMING AGENTS

In section 2.3, an example of a dynamic landscape object was presented, which may or may not provide a realistic model of wildfire propagation. However, the software object presented in section 2 gives a way to model diffusion processes using GIS datasets and simple intercell rules. While wildfire was used as the example, it is hoped that the reader will see the applicability to such a modeling process to other diffusion processes such as vegetation encroachment and groundwater flow.

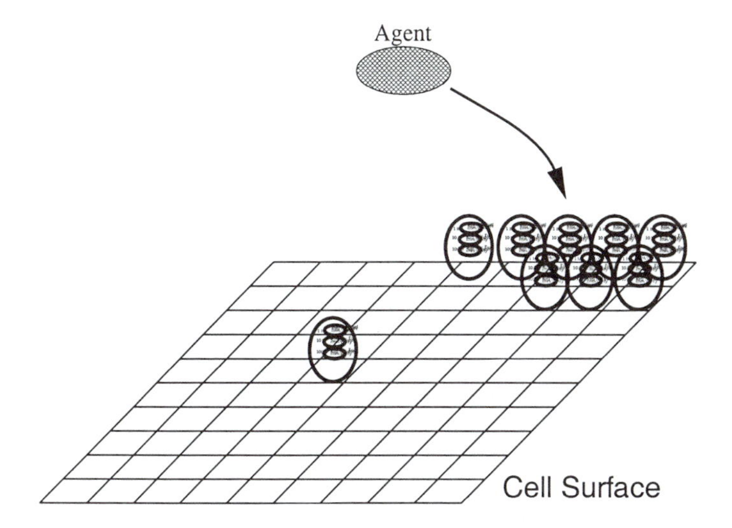

FIGURE 6 Agent in cell surface.

It is conceivable that one would wish to explore situations where free-roaming agents would inhabit this dynamic landscape, who are capable of modifying the landscape while the landscape is modifying their behavior. Possible applications would be grazing animals who graze the vegetation in an area, leaving the selectively cropped vegetation to grow back by its own rules; recreational users in an area who use the most attractive areas, but whos collective impact reduce the area's attractiveness; or addition of firefighters in the fire model description described earlier, who attempt to contain an existing burn by increasing moisture or decreasing burnable biomass in cells, one cell at a time.

8 REFERENCING AN AGENT TO THE GIS-CA

In the Swarm libraries, the spatial information of an agent (such as location and proximity to other agents) is associated with an object called a *Discrete2D*, which is at its core a two-dimensional array. The coordinates inside the two dimensions represent the location of a grid cell in a raster world. An agent is associated with this two-dimensional object by the *Discrete2D* storing a pointer to the agent at the coordinates corresponding to the agent's location in the world.

In the GIS-CA, every location is occupied by a *cell object*, which has already been described in some detail. When a free-roaming agent is associated with the GIS-CA surface, it in fact is associated with the cell object that occupies the coordinates on the cell surface (fig. 6); it is the cell object that contains the pointer to the free-roaming agent. This gives the free-roaming agent immediate access to all of the cell-object's attributes, including methods for modifying the cell's attributes. Thus, if the free-roaming agent were a cow, and the cell a piece of ground, the cow may send messages to the cell informing it that it has eaten some of its grass, trampled other grass, compacted its soil, etc. The messaging can work both way; the cell is perfectly capable of sending messages to the cow along the lines that it is a hole and the cow must now fall in.

Once agents are added to the cell surface, the number of potential messages between objects increases dramatically. Agents must communicate with other agents, as well as with the cells on which they "stand." Cells must communicate with the agents "on top" of them, with their neighboring cells, with their cell surface, and possibly with any embedded models that they may contain. If more than one kind of agent inhabits this world, then the number of potential messages being passed goes up again, possibly exponentially. As one contemplates this, the advantages of using established software libraries of proven stability for list management and message passing become more apparent.

9 SAMPLE APPLICATIONS: ADDING FIREFIGHTERS

This next section describes the potential addition of firefighters into the wildfire model described in section 2.3. While the landscape is acting by its own set of dynamic rules, it is possible that agents acting within the landscape will have a significant impact on the landscape dynamics; in the case of firefighters, influencing the course of the burn is the explicit goal of the agents. The burn itself is a dynamic process that firefighters often refer to in terms of a living thing. While there may be an infinite number of possible ways to stop a fire, the actual containment of the fire is a process that will be severely constrained by the numbers of firefighters, their available tools, and their ability to move around the landscape and coordinate with each other. Even with the best conceivable combination of firefighters and equipment, the firefighters are still at the mercy of environmental conditions and random chance. Given all of the disadvantages to the firefighters, the behavior of the burn is still highly influenced by the behavior of the firefighters.

This is not intended to be a detailed discussion of firefighting models. A few basic assumptions will be presented about the firefighting agents, with enough information that the reader will understand how such agents could be added into the cellular model presented earlier. Discussions of how to create "intelligent" agents who navigate in a GIS-generated landscape are presented elsewhere in this book, and will not be covered in detail here. The main purpose of the following description is to establish the principles for a dynamic agent/environment interaction model.

10 ASSUMPTIONS IN DESIGN OF AGENTS

For an agent to roam the landscape fighting fires in the model described above, the agent needs minimally to

- be able to traverse the landscape from cell to cell,
- be able to modify the state of a cell (increase fuel moisture or remove burnable material), and
- survive (avoid being burned or breathing smoke).

With these minimal abilities, the agents could roam the landscape like ants, spraying water or fire retardant on vegetation and cutting underbrush wherever they happen to be. There obviously needs to be other abilities to a firefighter, like some degree of intelligence to know its destination, strategy, and contribution to group goals; these will be addressed after these minimal requirements have been discussed.

11 THE AGENT'S COMMUNICATION WITH THEIR WORLD

As an agent is standing on a plot of ground (cell), and wishes to move to a neighboring cell, it will need to know things about the area that it's traversing. Some things, like difficulty of traversing, can be deduced by knowledge of the slope and vegetation type; a grassy field is easier to cross than a fir stand or a lake, and traveling along or down a slope is easer than going up a slope. Additionally, traveling through a burning or recently burned area may be impossible. All of these attributes of the cell can be retrieved through a simple query from the firefighter agent. The translating of landscape attributes into costs of traversal is a well-established technique in spatial models.

In the GIS-CA libraries, since the agent is directly associated with the cell on which s/he is "standing," querying the state of the cell or the neighboring cells is a trivial matter. Through the hierarchy described above, every object knows its relationship to other objects through embedded functions. While it's technically possible to code an agent that carries around a list of the cells upon which it's standing and the immediate surroundings, there is no need. If the agent wishes to know the state of a fuel component in the cell to its north, it simply passes the request to "its" cell, the cell passes the request to the cell surface to find its northern neighbor, the cell surface passes the request to the neighboring cell, the cell passes the request to its fuel component, and the fuel component responds. The chain of communication is hidden from the agent making the request (and to the modeler writing the code for the request); all the agent knows is that it's getting information from "its" north. A rudimentary action that would be a nightmare of subroutines in Fortran or C becomes a simple request in an object-oriented structure.

For the agent to modify the state of a cell (its "own" cell or a neighboring cell), the actions are similar. The agent passes a request to the object for it to modify itself, and the cell complies (or doesn't, as per its internal state). Thus, to simulate an agent with a firehose, one would have the agent send a message to a neighboring cell requesting that it increase its moisture corresponding to the amount of water that is being sprayed on it.

The firefighters in this simulation are managed by a linked list. At each time step, a message is sent to every member of the list to "step." To have firefighters and their landscape coupled into a single virtual world, one only needs to associate the firefighters with locations on the world. As the simulation runs, one needs to ensure that the firefighters and the cell objects are all receiving instructions to update themselves (do things) at the appropriate time intervals. As this typically involves tens of firefighters and tens of thousands of cells acting in (virtual) parallel, the need for proper list management and scheduling is crucial. These aspects are handled directly from Swarm's list managers and schedulers. Lists are created, and the scheduler is simply told to send a "step" request to the landscape (who forwards the requests to its cells), and to every member of the firefighter list.

12 ON ADDING INTELLIGENT BEHAVIOR TO AGENTS

The behavior described in the previous paragraphs can account for the immediate needs of the firefighters. Firefighters require some level of intelligence in their actions, including abilities to read the landscape, anticipate future behavior of the fire, pick optimal paths to firefighting locations, coordinate activities with other firefighters, and evaluate the success of their strategy on the fly. Techniques for simulating these kinds of decisions is a subject beyond the scope of this chapter. There is an extensive body of literature in adding intelligence to agents; for a discussion of some of the practical considerations of this subject, see Deadman and Gimblett [9] and Gimblett et al. [15, 16].

What has been presented here is an appropriate set of inputs that such an "intelligent" agent would require if turned loose on the cell surface. However, even with a reasonable artificial intelligence engine attached to the firefighters, the simulation described will take on an unpredictable quality, where the outcome will be unable to be determined without actually running the simulation. This unpredictable aspect, even in the face of predictable behavior by any one agent, provides for the most exciting components of such a simulation.

13 ON EXPLORING AGENT-ENVIRONMENT INTERACTION

By elevating the landscape to an equal player in agent-based simulations, a new class of scenarios open for exploration via computer modeling. It has long been known that populations affect their environments, and that changes in the environment affect how the population acts. For reasons mentioned in section 1, it has not been practical until recently to explore these interactions via computer simulation.

Reusable software packages such as Swarm have dramatically increased the availability of complex system simulation exploration to researchers who are not experienced computer programmers. The ideas of dynamic populations and dynamic environments is certainly not new, and ingredients to create such simulations have been around for a while. Computer technology has been cheap and available for more than a generation of researchers. Newer theories about complexity and complex systems such as chaos theory, fractal geometry, and self-organized criticality are not quite new any more. Agent-based simulations and cellular automata have been well-used tools for simulating dynamic populations and landscapes. Their lack of application to agent/environment interaction has certainly not been for lack of tools or lack of talent on the researchers.

Though all of these ingredients have been around for some time, combining them into an integrated simulation has been difficult. The availability of software libraries such as Swarm may well be the catalyst that allows for creative recombination of research ideas into new packages, where the strengths of each of the components creates a picture that is greater than the sum of its constituent parts, and distinct from any of its parent disciplines. The emergence of a new class of problems may well mirror the emergence of new disciplines similar to what occurred with the coupling of relational databases and computer cartography to create geographic information systems, or the combining of statistics and economics to create econometrics.

It is anticipated that this modeling tool will allow access to a number of questions that wildland ecologists will be able to explore via simulations. While it's not possible to name or even to anticipate the number of applications, some areas that might be addressed are:

- What are stable plant communities, given an area inhabited with animals of given habits and densities? Plant communities inside exclosures (simple fences in wild areas that keep deer, rabbits, etc., from grazing inside) often give rise to plant communities that are surprisingly different from the world outside the fence, and introductions of different grazers can have dramatic and unanticipated effects on the makeup of the plant community.
- What is the viability of a predator who relies on certain kinds of brush for cover, if there are minor disturbances to that cover?
- How resilient is a plant community in an area that has people frequently walking through it? Any one person has an impact that is barely measurable, but sustained foot and bicycle traffic may have devastating effects.

While all these questions have been partially answered by existing simulation technologies, there are other questions that are difficult to address without a suitably dynamic representation of agent/environment interaction.

REFERENCES

[1] Agha, G. A. *Actors: A Model of Concurrent Computation in Distributed Systems.* Cambridge, MA: MIT Press, 1987.

[2] Andrews, P. L. "BEHAVE: Fire Behavior Prediction and Fuel Modeling System—BURN Subsystem, Part 1." Technical Report INT-194, U.S.F.S. Intermountain Forest Research Station, Ogden, UT, 1986.

[3] Bevan, K., ed. *Distributed Modeling in Hydrology: Application of the TOPMODEL Concept.* New York: Wiley, 1997.

[4] Bevins, C. D. "Firelib C Function Library." March 1999. U.S.D.A. Forest Service, Rocky Mountain Research Station. ⟨http://www.fire.org/firelib/firelib.html⟩.

[5] Box, P. W. "The Swarm GIS-CA Libraries." June 1999. Kenge Development Project, Department of Geography and Earth Sciences, Utah State University. ⟨http://www.nr.usu.edu/swarm⟩.

[6] Briggs, D., J. Westervelt, S. Levi, and S. Harper. "A Desert Tortoise Spatially Explicit Population Model." In *Proceedings, Third International Conference/Workshop on Integrating GIS and Environmental Modeling*, Santa Fe, NM, January 21–26, 1996. Santa Barbara, CA: National Center for Geographic Information and Analysis. ⟨http://www.ncgia.ucsb.edu/conf/SANTA_FE_CD-ROM/main.html⟩.

[7] Clarke, K. C., J. A. Brass, and P. J. Riggan. "A Cellular Automaton Model of Wildfire Propagation and Extinction." *Photogram. Eng. & Remote Sensing* **60(11)** (1994): 1355–1367.

[8] Clarke, K. C., S. Hoppen, and L. Gaydos. "A Self-Modifying Cellular Automaton Model of Historical Urbanization in the San Francisco Bay Area." *Envir. & Plan. B: Plan. & Des.* **24** (1997): 247–261.

[9] Deadman, P., and H. R. Gimblett. "The Role of Goal-Oriented Autonomous Agents in Modeling People-Environment Interactions in Forest Recreation." *Math. & Comp. Model.* **20(8)** (1994).

[10] Dean, J. S., G. J. Gummerman, J. M. Epstein, R. Axtell, A. C. Swedlund, M. T. Parker, and S. McCarroll. "Understanding Anasazi Culture Change through Agent-Based Modeling." Working Paper 98-10-094, Santa Fe Institute, Santa Fe, New Mexico, 1998.

[11] Drogin, E. B. "Evaluating the Boater Experience: The Interrelationship of Recreational Use, User Contacts, Experiential Impacts, Satisfaction and Displacement." Ph.D. diss., College of Health and Human Development, Pennsylvania State University, 1991.

[12] Edwards, T. C., C. G. Homer, S. D. Bassett, A. Falconer, R. D. Ramsey, and D. W. Wight. "Utah GAP Analysis: An Environmental Information System." Technical Report 95-1, National Biological Service, Utah Cooperative Fish and Wildlife Research Unit, Utah State Univeristy, Logan, Utah, 1995.

[13] Forrester, J. W. *Urban Dynamics.* Cambridge, MA: MIT Press, 1969.

[14] Fried, J. S., and B. D. Fried. "Simulating Wildfire Containment with Realistic Tactics." *Forest Sci.* **42(3)** (1996): 267–281.

[15] Gimblett, H. R., B. Durnota, and R. M. Itami. "Spatially-Explicit Autonomous Agents for Modeling Recreation Use in Wilderness." *Complex. Intl. J.* **3** (1996).

[16] Gimblett, H. R., R. M. Itami, and B. Durnota. "Some Practical Issues in Designing and Calibrating Artificial Human-Recreator Agents in GIS-Based Simulated Worlds." *Complex. Intl. J.* **3** (1996).

[17] Hammon, G. A., H. K. Cordell, L. W. Moncrief, M. R. Warren, R. A. Crysdale, and J. Graham. "Capacity of Water-Based Recreation Systems, Part II: A Systems Approach to Capacity Analysis." Technical Report 90, Water Research Institute of the University of North Carolina, 1974.

[18] Kohler, T. A., C. R. V. West, E. P. Carr, and C. G. Langton. "Agent-Based Modeling of Prehistoric Settlement Systems in the Northern American Southwest." In *Proceedings, Third International Conference/Workshop on Integrating GIS and Environmental Modeling*, Santa Fe, NM, January 21–26, 1996. Santa Barbara, CA: National Center for Geographic Information and Analysis. ⟨http://www.ncgia.ucsb.edu/conf/SANTA_FE_CD-ROM/main.html⟩.

[19] Luo, W., R. E. Arvidson, and M. Sultan. "Ground-Water Sapping Processes, Western Desert, Egypt." *Geol. Soc. Am. Bull.* **109** (1997): 43–62.

[20] Minar, N., R. Burkhart, C. Langton, and M. Askenazi. "The Swarm Simulation System: A Toolkit for Building Multiagent Simulations." 1998. Santa Fe, NM: Santa Fe Institute. ⟨http://www.santafe.edu/projects/swarm/overview⟩.

[21] Miyamoto, H. "Simulating Lava Flows by an Improved Cellular Automata Method." *Comp. & Geosci.* **23** (1997): 283–292.

[22] National Center for Geographic Information and Analysis, ed. *Third International Conference/Workshop on Integrating GIS and Environmental Modeling*, Santa Fe, NM, January 21–26, 1996. Santa sBarbara, CA: National Center for Geographic Information and Analysis. ⟨http://www.ncgia.ucsb.edu/conf/SANTA_FE_CD-ROM/main.html⟩.

[23] Pahl-Wostl, C. *The Dynamic Nature of Ecosystems.* New York: John Wiley & Sons, 1995.

[24] Provenza, F., J. J. VVillalba, and M. Augner. "The Physics of Foraging." In *Proceedings of the XVIII International Grasslands Congress III*, edited by J. G. Buchanan-Smith, L. D. Bailey, and P. McCaughey, 99–107. Calgary, Canada: Association Management Centre, 1999.

[25] Provenza, F., J. Villaba, M. Werner, and M. Augner. "Self-Organization of Behavior: From Simplicity to Complexity without Goals." *Nutritional Resh. Rev.* **11** (1998): 1–24.

[26] Pyne, S. J. *Introduction to Wildland Fire.* New York: John Wiley & Sons, 1984.

[27] Remm, M. "Robot Auto Racing Simulation Project." March 1997. Estonia: Institute of Molecular and Cell Biology, Tartu University. 1998. ⟨http://www.ebc.ee/~mremm/rars/main.htm⟩.

[28] Resnick, M. "Learning About Life." In *Artificial Life: An Overview*, edited by C. G. Langton, 221–241. Cambridge, MA: MIT Press, 1996.

[29] Savage, M. and M. Askenazi. "Arborscapes: A Swarm-Based Multi-Agent Ecological Disturbance Model." *Geograph. & Envir. Model.* (2000): submitted. Also Santa Fe Institute Working Paper 98-06-056.

[30] U.S. Geological Survey. "7.5 Minute Digital Elevation Models." Digital Elevation Data Archives, U.S. Geological Survey, Reston, VA, 1979.

[31] Wesseling, C. G., D. Karsenberg, W. P. A. van Deursing, and P. A. Burrough. "Integrating Dynamic Environmental Models in GIS: The Development of a Dynamic Modelling Language." *Trans. GIS* **1** (1996): 40–48.

Geographic Information Systems and Agent-Based Modeling

James D. Westervelt

1 INTRODUCTION

As we enter the twenty-first century, decreasing computer costs continue to result in the development of new generations of computation-based land management tools. Geographical information systems (GIS) emerged from laboratories in the mid-1970s [10] and became, by the mid-1990s, a mainstay of the land manager's toolbox. During this time, GIS technicians moved from a role as the sole GIS operator to one that includes the design and development of decision support systems (DSS) developed on a foundation of GIS technologies. The land manager is now provided with interactive environments that provide limited, but directed, manipulation and querying of GIS data files. GIS, now a mature technology, is an incomplete tool. It allows for the capture and analysis of landscape system state information. Historic maps, satellite and high-altitude photography, survey data, communication networks, and other sources now provide a rich record of historic and present conditions. Our culture has now embraced the idea that ideas about the state of our landscapes should be formally captured. Ideas about how landscape works, however, remain concepts in the minds of land managers. These ideas develop through formal education, continual review of the literature, and perhaps most importantly, personal experiences during one's career. While applications of

GIS technologies reflect land manager knowledge about the land processes, the GIS does not easily allow that knowledge to be formally captured.

The scientific community has now developed a rich array of formalized landscape processes in the form of computer simulations. Hydrologic engineers have a wealth of groundwater, surface water, and overland water flow simulation models [2, 3]. Regional planners offer simulations of urban growth [4] and traffic flow. Ecologists are developing plant and community succession models [5, 15], and models of habitat responses to land use patterns [18]. Working with scientists, land managers can apply such models to their land management decision processes. However, each model focuses on only part of the landscape system. Full use of the scientific models to affect land management decisions will not occur until the models are integrated as components of a simulation-based geographic modeling system (GMS). Over the next decade experimental systems are expected to result in the release of commercially viable land simulation modeling environments. As we currently capture our understandings of state information in GISs, we will formalize our understandings regarding the dynamics of the landscape in GMSs.

In a GIS, digitally captured maps are most useful when they can be analyzed with respect to other digital maps. Similarly, today's stand-alone simulation models that represent parts of the landscape processes must be combined and analyzed with respect to other digital models. This chapter explores current opportunities for linking landscape simulation models into prototype GMSs. We begin by categorizing integration approaches into a small number of approaches and exploring the costs and benefits of each approach. An example of how to integrate a simulation modeling environment with a GIS is then presented in some detail. The integration approach is identical to the approach currently used to link programs into a common GIS. An exploration and analysis of the shortcomings of this approach lead to a discussion of anticipated improvements.

2 THEORY

Most substantial software products available today differ dramatically from software of the past. Perhaps the most important difference is the sheer size of the computer programs. The principal commands found in DOS or earlier versions of UNIX required on the order or 100 Kbytes of computer memory to run. Early versions of UNIX shipped in 1980 on computers with one to four megabytes of memory. Today it is not uncommon to run programs that require tens of megabytes of memory. The class of programs of concern in this chapter can require data spaces that accommodate a number of large map layers that are changing over time. A dozen 1000-by-1000 digital raster map layers can require 12 Mbytes of main memory if the category values are integers in the range of 0 to 255. Floating-point raster maps will require up to eight times that requirement. The programs and data space requirements have grown enormously.

To develop today's new software, many software programmers now must collaborate, directly or indirectly. Software libraries, or application programmer interfaces (API), provide an indirect collaboration; they are developed to be maximally useful to a broad range of applications. These are then provided "as is" to other programmers. APIs are now available to provide database management system (DBMS) functions, low- to high-level graphics, access to operating system resources, interprocess communications, cross-platform communications, cross-language communication, etc. It is now common to develop software programs of several thousand lines, which rely on the correct operation of a number of APIs, each of which represent thousands of lines of code. Collaboration can also be direct and immediate. To turn out a product in time to meet market demands, software design teams must collaboratively settle on an overall design that allows individual programmers to focus on a part of the design. Each programmer toils in anticipation that the rest of the team will generate all of the other components necessary to create a final product.

While APIs are excellent, powerful, and necessary to support software development in the future, integration through the API concept is not always appropriate. APIs, by definition, are collections of software intended to be subsumed inside end-user-oriented software products. Challenges that are not solved through APIs emerge when it becomes important to combine two or more existing end-user products. This challenge is arguably the most critical obstacle facing the development of discipline-specific dynamic landscape simulation modeling into land-management-oriented DSSs.

Integrating discipline-specific modeling must occur before the models can have full impact on land management decision processes. A number of excellent efforts have resulted in our ability to integrate discipline-specific models. The modular modeling system (MMS) [14] provides a graphical interface environment that allows managers and scientists to assemble computer subroutines into location-specific landscape simulation models. A growing number of subroutines for hydrology, weather, insulation, transpiration, snow melt, and plant growth can be linked together by scientists without direct and immediate programmer support. The spatial modeling environment (SME) [16, 17] uses a modular modeling language (MML) to establish libraries of integratable submodels. A new collaborative modeling environment (CME) [22] provides a graphical user interface (GUI) to SME and potentially to other modeling environments. Model developers are allowed to probe MML libraries for salient submodel components, link those components, and run simulations. The United States military has contributed to the development of simulation modeling integration environments such as the high-level architecture (HLA) [8], distributed interactive simulation (DIS), and the dynamic information architecture system (DIAS) [7]. HLA is an architecture that is used by "federations" of separately running models to interact at run time over a network and is intended to replace the earlier DIS approach. DIAS was developed also as an improvement to DIS and provides a unique approach to capturing existing stand-alone software models into collaborative participation with other software. The across-trophic-level system simulation (ATLSS) environment [6] was developed to support a complex multiresolution landscape model of the natural components of the Everglades system.

Each of these simulation modeling environment examples has addressed the challenge of simulation model integration. Several different strategies are represented including file sharing, direct communication, compilation of all processes into single programs, and interprocess communication through network protocols. All solutions seek to find a balance between a number of competing linkage objectives. These objectives include:

- *Integration Speed*: The programmer time involved in linking the programs.
- *Programmer Expertise*: The required level of software development expertise.
- *Avoidance of Multiple Authorship*: Must a software product developed by one programmer be modified by an integrating programmer? If so, how much? Any modification damages the ownership link. Significant modification breaks the link. Loss of the ownership link results in the loss of support by the original author(s).
- *Execution Speed*: How rapidly does the integrated software execute?
- *Simultaneous Execution*: Can components of the system run simultaneously and communicate with one another? Can the components operate on separate platforms?
- *Debugging*: How difficult is it to track down execution errors in the integrated system?

Solutions to the balance between the needs of different projects might be fit into three general categories: loose, moderate, and tight. A description of each approach is developed below followed by a review of the approach with respect to the above list of objectives. Two other "add-on" approaches are also considered: distributed processing and object oriented. They can be adopted in addition to the loose, moderate, and tight approaches. Table 1 summarizes this information for quick reference.

3 APPROACH: LOOSE

A loose connection between computer programs typically involves (1) the asynchronous operation of each model by itself, and (2) the exchange of information using simple ASCII files. This approach is extremely common and is used, for example, in UNIX shell script and DOS batch file operations. Typically output files from one program must be massaged with standard DOS or UNIX operations to extract, reformat, and otherwise edit the data.

TABLE 1 Quick comparison of three approaches to software integration. Boxed blocks indicate most desirable characteristics.

	Loose	Moderate	Tight
Integration speed	Fast	Medium	Slow
Programmer expertise	Low	High	High
Avoidance of multiple authorship	High	High	low
Execution speed	Slow	Medium	Fast
Simultaneous execution	Low	Low	High
Debugging	Easy	Moderate	Hard

Occasionally, specific new programs must be developed to perform format modifications.

Integration Speed. Integration can be medium to fast depending on the intricacies involved in extracting and reformatting data. The approach may involve modifying existing programs to put data out in a usable ASCII format.

Programmer Expertise. Assuming that no programmer expertise is required to modify existing programs, the skill level required will be that of an expert computer user. Scripting languages like AppleScript, UNIX shell scripts, DOS batch files, or cross-platform PERL might be required to manage input and output files.

Avoidance of Multiple Authorship. Avoidance is high. Any modifications to existing programs to generate ASCII output will be typically minor and easily handled by program authors.

Execution Speed. Slow. The process of converting binary data from an internal program format to ASCII, storing that data to disk, and then reading that data into binary data storage in a second program is very slow.

Simultaneous Execution. Low. In this approach, each process is run sequentially. Using UNIX pipes, it is possible for several processes to be running simultaneously and across a number of different processors on one or more machines. Pipes allow the output of one program to be fed directly into the input stream of another program without the interim requirement for storing the transient information. The flow of information is one way.

Debugging. Easy. Because each program runs as it was originally designed and built. The debugging requirements added through integration only involve the proper manipulation of the ASCII files.

4 APPROACH: MODERATE

The moderate approach is achieved through common files in standard file formats. Files are accessed through standardized library functions. Ideally, file formats are binary to save file storage requirements, and to speed the reading and writing of data. Virtually all commercial GIS systems use this approach. For example, the geographic resources analysis support system (GRASS) [23] stores GIS data primarily in raster, vector, and point files. All GRASS-compliant programs access data in these files through standard GRASS API subroutine calls [21]. GRASS programs work together through these files. A modern suite of water simulation software that includes the groundwater modeling system (GMS), the surfacewater modeling system (SMS), and the watershed modeling system (WMS) [2, 3, 18] uses this efficient approach. A variant of the moderate approach utilizes an early invention in the UNIX world: pipes. A number of sequentially operating processes communicate directly with one another by sending output from one to the input stream of the next sequential operation. Any number of processes can be strung together and hidden behind the scenes in a shell script (batch file) providing an end-user with the impression that there is only a single step need to convert an input file to a required output. The Poskanzer suite of image input/output and processing programs used this approach [20].

Some commercial vendors have replaced and extended the scripting approach described above with much more sophisticated capabilities. ESRI still offers the arc macro language (AML) scripting approach, but in recent years has offered a new language, Avenue, that perhaps lies somewhere between traditional scripting and computer programming. Both are cross-platform languages that avoid making use of operating system specific capabilities. Other approaches have attempted to be more open. The GRASS community offered Xgen [1], which allowed sophisticated shell programmers to generate UNIX X-windows user interfaces that integrated any available UNIX commands—including the GRASS extensions to UNIX. GRASSLAND 1.1 is programmed in the freely distributed Tcl/Tk application generation language—a much more sophisticated and complete cousin to Xgen. As Arcview users can modify their interface using the Avenue environment, so too GRASSLAND users modify theirs by writing Tcl/Tk scripts.

Integration Speed. Rapid. Because no file manipulations are required, the user need only concentrate on the appropriate sequential linking of the processes. In addition to the scripting approaches described above in the "Loose" approach, many integrated environments offer specialized GUIs. The ERDAS image-processing package [9], the KHOROS image-processing system [13], and the GRASSLAND GIS software [11] all provide graphical canvases on which users can visually link data and analysis processes together. The resulting defined model can then be saved for sharing with others and for later recall and modification.

Programmer Expertise. Low to medium. Programmer expertise is typically not required at all to link the software capabilities of existing programs. This has been the primary selling point for modern commercial GIS and image-processing packages. As the sophistication of user interfaces has improved, the sophistication of users to modify these environments has increased.

Avoidance of Multiple Authorship. High. The basic programs are designed to run "as is" with no option for integration specialists to modify the programs. The integration software, in the form of scripts or Tcl/Tk and Avenue types of programs, is a separate piece of software.

Execution Speed. Moderate. Individual programs are communicating through files requiring significant disk I/O.

Simultaneous Execution. Low. Programs are still running sequentially, or possibly iteratively.

Debugging. Easy. There is no significant complications associated with process integration as the processes can be debugged individually.

5 APPROACH: TIGHT

Tight interactions are characterized by a simultaneous operation of multiple models that allows intermodel communication during model runs. One variant, covered by this approach, is to compile all of the different modeling components together into a single program. A second is to facilitate intermodel communication across different process that may be running on one or more networked computers. This approach is covered separately below in the "Distributed Processing" approach. The ability to develop single programs by linking multiple components is not typically supported in commercial software packages—except in support of traditional programmers. The integration work described in detail in this chapter fits this category. Two disparate landscape simulation model approaches resulted in two very different models. A tightly linking strategy was then used to combine the separate models. A landscape simulation modeling package, called the modular modeling system (MMS) [14], has been developed by a consortium of government-sponsored research. MMS offers a graphical user interface through which watershed management scientists can graphically connect the inputs and outputs of a wide variety of landscape simulation components. Once all interconnections are satisfied, the system compiles together all of the identified components into a single tightly connected dynamic landscape simulation model. By uniquely parametrizing the model modules with information specific to the watershed, the resulting program represents that watershed.

Integration Speed. Very slow. Linking two or more simulation models in this intimate fashion requires programmer agreement on fundamental programming approaches, data interchange formats, a common data dictionary, and (sometimes) programming languages. The conversion of an existing simulation model to conform to the standards, guidelines, and definitions can often result in a completely different piece of software. Often, existing software models, stripped of some or all data I/O and user interface, will be encapsulated in software that is conformant with the agreed standards. Ideally, this leaves a core of relatively untouched original algorithms often in the original language.

Programmer Expertise. Significant. Intimate knowledge of one or more programming languages as well as the algorithmic intent of the model being adapted is required.

Avoidance of Multiple Authorship. Medium to low. Unless it is clearly in the best interest of the original programmer to dramatically rewrite or rework the software, it is likely that a different programmer will be assigned the task of adapting existing code. This breaks the original ownership connection leaving the second programmer with all responsibilities for debugging any end result.

Execution Speed. Execution speeds are extremely fast, but generally limited to execution on a single processor. Because data will be shared in a native binary format, data I/O requirements between models using pipes or files is eliminated. If insufficient core memory is available in the computer, paging to disk will dramatically decrease execution speeds and can render the model even slower than loosely connected counterparts.

Simultaneous Execution. Excellent. This is the major benefit of this approach.

Debugging. Some extra debugging challenges are presented. Individual submodels should be debugged individually. Then, collections of submodels should be debugged until finally the entire model is tested as a whole. As interactions become more complex, the potential for problems with numerical chaos grow and the potential for submodels running in environments outside of the range for which they have been designed/validated increases.

6 APPROACH: DISTRIBUTED PROCESSING

This is a variant of the "tight" approach. The principal difference is that instead of compiling all of the model components into a single program, model components run as separate programs that communicate with one another during simulation runs. Communication channels can make use of fifo (first-in-first-out) files, sockets, or shared memory. An array of software libraries support interprocess communication such as Microsoft's COM/OLE, parallel virtual memory (PVM), REF, and the common object request broker architecture (CORBA). The Department of Defense's high-level architecture (HLA) [8] defines an architecture that can be realized through interprocess communication for collaborative interdisciplinary battlefield and war simulation models. Argonne National Lab's DIAS [7] also supports collaborative simulation modeling by supporting distributed processing. To become compliant, an existing stand-alone simulation model must typically be stripped of all of its user interface and data I/O operations. The stripped model is then encapsulated in software routines (specific to the integrating environment being employed) that help mediate the user interface and data I/O requirements. This allows for tight model integration with the option of distributing the simulation programs across multiple programs and, optionally, across multiple machines.

Integration Speed. Slow but, if the encapsulating mechanisms are predefined, the integration effort can be accomplished with minimal coordination with other programmers.

Programmer Expertise. High to very high. Intimate knowledge of the intricacies and implications of interprocess communication software and approaches are required.

Avoidance of Multiple Authorship. Low. Like the "tight" approach, it is most likely that the integration effort will not be carried out by the original programmer of the models of interest.

Execution Speed. High to very high. Because the individual submodel components run as separate processes, modern standard operating systems that support multiple processors can easily balance processing loads among available processors. The ability to run processes on separate platforms across a network further increases the available CPU power that can be simultaneously harnessed.

Simultaneous Execution. Very high. This is the primary advantage of this approach.

Debugging. Very difficult. Besides the difficulties associated with the "tight" approach, the extra software associated with maintaining synchrony and communication between very different processes spread, potentially, across a number of platforms increases the potential for problems.

7 ADD-ON APPROACH: OBJECT ORIENTED

Here, object-oriented programming is considered an "add-on approach" as it can be used in any of the above-described approaches. While most commercial

GIS software is based on non-"object-oriented" software development, we will continue to see more software developed with this approach. Object-oriented programming languages in use today include C++, JAVA, and Objective-C. Products already developed with nonobjective languages (e.g., Fortran, C, and Pascal) can be encapsulated within objects. This provides a mechanism to rapidly develop object-oriented APIs for existing capabilities without completely rewriting software.

Integration Speed. Increases. Objects in object-oriented programming are typically more self-contained than subroutines and have increased potential for rapid integration with other object-oriented software modules. That is, they are more easily extracted from the application for which they were originally developed and reused in other applications. This is not absolutely true, for objects may be developed that only make sense within the context of certain other objects developed for the original application. The reliance of objects on other objects can be difficult to trace when trying to reuse an object.

Programmer Expertise. Little or no change. While object-oriented programming is conceptually different from subroutine programming, there is no added complexity. Today most new programmers are being trained to think in object-oriented terms.

Avoidance of Multiple Authorship. Increases. Objects, being designed to be self-contained can be more easily used and "modified" by others without touching the original object code. Modification is accomplished through the process of inheritance of super-class characteristics without changing the original objects.

Execution Speed. Decreases—sometimes dramatically. Modern computers are designed to execute instructions within the context of certain information currently in memory. Calling another subroutine or another method requires that the current context be switched out for the next context. Stacks of these contexts can be maintained in fast cache. Compare these two approaches to accessing a value in a raster GIS array. In traditional GIS software the array is likely maintained in a single block of contiguous memory. A subroutine, knowing the starting memory address for the array can calculate the memory location of the desired data based on its row and column. The subroutine makes a call to this memory location and loads the result into CPU memory. In an object-oriented environment, the grid cell could be an object, the value in that cell could be stored as part of an object, and the full raster could be an object. Code requiring the value at a given location must ask the full raster object for the value at the given row and column. This requires a context switch. The raster object identifies the responsible cell object and asks it to return its value—context switch. The value object, which could contain any number of different values, is asked to return the integer value of its contents—for another context switch. The value is then returned to the cell object, to the raster object, and finally to the original calling object requiring another three context switches. The whole process requires six context switches plus computations within each of the steps resulting in a much slower overall process. The dramatic increases in processor speeds have helped mask object-oriented programming CPU requirements.

Simultaneous Execution. Unaffected.

Debugging. Improved efficiency. Objects are, by definition, self-contained, making it is easier to test the objects individually.

The modeling effort described below adopted a moderate integration approach using object-oriented programming. Our goal was to link a simulation model of agents interacting with one another to digital maps managed and processed by a geographic information system. The agents required rapid access to landscape characteristics and generated information appropriately captured as digital maps. A tight integration was not required because the processing of the digital maps and the interactions of the agents did not require, nor would they benefit, from simultaneous operations. Hence, the added burden involved in developing a tight interaction would not be rewarded with tangible benefits. Object-orient programming was adopted primarily because the agent-based simulation software was completely developed in Objective-C, an object-oriented version of the popular C language.

8 EXAMPLE MODERATE LINKAGE OF GIS WITH ENTITY-BASED SIMULATION

Landscape simulation modeling today has the great advantage in the growing development of digital landscape information and the geographical information systems that manage and manipulate this information. The challenge addressed in this case study is how to efficiently utilize existing GIS data in the operation of entity-based landscape simulation models. The approach described here was originally used in the integration of the GRASS GIS with the IMPORT/DOME programming language [24]. IMPORT/DOME is an object-oriented language designed to support agent-based simulation modeling [19]. The integration was accomplished to easily provide agents with digital landscapes recorded through measurement and observation in the GRASS GIS. IMPORT/DOME language based software encapsulated GRASS GIS map input and output functions that allowed IMPORT/DOME simulation models to access GRASS GIS data. The success of that effort was repeated in the integration of the GRASS GIS with the Swarm programming environment. Details for only the later effort are presented, although the driving objectives and conceptual approach were identical.

IMPORT/DOME and Swarm are both agent-based simulation environments that provide common environments for managing the passage of simulation time while objects interact with one another. The ability for objects to operate on synthetic environments associated with real landscapes required three important components: (1) the ability to read and write to files containing representations of the landscape, (2) coordinate systems associated with the landscape, and (3) the ability for entities to rapidly identify which objects in the simulation space were within a perceptual range. Only the first of these additions is discussed here because the others are not directly involved in the challenge of software model integration. Because the available databases for the areas associated with the developing simulations were in GRASS format and the GRASS GIS has an open API allowing easy access to the GRASS format data files, a "moderate" approach was selected for integrating the GIS and the modeling environments. GRASS, like all modern GISs, is built on the "moderate" approach. Therefore, this was the most integrated approach easily accomplished.

GRASS was written in C and Swarm was developed with Objective-C. The GNU g++ compiler is able to compile and link both programming languages. IMPORT/DOME is C++ based, and here, too, the GNU g++ compiler was employed to compile and link it with GRASS. For the Swarm effort, a new Swarm object was created and called GrassDiscrete2d. This object is a subclass of the Swarm Discrete2d class, thereby allowing programmers to access all of the capabilities of Discrete2d—including the ability to do run-time displays. Discrete2d is a fundamental Swarm object for managing two-dimensional arrays of numbers or objects. It provides the ability to set and retrieve integer values or object pointers. GrassDiscrete2d adds the following methods to its Discrete2d superclass:

CREATION

```
- createBegin
    Initialize the Discrete2d superclass.
- initialize
    Read the current GRASS "window" to calculate the number of
    rows and columns.
- createEnd
```

READ/WRITE/QUERY RASTER DATA

```
-(void) readMap: (char *) name ;
    Reads the map into the Discrete2d data space (array of ids).
-(void) writeMap: (char *) name ;
    Writes a map out to a GRASS raster file.
-(int) getValueAtN: (float) n E: (float) e ;
    Return the category value associated with the location
    defined by the real-world N and E coordinates.
-(void) putValue: (int) val AtN: (float) n E: (float) e ;
    Save the category value associated with the location
    defined by the real-world N and E coordinates.
-getColor: (int) c R: (float *) r G: (float *) g B: (float *) b ;
    Return red, green, and blue color intensities (range 0-1.0)
    for the category.
```

QUERYING GRASSDISCRETE2D STATE

```
-(int) getMax
    Return the maximum value found in the GRASS raster map.
-(int) getMin
    Return the minimum value found in the GRASS raster map.
-(float) getE ;
    Return the coordinate of east edge of the map.
-(float) getW ;
    Return the coordinate of east edge of the map.
-(float) getS ;
    Return the coordinate of east edge of the map.
-(float) getN ;
    Return the coordinate of east edge of the map.
-(float) getEWres ;
    Return the east-west gridcell resolution of the map.
-(float) getNSres ;
    Return the north-south gridcell resolution of the map.
-(void) setCoorTranslation: ct ;
    Communicate region resolution and extent information to a
    CoorTranslation object.
```

From a programming standpoint, a direct mechanism to allow the Objective-C code supporting the above design to call GRASS C subroutines was not found. Objective-C could, however, call C++ objects that, in turn, could call the C subroutines. Therefore, the GRASS subroutines were encapsulated in a C++ class object that was then called by an Objective-C class object.

The GRASS GIS programs are "moderately" linked because they share common data formats in disk files through the adoption of a suite of digital map access subroutines. By sharing these same subroutines, we were able to "moderately" link agent-based simulation models developed in Objective-C through the libraries of C++ object classes.

9 SAMPLE APPLICATION

We developed a linkage between Swarm and GRASS to create the individual-based cowbird behavior model (ICBM) for the Fort Hood, Texas, military installation's environmental office [12]. The model provides the installation managers with anticipated performance feedback on alternative cowbird trapping scenarios. Cowbirds are obligate nest parasites (they must lay their eggs in other bird's nests) and pose a significant threat to the population viability of two threatened bird species: the Golden-Cheeked Warbler and the Black-Capped Vireo. The model combines a trapping scenario provided at run time with information about the current state of the landscape including densities of Cowbirds, attractiveness of the landscape for cattle grazing, location of cattle herds and corrals, and distribution of habitat for the two threatened birds. Interface with the GIS is both "loose" and "moderate." To provide information about the suitability of cattle grazing areas, the GRASS GIS is used to segment the landscape into 750-square-meter areas and then analyze each of those areas for grazing suitability. This analysis is based on proximity to corrals (supplemental feeding and salt licks) and water and on the density of grasses upon which the cattle feed. A similar analysis identifies suitable habitat for the threatened species. Information about the feeding areas and cowbird, corral, and herd initial locations are then stored in ASCII files that are read at run time by the Swarm ICBM model. The "moderate" approach is used in two ways. First, at run time, a GRASS map is read directly into ICBM model to provide a graphical backdrop to the visual display of movement of the cattle, cowbirds, and cowbird traps. Second, as the simulation runs, such information as cattle herd visits, cowbird visits, and trap success are stored in arrays that are occasionally output to GRASS GIS raster data layers. Postprocessing and visualization of these layers helps the user evaluate the effectiveness of the alternative trapping scenarios.

10 NEXT STEPS

10.1 CHALLENGE

Our success in linking a standard research and development-based GIS, GRASS, with an agent-based simulation-modeling environment, Swarm, has proven useful. The success is substantial, resulting in an integration identical to that between standard GIS programs. The link is in the sharing of common data format files. Using the operating system as a programming language, or such languages as Perl, Tcl/Tk, and Avenue, Swarm-GRASS programs can be run asynchronously with other GRASS programs. Iterative programming makes it possible to update portions of the simulation space resulting in a powerful interchange between GIS and agent-based processes. For example, using the GIS, a plant community model might update the landscape every few months of simulation time while a Swarm-GRASS agent simulation updates the location of individual animals at a daily time step. Each program reads in the current state, processes that state, and writes the new state out to disk to allow the other process then to proceed likewise. Alternating between the two processes allows for the development of integrated GIS-based and agent-based simulation models.

There are two fundamental requirements that are not met through the "moderate" integration. First, the process of exchanging information through disk files is very slow. Depending on the frequency with which two different operations must be performed, the disk I/O can dominate the wall clock time expended during a simulation run. This is not a drawback of only the Swarm-GRASS link, but of the link between GIS programs in general. If a particular analysis will be performed only once, then the disk I/O overhead is minimal and easily tolerated, as it is minor compared to the wall clock time expended to support simple user interfaces. The second unmet requirement is the difficulty in compiling separately developed models together into single programs. While successes have been made linking two separately developed programs together (the author has accomplished this on many occasions), the complexity increases dramatically as one attempts to draw together more programs into a single executable program. And, a primary drawback of linking models in this way is that the original authors of the individual programs become very disinterested in assisting in the process of debugging the combined codes. The ownership thread is broken.

10.2 CAPTURE GIS CAPABILITIES BEHIND OBJECT-ORIENTED SOFTWARE

To address the first challenge, it will be necessary to change the fundamental way that GIS is structured and released. Today, each GIS function is captured in a separate executable program. UNIX programmers have long understood that the UNIX shells offer powerful and easy-to-use programming environments that use individual executables as subroutines. Shell scripting is used to rapidly develop new functionality—built by combining the several hundred different UNIX programs. If the scripts prove to be too slow or frequently used, the UNIX programmer then employs a lower-level programming language such as C or C++ to replace the shell script with more efficient compiled software. Similarly, to improve the execution time of scripts that use GRASS programs, the functionality of a number of GRASS programs must be packaged into a single compiled program. An effort is currently underway at the University of Illinois' geographic modeling systems laboratory (GMSLab) to create a library of GIS analysis functions based on the GRASS public-domain software. These functions will be easily embedded in a compiled program rather than used from the Unix shell as in current versions of GRASS. The completed subroutine library can then be used as the GIS engine behind a next-generation object-oriented GIS API based on Java, Objective-C and C++. This API will directly facilitate the "tight" integration of GIS functions and other systems (e.g., Swarm).

10.3 INTEGRATE LAND SIMULATION MODELS: APPROACHES

The second challenge is to facilitate the integration of land simulation models that individually represent part of the landscape process. Our area of focus is in the development of software to support land management decision making. At the turn of the century, individual R&D academic efforts across civil engineering, landscape architecture, geography, crop sciences, economics, regional planning, and sociology have generated a very large array of models focused on different parts of the landscape processes. Through these models scientists, economists, and engineers have formally captured their understandings of the landscape processes. Arguments based on professional judgment have been enhanced with formalized representations of the conceptual models that support these judgments. Unfortunately, the academic field-centered models are typically very well developed only with respect to the represented field. Intricate and detailed interactions represent the understandings of one discipline, while representations of subject matter from other disciplines often remain cursory. For example, a hydrologic model may represent the associated vegetation as a simple static array of flow-resistance values. Scientists working on plant and community succession models based on intimate exchanges and competition for resources might represent the critical water availability with static representations. Of course, a landscape manager works with a system that combines a dynamic hydrologic regime with complex plant and community succession processes.

While the scientific models are individually excellent at formalizing the processes associated with their respective scientific disciplines, they must be linked together to become useful in evaluating alternative land management options. There are two fundamental approaches to facilitating this required step. Call the first the "closed approach." A small group taking full responsibility for the required integration characterizes it. Alternatively, a consortium of separate organizations developing components of the final system through open standards characterizes an "open approach." Each approach has costs and benefits and each has been chosen as the model for ongoing efforts in land management system technology development. The greatest difference between the "closed" and "open" approaches are the number of parties involved in defining and applying the approach.

In both cases a set of guidelines, requirements, and/or protocols must be established. For integrating landscape simulation models, decisions in the following area must be made:

Hardware Supported. What is the fundamental computer system environment to be adopted? Will it be multiple systems?

Operating System Supported. While Microsoft dominates the personal computer market, Mac OS, Linux, and others are available for PC-class machines. The workstation market is associated with NT and various flavors of UNIX. Which will be supported?

Software to be Iintegrated. Which of the many versions of GIS, graphics, human-computer interfaces, spreadsheets, statistical packages, text processing and DBMS environments will be supported?

Software Integration Approach. How will disparate software capabilities be allowed to communicate? Will communication use files, shared memory, and/or distributed processing?

Programming Language(s). Java, C, C++, Fortran, Pascal, Smalltalk, etc.?

In the "closed approach," a few individuals are free to rapidly design, prototype, and apply a set of guidelines, protocols, and standards. Design development is rapid because it is not typically encumbered by a wide array of objectives. Application of the standards can be very efficient, as the individuals intimately involved fully understand the process. Also, this small set of software developers can insure that the application of the protocols is consistent. This allows a future user to work with a single contact that can verify compliance. Finally, any shortcomings of the compliance design and protocols can be rapidly improved or corrected. A single one-stop vendor can then market the final integrated product. This is perhaps the dominant model used for commercial software development. There are a few drawbacks. In particular, any large development effort of this kind must be funded from a central source. If a large number of funding sources is required, the coordination of efforts required to line up the funding will likely involve numerous reviews of the planned efforts, and thus undermine the concerted effort to create integrated software. Secondly, competing efforts will result in capabilities that cannot be easily combined—resulting in the inability to cooperate across an industry as the technology matures.

A number of different projects and programs have used the "closed approach" with good success. The GRASS GIS, though eventually an open system, had a core library that was under the sole control of a few individuals working at a single location. A programmer's manual [21] enabled others to add new capabilities to the package, but the fundamental database and

GRASS API were developed with the "closed approach." Efforts resulting in the design and development of the set of hydrologic simulation packages (SMS, GMS, WMS), designed with the "moderate" linking approach of GISs, also took a "closed approach." The fundamental API was designed and developed with a small team consisting of government personnel working with programmers at a single university. For these projects, the use of the API has been accomplished in a closed fashion as well. DIAS development and application has taken a similar route at Argonne National Laboratories.

The "open approach" can be superior under certain situations. If a major goal of a software development effort is to achieve adoption by a broad audience that has strong design requirements, then the "closed approach" may not yield a politically acceptable result, nor one sufficiently flexible. For example, while DIAS was being designed, developed, and demonstrated, the defense modeling simulation office (DMSO) sponsored the design and testing of the high-level architecture (HLA). Although DIAS and HLA both offer solutions to the problem of integrating disparate simulation models, only HLA involved the participation of a broad consortium with top-level support for design and implementation. DIAS is arguably better in many technical respects. Because DMSO strongly supported and funded HLA, many other agencies and commercial firms contributed time and funds to leverage the DMSO funding. Some of the advantages of the "closed approach" are the disadvantages of this approach. The process of finding solutions that equally benefit all potential partners can consume years of political effort—while "closed" solutions are developed and demonstrated, changing the technical foundation upon which the "open" efforts are based. "Open" approaches typically result in large and potentially ungainly solutions as all requirements for all participants must be considered.

Regardless of the approach, the end goal is to develop capabilities that will be used. At the cutting edge of technological endeavors the "closed" approach is favored as it provides a competitive impetus to the generation of new knowledge. "Open" approaches can favor the consumer and the less-dominant market participants. It does so by providing a common environment within which lesser corporations have an opportunity at excelling by focusing limited resources at a small part of the domain—thereby successfully competing with the larger corporations forced to spread their investment over a broader range of development possibilities. With more competition in the market the consumer benefits with better technology and lower prices.

Several different models for taking the next steps are suggested here. They are put in the context of loosely defined technology stages. The first stage, of course, is the pure research. The applied research of the second stage has been well underway in the area of landscape simulation modeling—resulting in the current wealth of single-discipline land simulation models. There are two fundamental approaches to the third phase, both of which will result in the addition of landscape simulation models to land management decision support systems (DSS). If (1) there is a sufficient commercial market, and (2) a commercial vendor perceives this market, the third phase will be dominated by the commercial industries. If the market is insufficient to generate a profit, a continued second stage will generate the landscape simulation model-based DSS prototypes. Currently, DSS are being uniquely created for each management need.

11 CONCLUSION

We have identified four levels of integration: "loose," "moderate," "tight," and "distributed processing," and recognized the optional addition of object-oriented programming to these approaches. Current commercial GISs are composed of stand-alone programs that communicate through commonly understood data formats—"moderate integration." This approach has been successful to date and will continue to provide the primary method of integration into the next decades. An example of how this approach was readily used to integrate the Swarm simulation modeling environment with the GRASS GIS was discussed. Land management communities that have adopted GIS will be demanding a new class of geographic modeling systems. GMSs will add formalizations of our understandings about the dynamics of the landscape to our current formalized GIS map layers. Early demonstrations of the capabilities of GMS are already combining dynamic simulations with GIS static formalizations. We explored such an example: the linking of the Swarm agent-based simulation modeling environment with the GRASS GIS. Using the GRASS map data interface libraries, we demonstrated the use of "moderate" integration, which is the foundation of GIS. Evaluating this approach we found that this level of integration is adequate if GIS data is only read once and model output stored as GIS data is only saved. Like all GIS programs "moderately" integrated, maps are read, an analysis is completed, and optionally, maps are written out. We are now embarking on a next level of integration in which file I/O (and any file reformatting) is avoided. GIS operations will be handled through subroutine calls and object methods rather than system calls. This will allow GIS operations to be conducted on digital maps already in computer memory. However, any software developer must evaluate the "loose," "moderate," and "tight" options for software linking with respect to each application.

REFERENCES

[1] Buehler, K. "The Xgen Application." *The X Resource* 4 (1992).
[2] Coastal and Hydraulics Lab. "Groundwater Modeling System." U.S. Army Corps of Engineers, Vicksburg, MS, 1998.
[3] Coastal and Hydraulics Lab. "Watershed Modeling System." U.S. Army Corps of Engineers, Vicksburg, MS, 1998.

[4] Clarke, K. C., and L. Gaydos. "Long-Term Urban Growth Prediction for a Cellular Automaton Model and GIS." *Intl. J. GIS* **12(7)** (1998): 699–714.

[5] Costanza, R., R. F. Sklar, et al. "Modeling Spatial and Temporal Succession in the Atchafalaya/Terrebonne Marsh/Estuarine Complex in South Louisiana." In *Estuarine Variability*, 387–404. New York: Academic Press, 1986.

[6] DeAngelis, D. L., L. J. Gross, J. W. Day, Jr. "Landscape Modeling for Everglades Ecosystem Restoration." *Ecosystems* **1** (1998): 64–65.

[7] Christiansen. "A Flexible Object-Oriented Software Framework for Modeling Complex Systems with Interacting Natural and Societal Processes." In *Proceedings of the 4th International Conference on Integrating GIS and Environmental Modeling*, edited by D. A. Wolfe. September 2–8, 2000. Alberta, Canada, 2000.

[8] DMSO. "DoD High-Level Architecture (HLA) for Simulations." USD(A&T) memorandum, September 1996. Defense Modeling and Simulation Office (DMSO), Alexandria, Virginia, USA. June 2000. ⟨http://www.dmso.mil/projects/hla/⟩.

[9] ERDAS Image Processing Software Package. March 2000. ERDAS, Atlanta, Georgia, USA. June 2000. ⟨http://www.erdas.com⟩.

[10] Foresman, T. W., ed. *The History of Geographic Information Systems: Perspectives from the Pioneers.* Prentice Hall Series in Geographic Information Science. Upper Saddle River, NJ: Prentice Hall, 1998.

[11] GRASSLAND GIS software. 1999. Global Geomatics Inc., Laval, Quebec, Canada. June 2000. ⟨http://www.globalgeo.com⟩.

[12] Harper, S. J., A. M. Trame, and J. D. Westervelt. "The Individual Cowbird Behavior Model." November 1998. Geographic Modeling Systems Lab (GMS), University of Illinois, Urbana-Champaign, IL, USA. June 2000. ⟨http://blizzard.gis.uiuc.edu/htmldocs/Cowbirds/⟩.

[13] KHOROS image processing system. 1999. Khoral Research Inc., Albuquerque, New Mexico, USA. June 2000. ⟨http://www.khoral.com/⟩.

[14] Leavesley, G. "The Modular Modeling System." U.S. Geological Survey, Denver, CO, 1996.

[15] Loh, D. K., and Y.-T. C. Hsieh. "Incorporating Rule-based Reasoning in the Spatial Modeling of Succession in a Savanna Landscape." *AI Applications* **9(1)** (1995): 29–40.

[16] Maxwell, T., and R. Costanza. "Distributed Modular Spatial Ecosystem Modeling." 1997. International Institute for Ecological Economics, Center for Environmental Science, University of Maryland System, Solomons, MD, USA. June 2000. ⟨http://kabir.umd.edu/SMP/MVD/⟩.

[17] Maxwell, T., and R. Costanza. "A Language for Modular Spatio-Temporal Simulation." *Ecol. Model.* **103(2-3)** (1997): 105–113.

[18] McLendon, T., W. M. Childress, and D. L. Price. "A Successional Dynamics Simulation Model as a Factor for Determination of Training Car-

rying Capacity of Military Lands." U.S. Army Corps of Engineers, CERL, Champaign, IL, 1998.
[19] Morrison, V. P. "Import/Dome Language Reference Manual." U.S. Army Corps of Engineers, CERL, Champaign, IL, 1995.
[20] Poskanzer, J. "Netpbm." 1989. ⟨http://wwwcn.cern.ch/dci/asis/products/X11/netpbm.html⟩. The manual pages are located at ⟨http://amp.nrl.navy.mil/code5595/online-software/netpbm.man.html⟩.
[21] Shapiro, M., J. D. Westervelt, et al. "GRASS 3.0 Programmer's Manual." U.S. Army Corps of Engineers, CERL, Champaign, IL, 1989.
[22] Villa, F. "Usage of the Collaborative Modelling Environment." Maryland International Institute for Ecological Economics, Solomons, MD, 1998.
[23] Westervelt, J. D., M. Shapiro, et al. "Geographic Resources Analysis Support System (GRASS) Version 4.0 User's Reference Manual." U.S. Army Corps of Engineers Construction Engineering Research Laboratory, 1992.
[24] Westervelt, J. D. "Simulating Mobile Objects in Dynamic Landscape Processes." Ph.D. diss., Department of Urban and Regional Planning, University of Illinois at Urbana-Champaign, IL, 1996.

Management Application of an Agent-Based Model: Control of Cowbirds at the Landscape Scale

Steven J. Harper
James D. Westervelt
Ann-Marie Shapiro

1 INTRODUCTION

Brood parasitism by brown-headed cowbirds (*Molothrus ater*) has negative impacts on a large number of songbird species. Cowbirds are obligate brood parasites, meaning that females lay their eggs in the nests of other species and do not provide care to their offspring. Parasitism by cowbirds often results in reduced reproductive success for the host, sometimes to the exclusion of fledging any of their own young. Clearly parasitism by cowbirds can have a substantial impact on the population dynamics of the host species [15]. Over 200 species of birds are known to be parasitized by cowbirds [11].

Cowbirds breed in shrublands and forests, and especially parasitize host nests located near ecotones, or borders between habitat types. Human land use in general may promote the success of cowbirds; landscapes with forest openings, clearcuts, small tracts of forests, and large amounts of habitat edge have higher parasitism rates than do landscapes with contiguous forest tracts [4, 12, 18]. Cowbirds readily forage in feedlots, overgrazed pastures, and grasslands, and the expansion of agricultural land use over the past century has provided abundant feeding habitat for cowbirds. Large increases in the numbers of cowbirds have been documented [4, 16] and this increase has been implicated as one factor responsible for the decline of a large number of passerines [19, 20]. Compounding their impact is the fact that cowbirds can affect host populations over broad spatial scales. Because they do not protect their young or a nest, they can range large distances in search of suitable feeding areas; researchers have reported maximum daily movements from 7 to 13 km for cowbirds (Rothstein et al. [21], Cook et al. [7], respectively).

At Fort Hood, a U.S. Army military installation located in central Texas, cowbirds parasitize the nests of numerous songbird species, including those of the black-capped vireo (*Vireo atricapillus*) and the golden-cheeked warbler (*Dendroica chrysoparia*), two federally endangered species. The black-capped vireo appears to be particularly vulnerable to parasitism. Once her nest is parasitized, a host female often abandons it. The female may then attempt to renest but, when cowbirds are abundant, this nest is also likely to be parasitized. As a consequence, the overall reproductive success of black-capped vireos that have been parasitized is greatly reduced [13]. Low productivity due to cowbird parasitism has been identified as a major reason for the endangered status of this species [26].

To reduce the negative impacts of cowbirds on endangered species and other songbirds, Fort Hood personnel are conducting an extensive cowbird control program. A large effort has been made since 1989 to trap and remove cowbirds. The control program has been tremendously successful, resulting in a large increase in the nesting success of black-capped vireos [13]. Over 90% of black-capped vireo nests were parasitized prior to initiation of the control program, while in recent years less than 15% of nests were parasitized [13].

Unfortunately, control of cowbirds must be conducted indefinitely for continued benefits to be realized. Land managers at Fort Hood fully expect that cessation of their control program would result in increases in cowbird densities, increases in parasitism levels, and decreases in the reproductive success of songbirds, especially endangered black-capped vireos. Therefore, one goal of land managers at Fort Hood is to optimize the placement of traps in space and time in order to capture the most cowbirds with the least effort. Herein is described a two-stage modeling approach used to assist managers in meeting this goal. First, a model was developed to predict the feeding locations of cowbirds for a given landscape. Because locations predicted to support large numbers of feeding cowbirds are the preferred sites for traps, results of this model may assist managers in determining the most suitable locations for their traps. Next, the model was extended to simulate the trapping of cowbirds and to estimate the number of cowbird captures that could result from a given trapping program. By providing an environment in which land managers can explore the costs and benefits of alternative strategies, it is hoped this model will assist land managers as they develop an effective and sustainable cowbird control program.

2 MODEL STRUCTURE

The model was developed using both geographic-information-system (GIS) and agent-based modeling software packages; Westervelt [27] provides a detailed description of how these environments can be integrated. GRASS (version 4.2) GIS was used to develop and process initialization maps, as well as to analyze and display output maps. Swarm (versions 1.0.4 and 1.4) was used to conduct simulations in which independent agents (i.e., individual cowbirds, cattle herds, and traps) interacted with one another in space and time. During a simulation each agent developed a unique history according to the rules assigned to its type of object. The Swarm model was written in Objective-C and consisted of an observer Swarm (interface control, display-animation, and time schedule) and a model Swarm (equations describing ecological and behavioral features of modeled agents). The model was developed on Sun UltraSparc and Linux-OS PC computers. Simulations were conducted locally via command-line statements and remotely via a WWW-based interface.

Interactions among modeled agents across the Fort Hood landscape and surrounding areas were simulated. Land managers indicated that cowbirds may breed at Fort Hood and fly over the installation boundary to feed on neighboring, private property; by extending the simulated landscape approximately 7 km past installation boundaries, such movements were incorporated. The total area modeled was 58 km × 58 km, or over 335,000 ha. Using a daily time step, the model simulated interactions among agents over a 100-day period, representing the breeding season (April–June) of the endangered bird species of concern.

3 COWBIRD MOVEMENT PATTERNS

Our first objective was to develop a model capable of simulating the daily movements of cowbirds as they travel from their breeding territories to feeding areas. By identifying feeding areas frequently visited by cowbirds for a given landscape, it was hoped that results of this modeling effort could be used by land managers to select trap locations most likely to capture cowbirds. Although the primary interest was cowbird movements, these movements were a function of the movements of cattle, which in turn were a function of the physical characteristics of the available feeding areas. Therefore the dynamics and interactions among three types of agents (described in detail below) were simulated: Feeding areas utilized by cattle and cowbirds, cattle herds, and cowbirds. The focus was on dynamics that occur within the breeding season of the endangered species (April–June), and thus seasonal and annual changes in variables such as habitat quality, number of breeding cowbirds, and territory boundaries were ignored. Because the primary focus was on movement patterns of breeding cowbirds, other components of cowbird biology, such as host selection and reproductive success, were beyond the scope of this model.

3.1 FEEDING AREAS

This object type was created to represent physical locations on the landscape. While feeding areas remained stationary, their attributes were dynamic. Feeding areas maintained information about the grazing suitability of a given location for cattle, distances among feeding areas, and the presence of other agents (i.e., cattle herds, cowbirds, and traps). Feeding areas can be envisioned as representing foraging sites for both cattle and cowbirds, because in our model feeding cowbirds always associated with cattle herds (see below).

A feeding area represented a 56 ha area (750 m × 750 m) on the landscape, and was considered to be homogenous with regard to all attributes throughout this area. For example, the grazing suitability for cattle across the entire 56 ha area was represented as an attribute of a single feeding area. This approach was used so that landscape-level features could be captured without the necessity of having detailed knowledge of microhabitat characteristics within a 56 ha cell. Further, this spatial scale approximates the level of detail needed to simulate the relatively long-distance movements (up to 13 km) of cowbirds, the focus of this model.

Grazing cattle at Fort Hood are not evenly distributed across the landscape, reflecting their foraging preferences. Several factors likely contribute to this patchy distribution. First, the quantity and quality of forage have been documented to influence movements by cattle (e.g., Hodder and Low [14]). Second, provisions of supplemental resources also likely influence the range use of cattle [17]. For example, cattle at Fort Hood tend to congregate at "corral" sites (typically former ranches or homesteads) where cattlemen provide supplemental food, water, and salt [8]. Finally, accessibility to water has been shown to influence the movements of cattle [10, 22] and overall patterns of range use [6, 23, 25].

A mathematical function was created to describe the grazing suitability for cattle on the simulated landscape. Grazing suitability was determined from four components: (1) amount of grassland habitat, (2) patchiness of grassland habitat, (3) distance from nearest corral, and (4) distance from permanent water. Each component was scaled to range from zero (unsuitable) to one (highly suitable), and then the overall suitability was determined as the product of these components. A multiplicative function was used to combine components to ensure that a feeding area that was unsuitable for any of the four components had an overall grazing suitability of zero. In general, grazing suitability was positively related to the amount of grassland habitat and negatively related to the other three components [24].

Calculation of distance from nearest corral was problematic; while the 30 corral sites at Fort Hood were mapped, those located outside the installation boundary were not. Assuming an equivalent density, an estimated 86 corrals were located off-post. These off-post corrals were distributed on our simulated landscape at random, stratified by an initial estimate of grazing suitability determined from the other three components (corrals at Fort Hood were found to be located at sites with high suitability for other components). Once off-post corrals were in position, the final grazing suitability was calculated from all four components.

3.2 CATTLE HERDS

This object type was created to represent cattle that form small (approximately 30 head), cohesive herds as they graze and move across the landscape. Fort Hood currently grants leases for approximately 3,600 head of cattle to graze on the installation. Using an equivalent stocking rate, an additional 10,320 head of cattle were estimated to be on the modeled lands surrounding Fort Hood. Assuming 30 head per cattle herd, 120 cattle herds were simulated on the installation and 344 on surrounding lands. Because cattle often aggregate at or near corral sites, 116 of these sites were chosen as the initial locations for the 464 cattle herds (four herds at each corral).

While the microhabitat foraging decisions of cattle have been well studied by range scientists, relatively few researchers have investigated the large-scale movements of free-ranging cattle, the types of movements that occur at Fort Hood. Nevertheless, research indicates that cattle may utilize spatial memory [1, 2] and employ a "win-switch" foraging strategy [3]. Under this strategy, cattle find a productive site (i.e., "win") but then routinely move (i.e., "switch") to another location rather than continuing to forage at the productive site until the expected net energy gain drops below that of other locations, as predicted by optimal foraging theory [5]. This strategy may promote rapid regeneration of vegetation at the productive site. Cattle also utilize spatial memory to avoid recently grazed areas [1, 2] and have been shown to avoid locations with depleted food resources for up to 8 d [2]. Field studies also documented that when cattle switched foraging areas, they tended to move to adjacent sites rather than traveling to more distant areas [3].

Movement rules were programmed for cattle herds in the model to simulate a win-switch strategy with spatial memory. Although cattle herds occurred on both sides of the installation boundary, we prevented movement across this fenced boundary. A grazing quality variable was created for each feeding area to describe its relative attractiveness to each cattle herd. Unlike grazing suitability (described above), grazing quality did not quantify static habitat features but rather represented the dynamic attractiveness of a feeding area to a given cattle herd based on its unique history. Grazing quality was a function of: (1) grazing suitability, (2) time since previous occupation by the cattle herd, and (3) distance from the current location of the cattle herd. To determine grazing quality, the grazing suitability value (see above) was adjusted on a daily basis to reflect the decreased preference of cattle for recently visited sites, and their increased preference for adjacent sites. The decrease in grazing quality due to previous occupation by a cattle herd declined over time, and was eliminated when 8 d passed since the occupation; this approximates the temporal extent of the spatial memory of cattle [2]. Once the grazing quality of each feeding area was updated, the movement rule for a cattle herd was simply to move to the feeding area with the greatest grazing quality.

3.3 COWBIRDS

This object type was created to represent female cowbirds breeding within the Fort Hood boundary, individuals that can directly impact resident endangered species via brood parasitism. While cowbirds that breed off-post may fly onto the installation to feed, these individuals do not parasitize resident endangered species and thus were not incorporated in our model. Our approach (described in detail below) was to distribute cowbirds in suitable breeding habitat, assign rules to direct their daily movements from breeding to feeding areas, and then quantify visitation rates to all feeding areas across the landscape.

A map was developed to quantify the breeding suitability for cowbirds at all areas within the Fort Hood boundary. While the exact breeding requirements for cowbirds at Fort Hood are unknown, we used an approach that ensured all habitat that could potentially support endangered species would also be considered suitable for breeding cowbirds. The vegetative composition of all 56 ha areas that fell within designated "endangered species protection areas" on Fort Hood was examined; because they support or are likely to support endangered species, military training is restricted and biological studies are focused within these areas. The minimum amount of aerial cover by shrubs and/or forests within endangered species protection areas was 30%, and this value was used to classify all 56 ha areas across the landscape as either suitable or unsuitable for breeding cowbirds. Distance from grassland edge was another important landscape feature that was measured to further refine the breeding quality of suitable areas. Brittingham and Temple [4] found that breeding cowbirds more heavily parasitized host nests within 100 m of the boundary between forested areas and grasslands (65%), compared to more interior forest areas (18%). Relative breeding quality values were assigned to reflect the preference of cowbirds for edge habitat. First, the average distance from shrub/forest habitat to grassland habitat within each 56 ha area was calculated from a vegetation map comprised of 50 m $\times$ 50 m cells. Next, areas with an average distance of less than 100 m were classified as high quality (3.6; determined as 65/18 from parasitism values reported above), interior areas (with an average distance of greater than 300 m) as low quality (1.0), and the remaining suitable areas as moderate quality (2.3; determined as the mean of 1.0 and 3.6).

No comprehensive survey has been conducted at Fort Hood to enumerate breeding female cowbirds. However, Cook et al. [7] found that female cowbirds maintained exclusive access to breeding habitat, and that each breeding territory was approximately 56 ha in size ($n = 7$ telemetered cowbirds, median = 50 ha, mean = 60 ha). Thus, each 56 ha cell on our breeding quality map could be considered to represent the breeding territory of an individual cowbird. This provided us with a coarse estimate of the maximum number of breeding cowbirds that the landscape could support (assuming there is no relationship between territory size and cowbird density). Using this approach, it was estimated that a maximum of 1,249 breeding female cowbirds could be supported on the Fort Hood landscape. To initiate a simulation, the number of cowbirds to be instantiated was specified as a percentage of this "saturation" density. Cowbirds were randomly distributed among suitable breeding areas, stratified by breeding habitat quality. To follow better the movements of those cowbirds most likely to parasitize endangered species, each cowbird was labeled as breeding in black-capped vireo protection areas, golden-cheeked warbler protection areas, or locations not within protection areas.

Feeding cowbirds at Fort Hood nearly always associate with small herds of cattle; it is estimated that over 90% of afternoon feeding bouts occur in the presence of at least 30 head of cattle [8]. To our knowledge, no research has elucidated how cowbirds locate feeding sites with cattle, areas that can be over 13 km from their breeding territories. Four alternative behavioral algorithms or movement rules, were developed, to specify how cowbirds might search for a suitable feeding area. Regardless of movement rule, the goal of each cowbird was to leave its breeding territory and locate a feeding area occupied by at least one cattle herd. First, a "next-nearest rule" directed each cowbird to move randomly to the nearest feeding area, then to the next-nearest feeding area, etc. until a cattle herd was located. Second, a "memory rule" directed each cowbird to move, in reverse chronological sequence, directly to each of the three most recently utilized feeding areas until a cattle herd was located. If no cattle herd was found at any of these previously successful locations, the cowbird continued to search using the next-nearest rule. Third, a "memory-with-perception rule" directed each cowbird to use the memory rule (above), but provided them with the ability to assess feeding areas en route to the three previously successful locations. For this rule, the perception distance was defined (i.e., distance away from the straight-line travel path) within which cowbirds could locate cattle herds. Finally, an "omniscient rule" directed each cowbird to travel directly from its breeding territory to the nearest feeding area with a cattle herd. Results of this rule represented the shortest possible distance that cowbirds could travel, and thus served as a benchmark for evaluating the results of the other three movement rules.

3.4 SIMULATIONS

Each simulation of the cowbird movement model produced two general types of output. Maps detailed the cumulative visits by cattle herds and cowbirds to feeding areas across the modeled landscape, and tables of information detailed the distances moved by cattle herds and cowbirds during a simulation. Thus, the resultant patterns produced by the model (maps) were examined as well as the underlying processes that generated these patterns (individual movements).

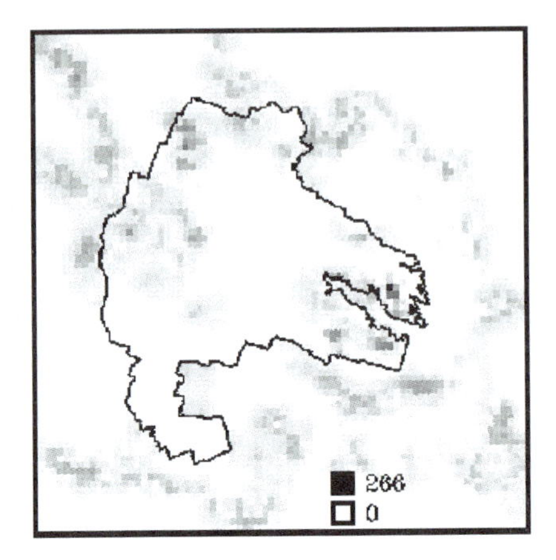

FIGURE 1 Map of the cumulative number of cattle herd visits to feeding areas across the simulated landscape. Darker shades of grey represent larger numbers, within the range of values indicated. The Fort Hood installation boundary is plotted for reference.

3.4.1 Cattle Herds. The movements of cattle herds on the simulated landscape were compared to those observed in a 248 ha pasture in Texas. Bailey et al. [3] examined movements of cattle among five large, relatively homogeneous sections of pasture (mean = 49.6 ha). Because the natural pasture sections approximated the size of our simulated feeding areas (56 ha), model results were qualitatively compared with those obtained from the field study. For observations made during March–June ($n = 17$), Bailey et al. [3] found that cattle remained in the same section of the pasture two mornings in a row for 29% of the observations. Cattle moved to an adjacent section 35% of the time, to more distance sections 29% of the time, and to the most distant sections only 6% of the time. Distances reported by Bailey et al. [3] correspond to movements across 0–3 contiguous feeding areas on our modeled landscape. Analysis of movements by 464 cattle herds during a 100 d simulation ($n = 46,000$) indicated that 17% remained in the same feeding area on consecutive days, 39% moved to an immediately adjacent feeding area, 31% moved two feeding areas, 13% moved three feeding areas, and none crossed more than three feeding areas. Thus, the model compared favorably to the field study for moderate distances, but underrepresented the number of herds that remained in the same feeding area and overrepresented those that traveled long distances. Despite these limitations, all modeled distances fell within the appropriate range, and they generally approximated the movements observed for cattle in the natural pasture.

One hundred simulations of the model were undertaken, and from these results a map was created detailing the average number of cumulative cattle herd visits to all feeding areas (fig. 1). This map was presented to Fort Hood biologists, and they indicated that locations predicted by the model to receive large numbers of cattle herd visits did, in fact, generally receive heavy use by cattle [9]. However, other areas at Fort Hood known to be used by cattle were not indicated on our output map. Small savannalike areas with mature trees and open understory, habitat readily used by cattle for shade and forage at Fort Hood, were not visited by modeled cattle herds as often as the biologists expected. This habitat feature was not specifically incorporated into our estimate of grazing suitability. Though a relatively small percentage of all habitat types, future modification of the model to include savannalike areas likely would enhance its ability to predict cattle feeding locations. Currently an effort is being undertaken to document the large-scale movements and distribution patterns of cattle at Fort Hood. By validating this part of the model with field data, it will be possible for us to quantify the accuracy of the model and also to identify other modifications that may be necessary.

3.4.2 Cowbirds. Encouraged by the results obtained for cattle herds, a series of simulations was conducted to evaluate the effects of the four movement rules on results for cowbirds. Again, the responses of the agents were examined as well as the overall pattern resulting from these responses.

Simulations with each of the four cowbird movement rules were run and cowbird responses were compared to results obtained from field studies conducted at Fort Hood. For each daily movement of a cowbird, the absolute distance (i.e., straight line) was recorded between its breeding territory and the selected feeding area as well as the total distance traveled during its search for that feeding area. Cook et al. [7] conducted a preliminary radio telemetry study of cowbird movements at Fort Hood, and results of their study allowed us to evaluate results of our model. Although these researchers did not track cowbirds continuously from breeding to feeding areas, and thus could not determine total search distances, they did document the absolute distances traveled by seven cowbirds.

It was found that the absolute distances of modeled cowbirds were influenced by the movement rule used (fig. 2a). Mean distances for omniscient and memory-with-perception rules approached those observed in the field, while those for the next-nearest and memory rules were larger. Mean search distances differed to an even greater extent among movement rules (fig. 2b), and decreased as cowbird agents were provided increasing knowledge about the landscape. From an energetic perspective, cowbirds should benefit by minimizing their total search distance, and selection may favor those cowbirds that are able to find suitable feeding sites with minimal search effort. A simple memory-with-perception rule, which minimized both the absolute distance and total search distance, appeared to best represent the behavioral decisions used by cowbirds. It seems plausible that cowbirds have the ability to remember locations where they had successfully foraged, as well as the ability to

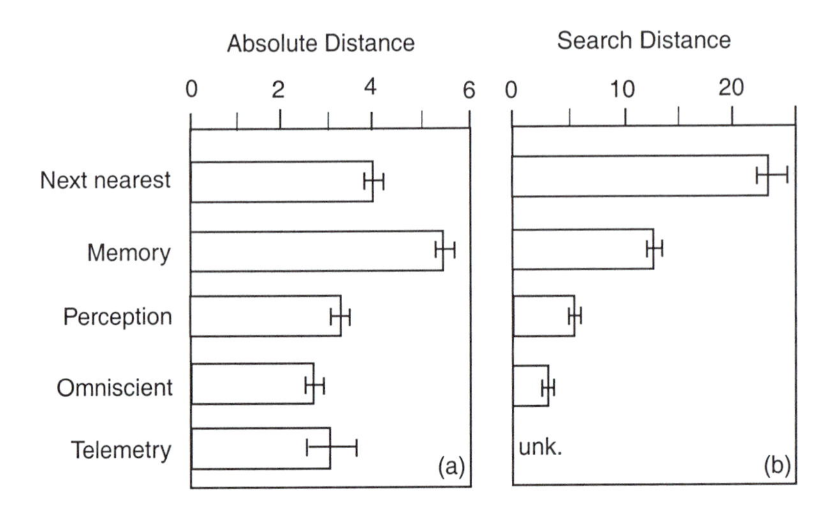

FIGURE 2 Results of simulations using four different cowbird movement rules with regard to (a) the absolute distance between breeding and feeding locations, and (b) the total search distance traveled to locate a suitable feeding area. Values are means (± 1 SD) of the daily movements of 312 cowbirds averaged over a 100 d simulation. Absolute distance data are also plotted for seven telemetered cowbirds studied at Fort Hood [7].

evaluate their surroundings as they travel to those sites. However, additional studies are needed to better understand the cognitive abilities of cowbirds.

As might be expected, the patterns of cowbird visits to feeding areas were dependent on the movement rule that was used (data not presented). Here, visitation patterns produced by the memory-with-perception rule are focused upon as this rule seemed to best simulate the daily movements of cowbirds. An examination of the visitation map for all cowbirds (fig. 3) revealed that cowbirds with breeding territories on Fort Hood traveled to feeding areas primarily within the installation boundary, though a few locations off-post received significant numbers of cowbird visits. While land managers currently do not trap cowbirds off-post, they have expressed a desire to enhance the trapping program by seeking cooperative agreements with owners of adjoining lands to trap cowbirds. Our results identify those locations where such an approach may be most effective.

The patterns of visits produced by cowbirds that had breeding territories in endangered species protection areas were examined. Patterns of visits to feeding areas by cowbirds breeding in black-capped vireo (fig. 4A) and golden-cheeked warbler (fig. 4B) protection areas differed from that of the general cowbird population (fig. 3). If these maps are representative of actual cowbird feeding patterns, this suggests that land managers may be able to target their trapping effort to capture those cowbirds most likely to parasitize nests of

FIGURE 3 Map of the cumulative number of visits to feeding areas across the simulated landscape made by all breeding cowbirds. Values are means generated from 100 simulations in which 312 cowbirds were instantiated and moved using the memory-with-perception rule. Darker shades of grey represent larger numbers, within the range of values indicated. The Fort Hood installation boundary is plotted for reference.

endangered species. A study has recently begun to document the visitation patterns of feeding cowbirds at Fort Hood. Results of this field study will be used to validate model predictions quantitatively and to identify possible modifications needed to the cowbird movement model.

4 TRAPPING OF COWBIRDS

Having successfully developed a model to simulate the movements of cowbirds, our next objective was to extend the model to simulate the trapping of cowbirds. By quantifying the success of alternative trapping approaches at capturing cowbirds, it was hoped that this model could be used by land managers to identify how the efficacy of their trapping program might be improved. Two additional types of agents, traps and a trap manager, were incorporated. As before, the focus was on the daily dynamics that occurred from April–June.

4.1 TRAPS

This object type was created to represent the physical attributes of traps that Fort Hood land managers use to capture cowbirds. Two trap types were represented, "fixed" traps which remained in the same location throughout a

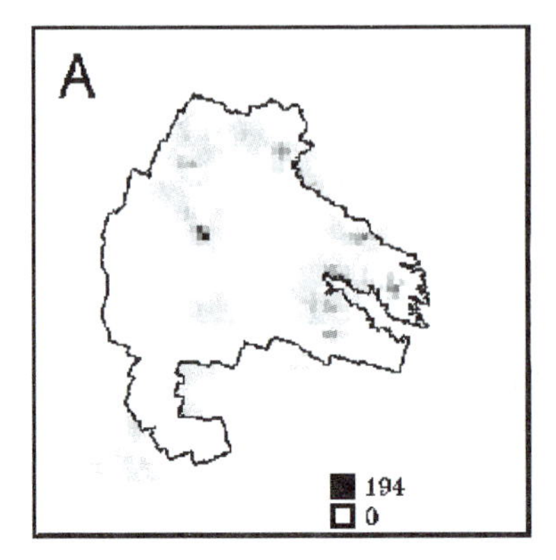

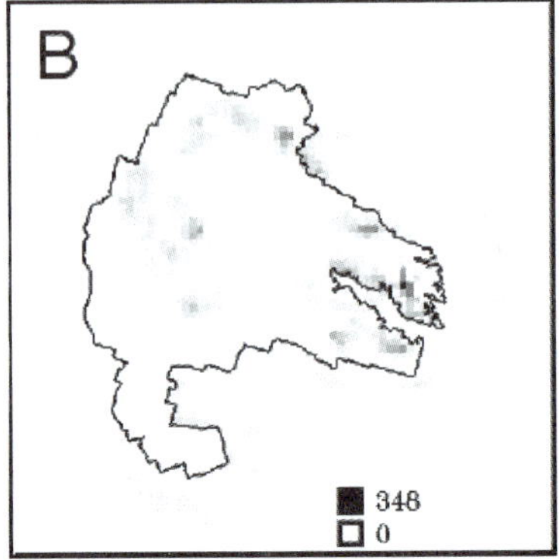

FIGURE 4 Maps of the cumulative number of visits to feeding areas across the simulated landscape made by cowbirds breeding in (A) black-capped vireo protection areas, and (B) golden-cheeked warbler protection areas. Values are means generated from 100 simulations in which 312 cowbirds were instantiated and moved using the memory-with-perception rule. Darker shades of grey represent larger numbers, within the range of values indicated. The Fort Hood installation boundary is plotted for reference.

simulation and "mobile" traps which could be moved among locations. Fixed traps, which are larger than mobile traps, tend to have higher capture rates. The efficiency of both trap types was specified as the probability that a cowbird visiting a feeding area would be captured. Using preliminary data collected at Fort Hood, trap efficiency values of 0.18 to mobile traps and 0.30 to fixed traps were assigned. The role of traps in these simulations was simply to identify those cowbirds visiting a feeding area that were captured.

4.2 TRAP MANAGER

This object type was created to simulate the typical decisions made by land managers at Fort Hood. It was designed to possess only information typically known to land managers, and did not have complete knowledge about the modeled system. For example, during a simulation the trap manager could access the number of cowbirds being captured at trap locations, but not the number of cowbirds visiting feeding areas across the simulated landscape. The primary role of the trap manger was to implement a trapping program, guided by information provided by the user. This involved determining the number of fixed and mobile traps to use, the initial locations of all traps, the frequency with which to move mobile traps, and the locations to receive relocated mobile traps. The trap manager visited each trap daily and counted and removed captured cowbirds from the simulation.

The trap manager selected locations that received traps from a list, provided by the user, of suitable feeding areas and their likelihood of being visited by cowbirds. While this list could be created in any number of ways, a reasonable approach was to use output generated by the cowbird movement model (above), which was developed specifically to provide this information. As the same rules governed movements of cattle herds and cowbirds in both models, results of the cowbird movement model obviously should provide the best possible estimate of cowbird visits for the trapping model. Again, 100 simulations were performed with the cowbird movement model and generated a map summarizing cowbird visitation patterns of all runs. For the trapping simulations described here, the visitation map for all cowbirds on Fort Hood (see fig. 3) was used; it would also have been possible to use visitation maps for those cowbirds breeding in endangered species protection areas (see fig. 4). This map was processed to generate a list of suitable feeding areas and their likelihood of being visited by cowbirds. First, those feeding areas that were visited by very few cowbirds were eliminated. Those feeding areas that were more than 750 m from a navigable road were subsequently eliminated. This reflects current management practices on Fort Hood; while remote sites are visited by cowbirds, traps placed in such locations cannot be checked and maintained on a regular basis. Finally, each remaining feeding area and its likelihood of being visited by cowbirds, which was defined as simply the number of visits obtained during simulations of the cowbird movement model, was listed.

The goal of the trap manager was to place traps in those locations most likely to capture cowbirds; therefore, locations predicted to have the largest numbers of cowbird visits preferentially received traps. However, some flexibility in the way that the trap manager selected trap locations was incorporated. By specifying that the trap manager should maintain some minimum distance between traps, dispersion of traps on the landscape could be influenced. To illustrate this effect, visualize the map of predicted cowbird visits (fig. 3) plotted three-dimensionally as an irregular surface with peaks and valleys. The largest values, or peaks, are surrounded by neighboring cells that also have large values. If the trap manager simply placed a given number of traps, say, 12 fixed traps, at the 12 locations predicted to have the largest numbers of cowbird visits, then all traps likely would be positioned in very close proximity to one another on the top and sides of a few peaks (i.e., in a clumped distribution). In contrast, if the trap manager required a minimum distance between trap locations of, say, 2 km, and then selected the 12 locations predicted to have the largest numbers of cowbird visits, the traps likely would be positioned across the landscape on different peaks (i.e., in a more uniform distribution). To change the dispersion of traps on the landscape, any value (including 0 m) could be specified as the minimum distance between traps.

Once traps were placed on the simulated landscape, the trap manager became responsible for the movement of mobile traps (recall that fixed traps remain in place throughout a simulation). The trap manager determined when a mobile trap was to be moved from a given location and where it was to be placed. Two different approaches were incorporated for specifying when mobile traps should be moved. In the first, all mobile traps were rotated after remaining in place for a specified duration. By moving all mobile traps simultaneously, this approach may satisfy the need of land managers to schedule labor requirements in advance. Using an alternative approach, mobile traps were individually moved once their capture rate dropped below some specified level. This approach satisfies the need of land managers to move ineffective traps to better locations, but requires regular calculation of capture rates and prevents advanced scheduling of labor. Following removal of a mobile trap, its new location was determined by the trap manager as the location, among those that had not yet received a trap, predicted to have the largest number of cowbird visits. As with initial locations of traps, the user could direct the trap manager to maintain some minimum distance between traps.

4.3 SIMULATIONS

Each simulation of the cowbird trapping model produced output detailing the number of cowbird captures over time, allowing the relative success of alternative trapping programs to be evaluated. First, the model was used to examine how numbers of cowbird captures were affected by the frequency of movement of mobile traps and the minimum distance maintained between traps. Next, the model was used to examine how the proportion of fixed and mobile traps influenced cowbird captures. Finally, simulations were conducted to determine how an actual trapping program used at Fort Hood fared on our simulated landscape.

4.3.1 Trap Movements. In this series of simulations the frequency with which mobile traps were moved was changed, and frequencies of 7, 28, and 56 d (i.e., weekly, monthly, and once per simulation, respectively) were used. Results of 100 simulations for each trapping program indicated that the number of cowbirds captured differed among frequencies ($F = 198.24$; $df = 2, 297$; $P < 0.0001$), with more cowbirds captured when traps were moved every 56 d (table 1). This response likely resulted from the fact that traps left in place continued to capture cowbirds during most of the 100 d simulation period. Early in a simulation, the trap manager placed traps at those locations most likely to capture cowbirds (i.e., sites predicted to have large numbers of cowbird visits); thus, when traps were scheduled to be moved at short time intervals, it resulted in them being removed from successful sites and placed at relatively poorer sites (i.e., those predicted to have fewer cowbird visits).

A series of simulations was conducted to explore the effect of changes to the minimum distance between traps, and values of 0 (i.e., no restriction), 1, 2, 3, and 4 km were used. Results of 100 simulations for each trapping program indicated that the number of cowbirds captured differed among minimum distances ($F = 21.99$; $df = 4, 495$; $P < 0.0001$), with more cowbirds captured when traps were forced to be at least 2 km apart from one another (table 1).

Traps placed very near one another likely competed for the same cowbirds; in contrast, when traps were forced to remain very far apart from one another, some areas with large numbers of cowbird visits probably did not receive traps. A distance of 2 km appeared to best balance this trade-off between competition among traps and utilization of the most suitable sites.

4.3.2 Trap Types. In this series of simulations the percentage of mobile traps in a total fleet of 12 traps was varied, and values of 0, 50, and 100% were used. Results of 100 simulations for each trapping program indicated that the number of cowbirds captured differed among proportions of mobile traps ($F = 197.91$; $df = 2,297$; $P < 0.0001$), with more cowbirds captured when all 12 traps were fixed (table 1). Because the efficiency of fixed traps was 1.7 times that of mobile traps, they caught more cowbirds for a given number of visits than did mobile traps. Nevertheless, it was expected that the use of mobile traps might result in a greater number of cowbird captures if they allowed areas that became effectively "trapped out" to be abandoned and new areas to be trapped. It appears that areas did not become devoid of cowbirds; as noted above, fixed traps continued to capture cowbirds throughout much of a simulation. While their predicted benefit was not realized in these simulations, mobile traps may still be useful to land managers at Fort Hood. For example, locations of mobile traps can be changed easily from one year to the next, allowing trapping programs to be modified in response to long-term changes in the landscape.

4.3.3 Trapping Strategies. In a more direct application of the model, a series of simulations was conducted to evaluate the effectiveness of an actual trapping program at capturing simulated cowbirds. Land managers at Fort Hood used 8 fixed and 36 mobile traps to capture cowbirds in 1995, and they recorded the locations of all traps. Trap agents were placed at these exact locations on our simulated landscape. Because land managers did not move mobile traps during the time period covered by our simulations, the trap manager was prevented from moving mobile traps during the simulation. It is important to note that land managers did not have access to the cowbird visitation map generated by the cowbird movement model, but rather based their placement decisions on experience gained during six years of trapping cowbirds. For comparison, simulations were run in which the same number of traps was used (8 fixed, 36 mobile) but the trap manager was allowed to select their locations. Again, mobile traps were not moved during the simulations.

The scenario in which traps were placed by the model captured about 42% more cowbirds than the scenario in which locations of traps in 1995 were used (table 1). Of course, the predicted numbers of cowbird visits used by the trap manager to select trap locations were generated by previous runs of the cowbird movement model, which served as the foundation for the trapping model. Therefore, it was not surprising that the trap manager could place simulated traps on the modeled landscape better than Fort Hood personnel could in 1995. More interesting to us was the fact that the locations used in 1995 resulted in the capture of a relatively large number of simulated cowbirds (table 1). Land managers in 1995 placed traps at those locations they expected to be regularly visited by cowbirds, and these same locations received relatively high visitation rates in our simulations (as evident by the number of cowbirds that were captured). This finding lends further support to our suggestion that the cowbird movement patterns generated by our model may be representative of natural patterns.

Finally, the model was used to test the effectiveness of a trapping program that land managers at Fort Hood might implement in the future. Because of the high cost of servicing and maintaining a large fleet of traps (typically over 40 traps are in service), biologists have indicated that a future management strategy may be to use a total of 12 fixed and 12 mobile traps. The team tested this trapping program on our simulated landscape, and found that it captured about 92% as many cowbirds as did the program using 8 fixed and 36 mobile traps, a fleet of traps nearly twice its size (table 1). Thus, the use of fewer traps placed at appropriate locations appears to be a very effective strategy, one that could greatly reduce the time and effort of the land manager while maintaining a high cowbird capture rate.

5 SUMMARY

The model developed to predict cowbird visitation rates across Fort Hood and the surrounding areas incorporated interactions among feeding areas, cattle herds, and cowbirds. It was found that movement rules of cattle herds resulted in their traveling distances consistent with those observed in a large pasture in Texas, and that these movements produced realistic patterns on the Fort Hood landscape. A simple memory-with-perception rule produced movement distances for cowbirds that approximated those observed for cowbirds at Fort Hood. By predicting the visitation patterns for cowbirds breeding within endangered species protection areas, results of our model may allow land managers to target their control program towards capturing those cowbirds that pose the greatest threat to endangered species. Cowbird visitation maps produced by this model should prove useful in assisting land managers as they identify locations for their traps.

The cowbird movement model was extended to allow simulation of the trapping of cowbirds. Through the use of a trap manager agent, the model incorporated the myriad of decisions that face the land managers of Fort Hood. By exploring alternative trapping strategies, it was found that frequent movement of mobile traps may decrease total cowbird captures, that a minimum distance of about 2 km should be maintained between traps, and that the benefit of being able to move mobile traps to previously untrapped areas may not compensate for their reduced efficiency. Simulation of an actual Fort Hood trapping program resulted in the capture of a relatively large number of cow-

TABLE 1 Mean (± 1 SD) numbers of cowbirds captured using different trapping programs. Results are based on 100 simulations of each trapping program. All simulations were instantiated with a total of 312 cowbirds on the landscape (i.e., 25% saturation density), which moved using the memory-with-perception rule. Lines divide simulations into comparisons discussed in the text.

Trapping Program				
No. Fixed	**No. Mobile**	**Frequency**	**Distance**	**Cowbirds Captured**
12	12	7	2	213.01 ± 12.23
12	12	28	2	236.74 ± 12.60
12	12	56	2	246.77 ± 12.11
12	12	56	0	234.81 ± 11.23
12	12	56	1	243.58 ± 11.81
12	12	56	2	246.77 ± 12.11
12	12	56	3	238.03 ± 11.12
12	12	56	4	233.91 ± 13.19
12	0	56	2	197.32 ± 13.01
6	6	56	2	176.98 ± 11.38
0	12	56	2	164.25 ± 11.09
8[1]	36	N/A[2]	N/A	185.09 ± 14.31
8	36	N/A	N/A	262.30 ± 13.03
12	12	N/A	N/A	240.91 ± 11.63

[1] Locations were from 1995 Fort Hood trapping program.

[2] Not applicable because mobile traps were not moved.

birds, though not as many as when the same number of traps were placed by the trap manager. Thus, it was the selection of appropriate locations for fixed traps, rather than the movement of mobile traps, that allowed large number of cowbirds to be captured. This suggests that land managers at Fort Hood could reduce the costs associated with their extensive trapping program, while still maintaining effective control of cowbirds, by using fewer, well-placed traps.

REFERENCES

[1] Bailey, D. W. "Daily Selection of Feeding Areas by Cattle in Homogeneous and Heterogeneous Environments." *App. An. Behav. Sci.* **45** (1995): 183–200.

[2] Bailey, D. W., J. E. Gross, E. A. Laca, L. R. Rittenhouse, M. B. Coughenour, D. M. Swift, and P. L. Sims. "Mechanisms that Result in Large Herbivore Grazing Distribution Patterns." *J. Range Mgmt.* **49** (1996): 386–400.

[3] Bailey, D. W., J. W. Walker, and R. L. Rittenhouse. "Sequential Analysis of Cattle Location: Day-to-Day Movement Patterns." *App. An. Behav. Sci.* **25** (1990): 137–148.

[4] Brittingham, M. C., and S. A. Temple. "Have Cowbirds Caused Forest Songbirds to Decline?" *Bioscience* **33** (1983): 31–35.

[5] Charnov, E. L. "Optimal Foraging: The Marginal Value Theorem." *Theor. Pop. Biol.* **9** (1976): 129–136.

[6] Cook, C. W. "Factors Affecting Utilization of Mountain Slopes by Cattle." *J. Range Mgmt.* **19** (1966): 200–204.

[7] Cook, T. L., J. A. Koloszar, and M. D. Goering. "1995 Annual Report: Behavior and Movement of the Brown-Headed Cowbird (*Molothrus ater*) on Fort Hood." Report of The Nature Conservancy of Texas, Texas, January, 1995.

[8] Cornelius, J. Personal communication, March 2, 1997.

[9] Cornelius, J., and T. Cook. Personal communication, December 16, 1997.

[10] Coughenour, M. B. "Spatial Components of Plant-Herbivore Interactions in Pastoral, Ranching, and Native Ungulate Ecosystems." *J. Range Mgmt.* **44** (1991): 530–542.

[11] Friedmann, H., and L. F. Kiff. "The Parasitic Cowbirds and Their Hosts." *Proc. Western Found. Vertebrate Zool.* **4** (1985): 225–302.

[12] Gates, J. E., and L. W. Gysel. "Avian Nest Dispersion and Fledgling Success in Field-Forest Ecotones." *Ecology* **5** (1978): 871–883.

[13] Hayden, T. J., D. J. Tazik, R. H. Melton, and J. D. Cornelius. "Cowbird Control Program on Fort Hood, Texas: Lessons for Mitigation of Cowbird Parasitism on a Landscape Scale." In *The Ecology and Management of Cowbirds*, edited by T. Cook, S. K. Robinson, S. I. Rothstein, S. Sealy, and J. Smith. Austin, TX: The University of Texas Press, in press.

[14] Hodder, R. M., and W. A. Low. "Grazing Distribution of Free-Ranging Cattle at Three Sites in the Alice Springs District, Central Australia." *Austral. Rangeland J.* **1** (1978): 95–105.

[15] May, R. M., and S. K. Robinson. "Population Dynamics of Avian Brood Parasitism." *Amer. Natur.* **4** (1985): 475–494.

[16] Mayfield, H. "The Brown-Headed Cowbird, with Old and New Hosts." *Living Bird* **4** (1965): 13–28.

[17] McDougald, N. K., W. E. Frost, and D. E. Jones. "Use of Supplemental Feeding Locations to Manage Cattle Use on Riparian Areas of Hardwood Rangelands." Forest Service General Technical Report, PSW-110, United States Department of Agriculture (USDA), 1989.

[18] Robinson, S. K., J. P. Hoover, R. Herkert, and R. Jack. "Cowbird Parasitism in a Fragmented Landscape: Effects of Tract Size, Habitat, and Abundance of Cowbirds and Hosts." In *The Ecology and Management of Cowbirds*, edited by T. Cook, S. K. Robinson, S. I. Rothstein, S. Sealy, and J. Smith. Austin, TX: The University of Texas Press, in press.

[19] Robinson, S. K., and D. S. Wilcove. "Forest Fragmentation in the Temperate Zone and Its Effects on Migratory Songbirds." *Bird Conserv. Intl.* **4** (1994): 233–249.
[20] Robinson, S. K., S. I. Rothstein, M. C. Brittingham, L. J. Petit, and J. A. Grzybowski. "Ecology and Behavior of Cowbirds and Their Impact on Host Populations." In *Ecology and Management of Neotropical Migratory Birds: A Synthesis and Review of Critical Issues*, edited by T. E. Martin and D. M. Finch. New York, NY: Oxford University Press, 1995.
[21] Rothstein, S. I., J. Verner, and E. Stevens. "Radio-Tracking Confirms a Unique Diurnal Pattern of Spatial Occurrence in the Parasitic Brown-Headed Cowbird." *Ecology* **65** (1984): 77–88.
[22] Senft, R. L., M. B. Coughenour, D. W. Bailey, L. R. Rittenhouse, O. E. Sala, and D. M. Swift. "Large Herbivore Foraging and Ecological Hierarchies." *Bioscience* **37** (1987): 789–799.
[23] Senft, R. L., L. R. Rittenhouse, and R. G. Woodmansee. "The Use of Regression Models to Predict Spatial Patterns of Cattle Behavior." *J. Range Mgmt.* **36** (1983): 553–557.
[24] Shapiro, A. M., S. J. Harper, and J. D. Westervelt. "Management of Cowbird Traps on the Landscape: An Individual-Based Modeling Approach for Fort Hood." Technical Report, No. 98/121, United States Army Construction Engineering Research Laboratories, Texas, 1998.
[25] Stafford-Smith, M. S. "Modeling: Three Approaches to Predicting How Herbivore Impact is Distributed in Rangelands." Agricultural Experiment Station Research Report, No. 628, New Mexico State University, Las Cruces, NM, 1988.
[26] United States Fish and Wildlife Service. "Black-Capped Vireo (*Vireo atricapillus*) Recovery Plan." U.S. Forest and Wildlife Service Publication, Austin, TX, 1991.
[27] Westervelt, J. D. "Computational Approach to Integrating GIS and Agent-Based Modeling." This volume.

Integrating Spatial Data into an Agent-Based Modeling System: Ideas and Lessons from the Development of the Across-Trophic-Level System Simulation

Scott M. Duke-Sylvester
Louis J. Gross

1 INTRODUCTION

Agent-based or individual-based models allow for variation in the state and behavior of the basic objects that interact within the model. Modeling each individual as a separate entity allows for spatially explicit components to be included so that the individuals can interact with a heterogeneous landscape and with each other. Realism is added to the models by incorporating spatially explicit data for the area of interest. Integrating spatial data into an agent-based system requires that a significant level of geographic information systems (GIS) functionality from traditional GIS be incorporated into the modeling system. This approach may seem redundant and costly, but current GIS systems do not offer a framework for building dynamic agent-based models.

The across-trophic-level system simulation (ATLSS) is characterized by the integration of several distinct agent-based models and spatially explicit data into a single modeling system [3]. One of the goals of the ATLSS (pronounced like "atlas") project is to investigate the relative response of various interconnected trophic levels of the South Florida (SF) Everglades to different hydrologic scenarios over a thirty-year planning horizon. The ATLSS approach consists of several distinct component models, each of which represents different biotic components of the Everglades system, linked together as a multimodel. Currently, ATLSS includes component models for the Florida Panther [2], the Cape Sable Sea Side Sparrow, white tail deer [2], fresh water fish [6], wading birds [11], the Snail kite, and vegetation biomass. Additional models for alligators and various reptiles and amphibians are in development and will be added as they are completed. The list above by no means enumerates all of the species in SF. It reflects instead the initial attempt to include key components that biologists with many years of field experience in SF believed were critical to include. It also reflects the time and funding limitations to carry out empirical studies and develop both the theory and software needed to model each system component. The Florida Panther, the Cape Sable Seaside Sparrow, the Snail Kite and several wading bird species of SF are all listed as endangered species. The fish, deer, and vegetation on the other hand are included since they are critical resources for the endangered species. The interaction between the models is a central feature of the ATLSS models and the objective of this chapter is to discuss the resultant methodology which handles these interactions.

Each model within ATLSS is agent-based, but different levels of aggregation are used for different models. The level of aggregation is a function of many considerations including computational efficiency, availability of empirical data to support a model, and the level of spatial and temporal resolution needed from the model. The component models for the panther, the deer, and the sparrow are all individual-based. In this type of model each individual is represented by a separate agent within the system. Each agent has its own separate set of state variable and behaviors. For example, each deer agent has values for the age, sex, and mass of the deer that are unique and particular to an agent. This level of resolution, where each individual in the population is modeled as a distinct entity, is necessary for these taxa. Variation in individual response creates population dynamics that can be quite different from the situation in which all individuals in the population are characterized by a single set of parameters and behaviors, which is typical of the classical population models in ecology.

In the freshwater fish and vegetation models, each agent represents a collection of individuals. For example, all individuals within a range of sizes are grouped together and the model dynamics represent the responses of that group. All of the individuals within that group may have a certain growth rate or reproductive rate. The fish model is broken into two classes that are defined by the maximum size of individuals of the fish species included. All large fish are represented by one group, all small fish by another. An agent, therefore, represents either the large fish or small fish and copies of the agent are distributed across the landscape to provide the spatial aspect of the model. The agent's properties in this case include information about the size distribution within each of the spatially distributed agents. These types of models are called functional group models. Other modeling approaches are currently used for wading birds and the Snail Kite, in which a simple index is calculated as a function of the dynamics of hydrology at a location. These index models are intended to provide simple assessments until individual-based models are available for these species groups. More details about all the ATLSS approaches are available at ⟨http://www.atlss.org⟩.

In addition to the component models, ATLSS uses a high-resolution topography model which provides high-resolution hydrology for the study area (fig. 1). This model is run as a preprocess and generates ground surface elevations for the study area at resolutions as small as 28.5×28.5 meters per pixel. The model generates topography based on a habitat cover map, hydrology data, and the relationship between habitat cover type and the hydroperiods (e.g., the length of time during a year for which a location has standing water) most commonly associated with those habitat types.

Each of the component models are run over a large portion of SF consisting of about 20,000 km^2. The boundaries for the model are defined by the intersection of the regions for which there are adequate spatial data. Currently this area is identical to the area covered by the South Florida Water Management Model (SFWMM) since this data set has the most restrictive boundaries (fig. 1). In addition to the hydrology data, the models use fine-resolution ground surface elevation and hydrology information as described above, a habitat cover map, and several other sources of spatial data. Each of these data sets are provided at a different resolution and registration. The differences between the data sets must be resolved for the models to operate smoothly and correctly. Additionally, the hydrology data is a time series, and the differences between the temporal extent and increments of the data and that of the models must also be resolved.

The models and the data are tied together in ATLSS by the Landscape classes. These are a collection of C++ classes, each designed to handle a difference facet of spatial data. Objects derived from some of these classes manage the data while the models are running, objects from other classes provide an interface to several input and output devices, others provide GIS-type functions which can be applied by the agents on the spatial data. Each spatial data set is associated with metadata that describes spatial features of the spatial data such as where it is located, how the data is structured, and the area covered by each cell. A set of classes has also been developed to handle this type of information and coordinate each metadata set with the appropriate spatial data set.

There are many aspects of integrating spatial data and GIS functionality into a system of agent-based models. In this chapter we will focus on three related issues fundamental to the task of integration. First, we will look at how spatial data and GIS functionality are provided to the models. Second, we will look at mechanisms for structuring the internal spatially explicit components of a particular system. Finally, we will look at mechanisms for communicating spatial data between the models. These three elements each comprise a major segment of the overall model construction. Similarities between these elements can be utilized to simplify many aspects of the models and the model

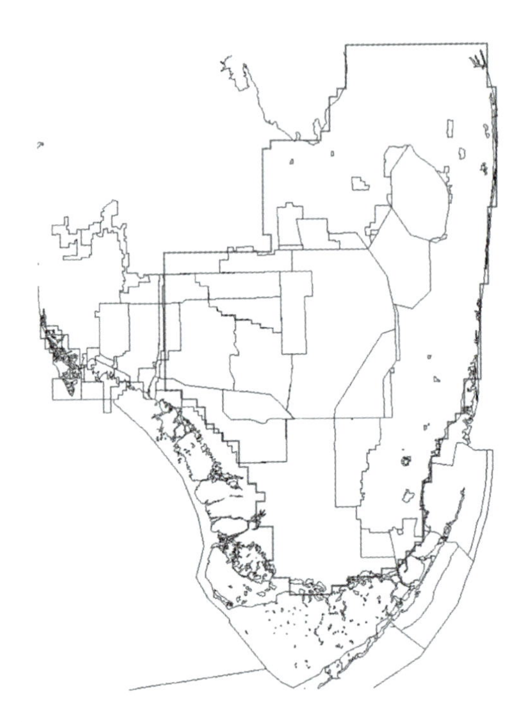

FIGURE 1 The outline of the South Florida water management model (SFWMM) study area with the boundaries of major management units and state boundaries for reference.

design process. The simplifications stem from using functionally identical programming elements throughout the models and model development and using these elements for a variety of related tasks. The similarities also allow us to unify the models and modeling process. Although many of the issues and solutions raised in this paper have been developed for the ATLSS project, the solutions we have developed are applicable to other similar modeling systems. The three major topics included in this chapter are common to any modeling effort where agent-based modeling incorporates spatially explicit data.

2 PROVIDING DATA TO THE MODELS

One of the primary tasks in spatially explicit modeling is making spatial data available to the models. A simple idea, but as with many simple ideas, one fraught with a multitude of annoying difficulties. In the ATLSS project some of the difficulties have been resolved by only using raster-based data. This type of data is appropriate for the models in ATLSS that also are based on a grid. It is possible to incorporate vector-based spatial data into a modeling system and, for some modeling systems, this may be appropriate. No vector-based data are used directly by any of the ATLSS models. Some raster maps are derived from vector-based information, but the models do not interact directly with the vector-based maps. This narrows the class of problems that we had to address in ATLSS. Accordingly we will not address here the issue of integrating vector-based data into agent-based modeling systems.

Some of the difficulties encountered in dealing with spatial data sets arise from differences in the spatial extent, spatial resolution, or registration of the data sets. These differences occur because the data sets are derived from different sources and stored in different formats. Data sets for the ATLSS project come from satellite telemetry, field observations, and models external to the project. Differences can also occur because data sets come from different agencies that have used different standards for storing spatial data.

Differences in spatial extent are a problem of intersection and computer memory. The models can only be run over the intersection of all the data sets. However, if the intersection represents only part of each spatial data set, there isn't much point in loading each map in its entirety. Only the region covered by the intersection should be loaded from each data set. This will save memory and can make the model run faster.

Different spatial resolutions and registration can cause sampling inconsistencies between the data sets. By registration we mean the location of the corners of the data, and consequently the location of the cell boundaries. Figure 2(a) shows two simple maps with different registrations. The sampling errors occur because cells of one data set straddle the boundaries of several cells from another data set. Figure 2(b) shows two simplified data sets with different resolutions. In this figure the sampling error occurs because the cell of one map straddles the border of cells from another map. For example, if the smaller cells are the grid cells of a model and the larger cells contain data for the model, which value should the model choose when the model cell intersects four data cells? None of the four values is completely appropriate since the model cell intersects with each of the four larger cells. There are several answers to this problem. The model could take a simple average of the four values, an average weighted by the amount each data cell intersects with the model cell, or the model could choose the cell with the largest intersection by area. There are several other possible answers to this problem. The correct answer depends on many factors. If computational efficiency is at a premium, then a weighted average would not be appropriate especially if there are a large number of cells in the data set. Another example is the type of data being utilized by the model. If the values in a spatial data set represent habitat type indices, then averaging the four values together does not yield a meaningful result. The point is that differences in registration and resolution need to be resolved to ensure the models operate in a timely and deterministic fashion. How the problem is resolved is application specific. Differences in spatial features of the data also presents a problem for correlating values. Values from different spatial data layers may represent features at the

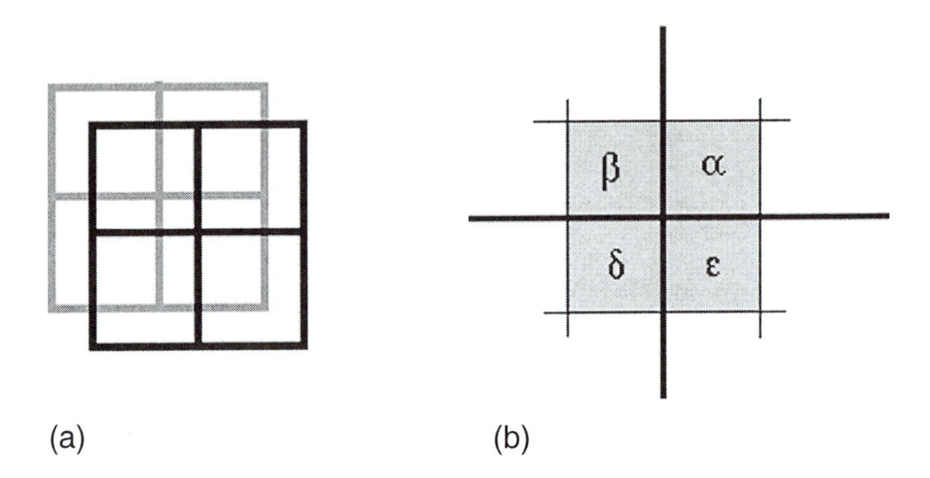

FIGURE 2 Illustrated the difference in (a) registration and (b) resolution. These types of differences can lead to sampling inconsistencies within the models.

same location in the landscape, but may have very different row and column indices.

Other difficulties in dealing with spatial data sets come from reading the data into the modeling system. Different data sets may store data values using different arrangements within a file. The type and arrangement of metadata can also be different from one file to the next. The arrangement of spatial data and metadata can also be the consequence of how the data was created and the agency from whom the data was obtained. Related to these sorts of problems are also the issues of exchanging data between the model system and outside programs such as databases and visualization packages. These third-party application may run concurrent with the model and provide valuable functionality. However, each of these packages uses different exchange format and mechanisms.

The basics of providing spatial data to the models is having a set of classes which store the data and which provide a single, common, and flexible interface. In the ATLSS project these classes are referred to as the Landscape classes. The Landscape classes can hold data sets of any spatial extent, at any resolution, and for any location. They preserve the spatial configuration of the data, and preserve georeferencing information for the entire data set, and each value within the data set. The ATLSS project currently uses both row-column indices and universal transverse mercator (utm) to allow component model to reference values from a spatial data set. Functions are also provided to translate between these referencing systems. Around this core functionality is built a host of other classes that can transform and translate the spatial data to and from other forms.

Input and output tasks are carried out by IODevice class objects. These objects encapsulate the details of translating spatial data between the outside world and the Landscape class objects. Translation is broken down into the format of the data and the source/destination of the data. The data format describes the sequence in which each value will be read from the source. It also includes the placement of metadata relative to the spatial data and the format of this metadata. Metadata are information about the spatial data that places the spatial data in context. A full description of what makes up metadata, its uses, and its importance is too lengthy a subject to discuss here and would be distracting. A host of literature is available to the interested reader [5, 9]. For the purposes of this paper, metadata will be used in a very restricted sense meaning the georeferencing information about the spatial data. This includes the location of the corners of data set, in utms, the size of the cells, the number of rows and columns, and other information defining the structure and location of the spatial data sets. Currently the ATLSS project uses a small number of data formats such as band sequential (BSQ) and a few other specialized formats inherited from government agencies from whom we receive spatial data. However, the IODevices have been designed to allow new data formats to be incorporated into the project without changing the code that depends on the IODevices. The source or destination of the data is also flexible. Data can be exchanged between the models and file stream or other applications such as databases or visualization systems. Flexibility in the format and source of the data is achieved by separating the interface for reading and writing the data, from the implementation which handles how the data are arranged for exchange and where it goes to or comes from.

Transforming the data between different resolutions and registrations is handled by a different collection of classes. The transforming class objects are configured by providing the size of a single cell in meters. A transforming object can also be given the exact number of rows and columns to which data sets should be resized. Spatial data sets can then be passed to the object, from which a new rescaled map is made and returned.

Subregions can also be extracted from any of the data sets using an object derived from the Poly class. This class accepts any number of vertices, specified in utms, which form a polygon. Subregions based on the shape and location of the polygon can then extracted from any spatial data. Because the polygon is defined in terms of utm coordinates, the same subregion can be extracted from data sets with different resolutions, registrations, or spatial extents.

There are some features of spatial data that the ATLSS library of class is not prepared to handle. As a result, differences between some types of data can not easily be resolved. Differences in projection and orientation are not handled by ATLSS data structures or tools. ATLSS has not had to deal with these issues since the data sets do not, in general, have a projection system associated with them. Many of the data sets are derived from other model systems that have assumed a flat earth over their regions of interest. For the level of detail available within our models and the current spatial extent of the models, these omissions do not present a serious problem. ATLSS has also been fortunate in that all the data sets have been delivered to us with the same orientation. However, both of these issues can further complicate integrating spatial data into a modeling system, by either having differences between the data sets or between the data sets and the models. These are both issues that should be kept in mind when integrating spatial data into a modeling system.

The temporal aspect of spatial data sets is also important. On the one hand, the data sets should represent features that are likely to have coexisted. The intersection of the time frames over which each data set is valid should be nonempty. However, there is little a set of programmatic tools can do to fix problems of this nature. The software can bring inconsistencies to the attention of a human, but techniques for solving the problem cannot be readily generalized or codified. Another consideration arises if a data set contains a time series of spatial data. For example, in the ATLSS project, hydrology is provided on a daily time step, each day being represented by a spatial data set that provides water depth at each location. While the time step of the data is daily, some of the models operate on a slower or faster time step. The ATLSS freshwater fish model, for example, operates on a five-day time step, while the deer and Panther models have a daily time step. In these cases the developer must make decisions about what will be provided to the model and how the data will be prepared. For the fish model it was decided that a five-day average of hydrology was appropriate. Other models may require different answers. Questions about the temporal resolution of the data have to be answered on a case-by-case basis and do not lend themselves to generalized answers.

3 STRUCTURING THE MODELS

All of the ATLSS models are spatially explicit. The individuals or functional groups are distributed over the landscape. The models do not all incorporate spatial distribution in the same manner, however. For reasons of computational efficiency, the agents which represent the panther and deer each explicitly track their own location by keeping explicit internal records of their positions on the map. In contrast, the ATLSS freshwater fish model arrays the agents in a Landscape object and the exact location of each agent is determined by referencing the container object. Many of the models make use of the Landscape classes in some fashion to provide spatial structure for internal components.

The internal spatial structure of many of the models is based upon a discrete partitioning of space into a rectilinear grid. This structure is identical to the structure of the spatial data sets. It makes sense, therefore, to reuse structures for holding input data to structure internal model elements. The major change from standard GIS functionality which we have developed concerns the types of data which can be placed within each cell of a Landscape object. Each spatial data set represents a single feature of the landscape and each cell contains a single value. The Landscape objects which hold these data sets while the model is running are simple two-dimensional arrays of variables. We would like the spatial structure of the models to be conceptually similar by having a model consist of a single layer. However, many variables may be needed at each location to model the state and dynamics of the system. For example, the fish model needs to track the numbers of fish in each of several size classes for each location over the landscape.

It makes sense for several reasons to combine all the features of a model, including both variables and functions, into a single package and place copies of the package in each cell of a container like a Landscape object. The difficulty is in creating a Landscape class that can use any type of object to fill cells. A simple way of doing this is by using C++ templates [8, 10]. Templates allows construction of generic classes where the basic structure of the class is described before compile time, but details concerning the type of data that they manage is defined at compile time. For example, a templated Landscape class describes a two-dimensional grid into which spatially explicit data can be placed. The structure of the data that are placed into the grid is not defined by the templated class. The spatial structure and the data that are to be included within this spatial structure are separate and independent of each other. One implication of this is that a model developer can focus on the details of a model at the cell level. The larger spatial structure of the model comes from combining the templated Landscape class and the model details.

Using the same data structure for both the spatial data set and the internal structure of a model is beneficial. Since they both have the same interface, they can be treated in an identical fashion. The same georeferencing system can be applied to both sets of spatial information, making it easier to coordinate the values between the two data sets. The same tools used for input and output can also be used with both types of Landscape structures. Similarity in structure also means the Landscape classes can be more rapidly developed and more easily maintained. This is particularly important when many developers from different institutions are involved in the project.

4 COMMUNICATING BETWEEN THE MODELS

One of the most significant uses of the Landscape classes is as a means of communicating between the component models. This allows several different models, representing different aspects of the system, to be joined together. The Landscape classes provide bidirectional communication between the models. They also provide a mechanism for transferring spatial data with the spatial aspects completely intact.

Where a data layer provides a relatively static view of the world, linkage between models provides temporal variability. Temporal variation in the spatial arrangement of values can significantly alter the outcome of a model and change the conclusions drawn from the model. Temporal variation also adds realism to the model.

In the ATLSS model there are many examples of linkage between models that make use of Landscape objects. In the ATLSS deer model the individual deer graze on the locally available biomass. These local biomass levels are generated by the ATLSS vegetation model. The vegetation model produces forage biomass in response to changing hydrologic conditions. Since the hydrology is changing from day to day, there is also temporal variation in biomass availability. This provides a temporally variable mosaic of available biomass to the deer model. This, in turn, changes the spatial distribution of the deer as they move to areas of high forage availability. The biomass values are communicated to the deer model from the vegetation model through a Landscape object. As the deer graze they lower the local biomass level. Thus deer grazing further alters the spatial distribution of forage availability. These lowered biomass levels are available to the vegetation model and are used as the starting point for regrowth. The same Landscape object used to communicate forage biomass level to the deer is also used to communicate the reduced biomass level back to the vegetation model. Landscape-class objects therefore provide bidirectional communication between the models.

All forms of communicating spatial data between models are carried out using the Landscape classes. Using the same set of tools for communicating information between models as are used for providing spatial data and internal structure simplifies and unifies the models. The same set of cell-referencing and georeferencing tools can be applied to all three parts of the modeling system. The same basic tools for manipulating the size and shape of spatially explicit data can be applied universally as well.

5 CONCLUSION

The ATLSS Landscape classes that we have discussed provide many different types of functionality that facilitate model development and which lead to significantly more unified models. The library of classes provide the basic data structures for sharing and accessing spatial data during the run time of the model system. They provide the methods of access such as row-column cell indexing and georeferencing through universal transect mercator (utm). They also provide commonly used functions such as converting data sets to other resolutions and other registrations or extracting subsregions. Other classes provide input and output between the Landscape classes and many different destinations such the file system, database, or visualization system, and provide compatibility with many different file formats. Many other functions and data classes could be added to the library. Additional data structures such as one-dimensional arrays or linked lists could be incorporated which would allow a better mapping between the real world and the model. Other functions could also be added, such as clustering algorithms. If a data structure or function is commonly used by the system, it makes sense to add these to the library. This allows a feature to be implemented once and then used many times which minimizes the reinvention of functions and data structures.

A collection of tools like the Landscape classes can also help with certain types of analysis of the models. Model sensitivity to different resolutions is important since the quantitative results may change with the resolution [4]. Run time also changes with resolution. The Landscape classes make it simpler to change the resolution of both the models and the data and removes some of the uncertainty and tedium of analyzing model sensitivity to spatial resolution. Yet to be addressed for these sets of models are issues of parallelization, and how run time might scale with different schemes to partition the workload across different processors. Though we have made some efforts to analyze these issues for the deer and Panther models, the indication is that considerable effort is required to efficiently parallelize agent-based models in which the agents have significant interactions with underlying landscape components [1, 7]. The simplest spatial partitioning methods can lead to tremendous communication loads depending upon the nature of the location information needed by the agents in order to ascertain their actions. Thus, agents that can move often between the spatial regions handled by each processor can create potential temporal conflicts between the model attributes being handled by each processor.

Agent-based modeling provides a mechanism for understanding larger spatial extent aggregate behavior of systems by modeling localized dynamics of component objects. The smaller units of individuals or functional groups can then readily respond to local conditions. A system of tools for dealing with the spatial structure and spatial operations that are common to many parts of the modeling system simplifies the design and development of such a system. These tools also unify the development and final implementation of the model system. This can result in shorter development times and faster running models. We have here described the object-oriented approach which we have taken to provide such a set of tools, though we are quite aware that additional functionality would be beneficial in order to make it feasible for these tools to be applied to other agent-based projects linked to GIS. Despite the present limited functionality, the issues of providing data to the models, structuring data handling within the models, and providing communication between the models all are required in a linked GIS agent-based system. Our described approach provides these basic functions and can readily be modified to provide enhanced options depending upon project needs.

REFERENCES

[1] Abbott, C. A., M. W. Berry, E. J. Comiskey, L. J. Gross and H.-K. Luh. "Computational Models of White-Tailed Deer in the Florida Everglades." *IEEE Comp. Sci. & Eng.* **4** (1997): 60–72.

[2] Comiskey, E. J., L. J. Gross, D. M. Fleming, M. A. Huston, O. L. Bass, H.-K. Luh, and Y. Wu. "A Spatially-Explicit Individual-Based Simulation Model for Florida Panther and White-Tailed Deer in the Everglades and Big Cypress Landscapes." In *Proceedings of the Florida Panther Conference*, held November 1–3, 1994, Ft. Myers, Fl, edited by Dennis Jordan, 494–503. Ft. Myers, FL: U.S. Fish and Wildlife Service, 1997.

[3] DeAngelis, D. A., L. J. Gross, M. A. Huston, W. F. Wolff, D. M. Fleming, E. J. Comiskey, and S. M. Sylvester. "Landscape Modeling for Everglades Ecosystem Restoration." *Ecosystems* **1** (1998): 64–75.

[4] Deutschman, D. H., S. A. Levin, C. Devine, and L. A. Buttel. "Scaling from Trees to Forests: Analysis of a Complex Simulation Model." 1997. Science online. July 10, 2000. ⟨http://171.66.122.53/feature/data/deutschman/⟩.

[5] Federal Geographic Data Committee. "Content Standard for Digital Geospatial Metadata." FGDC-STD-001-1998, 1998.

[6] Gaff, H., D. DeAngelis, L. J. Gross, R. Salinas, and M. Shorrosh. "A Dynamic Landscape Model for Fish in the Everglades and Its Application to Restoration." *Ecol. Mod.* **127** (2000): 33–52.

[7] Mellott, L. E., M. W. Berry, E. J. Comiskey, and L. J. Gross. "The Design and Implementation of an Individual-Based Predator-Prey Model for a Distributed Computing Environment." *Simul. Prac. & Theory* **7(1)** (1999): 47–70.

[8] Musser, D. R., and A. Saini. *STL Tutorial and Reference Guide*. New York: Addison-Wesley, 1996.

[9] Schweitzer, P. N. "Easy as ABC— Putting Metadata in Plain Language." *GIS-World* **11** (1998): 56–59

[10] Stroustrup, B. *The C++ Programming Language*, 3d ed. New York: Addison-Wesley, 1997.

[11] Wolff, W. F. (1994) "An Individual-Oriented Model of a Wading Bird Nesting Colony." *Ecol. Model.* **72** (1994): 75–114.

Models of Individual Decision Making in Agent-Based Simulation of Common-Pool-Resource Management Institutions

Peter J. Deadman
Edella Schlager

1 INTRODUCTION

Addressing the problems of natural resources management requires an understanding of the complex interactions between human and natural systems. Modeling and computer-based simulation has been utilized increasingly as a tool to facilitate this understanding. Numerous simulations of natural systems have been developed, from global-scale general circulation models to more localized models of watersheds or fisheries. Such simulations are useful in providing resource managers with an indication of how these systems behave under different conditions. But while considerable effort has been devoted to the simulation of natural systems, the amount of effort devoted to modeling human systems, and their interaction with natural systems, has been relatively small. Recently, researchers have outlined the importance of developing a discipline devoted to the modeling and simulation of human systems [39, 40]. Increased efforts are now being devoted to the simulation of social phenomena (see, for example, Doran and Gilbert [18]). The tools now exist to develop simulations that incorporate the behavior of both a natural resource and the human individuals or institutions that interact with the resource.

A considerable body of work exists devoted to understanding the behavior of the institutions that people have developed to manage natural resources.

Specifically, a large number of studies have been undertaken in an effort to understand how common pool resources (CPRs) have been managed in differing natural and institutional environments [8, 36, 43]. Numerous field studies and laboratory experiments using human subjects have supported the evolution of a theoretical foundation for the study of resource management institutions. But while field studies and experiments have been useful tools for exploring the management of natural resources, to date little effort has been devoted to exploring the potential role of modeling and computer-based simulation for understanding the behavior of resource management institutions.

This chapter seeks to combine the theoretical foundations of research on institutions for resource management, with recent advances in human systems modeling to outline a framework for modeling individual decision making in resource management environments. Starting with a brief review of social simulation and intelligent agent-based modeling formalisms, this chapter moves on to discuss models of individual decision making in the social sciences and in simulations. The discussion facilitates the construction of a theoretical framework for modeling individual action, drawing upon a series of simulations of intelligent agents interacting within the context of a common pool resource. This framework is built upon a number of theoretical tools for the study of institutions including, the grammar of institutions, and the institutional analysis and development framework developed by Elinor Ostrom and colleagues [16, 43].

1.1 AGENT-BASED MODELS AND SOCIAL SIMULATION

Creating simulations of social phenomena requires the utilization of an effective and appropriately structured modeling formalism. Although modeling formalisms are often associated with specific simulation languages, they are conceptually independent from any language [61]. No one formalism can best represent the variety of behaviors that exist in natural and human systems. Therefore, it is important to select a formalism that most naturally fits the objectives of the simulation, the framework adopted by the modeler, and the level of abstraction. A good fit will produce a simulation that is computationally efficient and well suited to analysis. Simulations of human-environment interactions may include a number of different connected models, expressed in different formalisms. For example, a simulation of human individuals interacting with a natural resource may require that intelligent agents represent the individuals while a cell space model such as a cellular automaton, or the spatially referenced database of a geographic information system (GIS), represents the natural system.

The formalism adopted for the simulations discussed in this chapter utilizes intelligent agents to represent human individuals interacting with their environment in an attempt to achieve certain goals. This agent-based approach to exploring social phenomena has received a great deal of attention in recent years. A detailed exploration of agent-based social simulation is beyond the scope of this chapter, but can be found in recent works by Holland [28], Epstein and Axtell [20], Conte et al. [14], Gilbert and Doran [23], Prietula et al. [48], and the chapters by Gimblett in this volume. Here, our representation of an intelligent agent is described as a tool to explore models of human decision making. Comparisons will be made between the simulations developed here and those of other authors to facilitate our exploration of decision making in individual agents.

Intelligent agents can be described as autonomous entities that exist within the virtual environment of a computer simulation. These agents have two basic components, a model of their environment, and a model of themselves [60]. The agent constructs the model of the environment as it interacts with the outside world of the simulation. The nature of this model, or the accuracy with which it reflects the conditions in the simulated outside world is influenced by the information that is available to the agent and the agent's capabilities for processing that information. This model influences an agent's actions. Reactive agents may simply react to incoming information as it is received. Intentional agents may attempt to predict future states or unknown conditions in the outside world and adjust their actions in anticipation of those future states. The agent's model of itself represents the knowledge base that it utilizes to govern its actions within the environment. Such a model may reflect the agents' individual preferences, norms, and selection criteria that are used to evaluate incoming information and determine possible future actions. In these systems, the knowledge base is frequently represented as a separate structure containing a collection of rules with a condition-action structure typically having the form "if such and such—then so and so" [28]. Agents who possess such a collection of rules may also be capable of adapting the knowledge base according to information received about the suitability of that knowledge relative to conditions in their environment. This adaptation frequently occurs within the context of an overall goal or set of goals.

Holland [28] has described the adaptive mechanisms employed by agents as having two basic forms, adaptation by credit assignment, and adaptation by rule discovery. As agents utilize the rules in their knowledge base, they keep track of how the rules performed, or how they might have performed, if they were utilized. Each rule, or collection of rules, has associated with it a score based on its actual or perceived performance during the simulation. Agents adapt by credit assignment when they compare these scores to select the best performing rule for use in future rounds. The agents in the simulations discussed in this chapter adapt by credit assignment, keeping track of the return earned by alternate strategies, and selecting the strategy that appears to yield the highest return for use in future rounds. Other agent-based simulations [1] have employed a similar adaptive mechanism.

When agents have the ability to create new rules, they are adapting by rule discovery [28]. With such a mechanism, agents break rules down into their constituent parts, and recombine these parts to produce new rules. The most widely utilized mechanisms for creating new rules include genetic algorithms [24, 31] and classifier systems [24, 30, 31]. Numerous examples exist in which evolutionary algorithms have been applied to social processes. Some of the potential benefits and pitfalls of using these mechanisms for simulations of social phenomena are outlined in Chattoe [12].

Recently, a number of examples have appeared in which agent-based simulations have been utilized as a tool for developing models of human behavior [17, 19, 21]. Computer simulation according to Conte and Castselfranchi [15] "is an appropriate methodology whenever a social phenomenon is not directly accessible, either because it no longer exists (as in archaeological studies) or because its structure or the effects of its structure, i.e., its behavior, are so complex that the observer cannot directly attain a clear picture of what is going on (as in some studies of world politics)." The authors observe that an important objective of this work is to realize, observe, and experiment with artificial societies in order to improve our knowledge and understanding, but through exploration rather than just through description [15]. These studies utilized a general model of multiagent simulations based on computation agents that represented individual organisms (or groups of organisms) in a one-to-one correspondence. These studies seek to understand the process of evolution in ecological and sociological systems.

Economists have also explored the utility of intelligent agents in simulating complex social interactions [27, 38]. Artificial economic agents have been designed to explore the complex interactions among themselves and their economic environment [1, 4, 29, 55]. Kephart et al. [32] explored the dynamics of a financial market in which different agents utilize different trading strategies (see also Sanglier et al. [51]). Caldas and Coelho [10] have explored the effects of strategic interactions. That is, interactions between agents in which other agent's behavior is taken into account in arriving at a decision between agents in oligopolistic markets. Bhargava and Branley [9] explored how belief systems may be simulated in agents. The interest in utilizing a modeling formalism for social simulation that is based on adaptive agents continues to grow.

Since it is likely that different decision-making contexts may result in different actions, the design of humanlike artificial agents may result in the creation of a multitude of different algorithms. To bound this modeling effort, Arthur [1, 4] has suggested that we calibrate the behavior of these artificial agents against real data. Thus the name of such agents—calibrated agents. The agents should then be able to be applied in new situations with a greater degree of confidence. Arthur's call for the development of calibrated agents highlights the need for more simulation efforts that explore the connections between specific models of individual decision making and the environment in which the individual agents are situated.

Many of the early agent-based social simulations were generic in scope and focused on agent architectures without exploring the settings for which that architecture would be appropriate. In the remaining sections of this chapter, we sketch one approach to the development of a framework for describing agent architectures and evaluating the relevance of those architectures to real-world settings. This framework has its theoretical roots in political science and economics. There are two reasons for selecting this approach. First, these fields have explored and debated alternative models of individual decision making. Second, these theoretical roots are closely connected to investigations that have explored how groups organize themselves around problems such as the management of a commonly held resource. The following section of this chapter explores individual models of decision making and their relevance to agent architectures in the exploration of simulations based on the commons dilemma.

1.2 MODELS OF INDIVIDUAL DECISION MAKING

Scholars devoted to the study of politics and economics have, for centuries, paid attention to individual decision making, or models of the individual. In the seventeenth century, Thomas Hobbes [26] developed his theory of the Leviathan based on a model of the individual in which people "sought power after power." In other words, individuals were assumed to act in ways that they believed would make themselves better off [46]. While Hobbes model was more descriptive, Machiavelli provided a prescriptive model of the individual, which was also decidedly more narrowly self-interested than Hobbes' individuals. If a Prince wanted to successfully gain and hold power, then he should be calculating, strategic, and opportunistic in pursuing his own personal gain [34]. In the eighteenth century, Madison, Hamilton, and Jefferson, grounded their theory of federalism on the assumption that individuals were self-interested [46]. In the nineteenth century, de Tocqueville, in explaining the operation of Federalism and the American form of democracy, grounded his analysis in the notion of "self-interest properly understood." By this, de Tocqueville meant that individuals do not narrowly attend to their own interests. Rather, they realize that their well-being depends on their own actions and choices, as well as the actions and choices of those residing in the same community. Thus, an individual who properly understands his own self-interest will "sacrifice some of his private interests to save the rest" [57, p. 527]. Each scholar shared a common goal—to use a model of individual decision making to explain the structure, function, and comparative performance of institutional arrangements.

Except for perhaps Machiavelli's prince, none of the models assumed narrowly self-interested decision making by individuals. Nor did the models assume isolated individuals cut off from community norms, values, and expectations. That has changed, and during the twentieth century a particular model of the individual held way over economics and political science. Neoclassical microeconomics is based on a model of the individual who makes choices so as to maximize his or her utility. The individual is quite capable of maximizing her utility because she exhibits perfect and boundless computational abilities, and possesses perfect and complete information. For analytical purposes, such an individual possesses a well-defined utility function, a complete and well-defined set of alternatives from which to choose, and a probability distribution over all possible states of the world. Armed with such information and computational power, the individual, when confronted with a choice, selects the alternative that maximizes her utility [53].

In neoclassical microeconomics, the model is used to examine, explain, and predict the outcomes produced by markets. Perfectly rational, utility-maximizing individuals, who are price takers, buy and sell until the marginal value of goods and services are equal. At that point, the market equilibrium, buying and selling ceases because all individuals' utilities are maximized. The focus is upon equilibrium outcomes. Exogenous shocks to the market affect prices, setting off another round of buying and selling, producing another equilibrium characterized by utility maximization. Utility is maximized because individuals can anticipate all other individuals' actions, assess every possible alternative, and select the voluntary exchanges that make them best off.

The utility maximizing model of human decision making is a form of a stimulus-response model. The stimulus is a change in price, the response is market transactions that make the parties to them better off. No actual consideration is given to the internal states of mind or thought processes of individuals. As Simon [53, p. 34] suggests, the model better describes the mind of God and not of man. Admittedly, the model does not describe human behavior, however, as Freedman argued in defense of neoclassical microeconomics, theories and models should be evaluated on the basis of how well they predict and explain, and not on the realism of their assumptions.

Precisely because the model does not describe human behavior, it has been subject to sustained criticism. Those who are highly sympathetic to the model make the most fruitful critique of it. Such a model usefully describes the outcomes achieved in competitive markets. In such settings, the outcomes achieved make it appear that humans act as if they were perfectly rational. Change the setting, however, and such a model of the individual poorly predicts outcomes. In other words, outside of competitive market settings, it is incapable of explaining important human actions and artifacts of central importance to social scientists. For instance, such a model of the individual, while useful in explaining and predicting competitive market equilibria, cannot explain the existence, let alone the variety and structure, of organizational arrangements found in competitive market settings [59]. Nor can such a model explain the emergence of cooperation among individuals to achieve a common end [42]. In the first case, perfectly rational individuals should not need to shield themselves from markets by bringing exchanges within the context of an organization. In the second case, individuals should not cooperate. That is, invariably they are better off if they free ride, according to the utility maximization model of decision making.

Instead of dismissing the model, scholars such as Williamson and Ostrom, recognize its value in explaining transactions in markets, yet suggest that other models of individual decision making be used, or developed, that better explain and predict behavior in diverse settings. For instance, Williamson [59] uses Simon's [54] boundedly rational model of the individual to develop a transaction cost theory of market organizations and contracts. A boundedly rational individual possesses both limited cognitive-processing capabilities and information. Such an individual is intendedly rational in the pursuit of instrumental goals, but is limited in that pursuit by her cognitive and information resources.

Ostrom [42] takes a similar approach in explaining why individuals cooperate to resolve social dilemmas. Perfectly rational individuals in a social dilemma setting should never cooperate to resolve the dilemma, unless the setting extends over an infinite length of time [56]. Yet, individuals do cooperate with one another in a wide variety of settings characterized by many different social dilemmas that extend over known and limited amounts of time. Boundedly rational individuals develop shared norms of trust and reciprocity that encourage cooperation and that allow individuals to achieve outcomes that make themselves better off [42]. Ostrom explains important social phenomena—cooperation, and institutional arrangements that emerge from cooperation—that allow individuals to resolve social dilemmas—by adopting a boundedly rational, behavioral model of individual decision making.

In contrast to the more deductive approaches of Ostrom and Williamson, who posit assumptions about individual decision making, and from those assumptions derive a set of testable hypotheses about human actions, other social scientists have taken a more inductive approach to individual decision making. The inductive approach takes two forms. One examines the myriad ways in which human decision making violates the perfectly rational model of the individual. The second attempts to develop a behaviorally grounded model (or models) of the individual. The two are not mutually exclusive, and often inform one another.

The literatures in economics, psychology, social psychology, and political science on how people systematically violate the presuppositions of the perfectly rational model of the individual are vast [11, 35, 58]. Individuals in experimental settings make choices that systematically violate the assumption that individuals possess well-defined, complete, and stable utility functions, and the assumption that individuals can and do assign consistent probability distributions to all future events [11, 58, p. 591].

In addition to examining how individuals' choices violate assumptions of perfect rationality, pyschologists, and to a lesser extent political scientists, have attempted to empirically examine and identify how people make choices, that is, the thought processes that guide decision making. Political scientists who study public opinion and voting using survey data, have long recognized that people do not possess well-constructed and coherent frameworks by which they assess information and reason about public affairs; that generally people base their opinions and votes on incomplete and incorrect information about events and candidates; and that people learn over time with experience [13]. Social psychologists have devoted considerable attention to identifying cognitive structures or schemas, and how those schemas affect the information individuals attend to, how they process information, and how they influence evaluations, judgements, and predictions in given situations [35]. Such work has been undertaken to determine how, in practice, individuals make decisions. Findings from this work have not been used to develop a model of individual decision making as simple and mathematically tractable as the model of perfect rationality.

What does all of this matter for the modeling of human decision making within complex adaptive systems? For such simulations to be interpretable, they must be grounded in theory. Theory informs the analyst of which variables to include in the model, and the expected relationships among them. One of the contributions of the social sciences is to make sense of, and to guide the simulation of individual decision making within complex adaptive systems. However, there remains a fundamental rift in the social sciences concerning appropriate models of the individual. Economists, in general, would argue that the model of perfect rationality should be used, especially in explanations of economic settings and processes, unless and until there is clear and convincing evidence that alternative models of individual decision making would perform better, i.e., provide better predictions and explanations of outcomes. On the other hand, many social scientists, who are not economists, would argue that the perfect rationality model should never be used because it fails to capture the essential features of individual decision making. And, a model of bounded rationality better explains particular types of behavior, even economic behavior, such as how people conduct job searches, or the role of reciprocity in the functioning of market organizations. In this chapter, we take the more ecumenical position of Ostrom [42]: settings and research questions should guide the selection of the model of the individual.

The model of the individual that should be used in a simulation depends on the questions to be answered by the simulation. For instance, if one were interested in how and to what effect individuals learn about the consequences of a particular policy, then one should use a model of the individual that allows for learning that passes through the filter of belief systems [50]. On the other hand, if one were interested in the emergence of social norms and their effects on coordination and cooperation among individuals, then one should use a model of the individual that allows for learning based on recognition of other individuals' likely strategies and experience [33, 42]. But, what would it mean to simulate such boundedly rational individuals? As the social sciences have demonstrated, a boundedly rational individual, i.e., any model that posits instrumental decision making short of perfect rationality, can take on a wide variety of characteristics. Furthermore, the social sciences, for the most part, provide only broad outlines of alternative models of individual decision making. These models work well for deducing hypotheses that are empirically testable, or that can be used to develop narrative accounts of particular events or processes [7, 52], but that are not easily transferable into a language interpretable by a computer.

Ostrom, Gardner, and Walker [43] posit a framework for organizing characteristics of individual decision making that may prove useful in making sense of the wealth of decision-making information provided by the social sciences, and that may provide a map of how to proceed in simulating individual decision making. In order to explain individual behavior, analysts must make assumptions about individuals': (1) preferences; (2) information-processing capabilities; (3) selection criteria; and (4) resources [43, p. 33–35]. We supplement this list with one additional category, norms/values, which is derived from Ostrom's [42] latest work. "The actor is, thus, the animating force that allows the analyst to generate predictions about likely outcomes given the structure of the situation" [43, p. 35].

Preferences reflect the tastes or desires of individuals over objects and outcomes. Norms and values are morally grounded and reflect "an internal valuation—positive or negative—to taking particular types of action" [42, p. 9]. Information-processing capabilities reflect both the information that individuals possess, or have access to, about a situation, and their ability to process it. Selection criteria determine individual choices among alternatives. Resources define individuals' abilities to take available actions. For instance, the model of perfect rationality assumes values for each of these characteristics. Individuals' preferences are assumed to be complete and stable over the outcomes that are available in a given setting, individuals are not guided by norms, individuals possess complete and perfect information which they are capable of infallibly processing, individuals select those alternatives that maximize their preferences, but individuals are constrained by limited budgets in so doing. Different models of the individual can be constructed by varying the values of one or more of these characteristics. The value of such a framework is that it allows for carefully and precisely changing these characteristics to determine the effects such changes might have on the choices and outcomes individuals make and achieve. The next section outlines a series of common-pool-resource experiments that were developed to explore theoretical predictions about individual behavior in commons dilemmas. In the following discussion we focus primarily on the characteristic of information processing, and to a lesser extent resources and norms and values. In our conclusions we will again return to a discussion of each of the characteristics.

1.3 COMMON-POOL-RESOURCE EXPERIMENTS

To explore the effects of different values of the characteristics of individuals outlined above on the choices that they make and the outcomes that groups collectively achieve, we turn to a series of laboratory experiments and simulations of those experiments. These experiments were motivated by a glaring contradiction between theoretically predicted outcomes and actual outcomes in common-pool-resource settings. Interest in the study of CPRs is fueled in part by the desire to understand how the apparent conflict between individual rationality and group rationality, referred to as a commons dilemma, can be avoided. CPR dilemmas have been illustrated in the past through a number of popular metaphors including the tragedy of the commons [25], and the Prisoner's Dilemma game. These metaphors have been used to argue that individuals who participate in the withdrawal of resource units from a CPR are likely to overappropriate the resource, resulting in suboptimal collective benefits.

However, Ostrom [44] challenged the universality of these metaphors, outlining numerous real-world examples in which individuals were able to organize their collective actions by establishing rules that facilitated a long-term improvement in joint outcomes. But despite the fact that we know that many CPR management institutions are able to function efficiently without depleting the resource, no theory exists which can explain how and why some appropriators are able to avoid CPR dilemmas while others are not [43]. One of the goals of researchers who endeavor to explain the behavior of CPR management institutions has been to understand what types of institutional and physical variables affect the likelihood of the successful resolution of CPR dilemmas [43]. Modeling and computer-based simulation promises to be a useful tool for exploring these theoretical questions. By developing simulations of the interactions between human institutions and natural resources, researchers could explore how variations in the parameters of the model influence the overall behavior of the simulation.

Researchers such as Ostrom et al. [43] developed a series of laboratory experiments in an effort to understand the degree to which predictions about individual and group behavior, and their resulting outcomes, derived from noncooperative game theory, are supported by empirical evidence. Built upon tight theoretical models of a CPR situation, laboratory experiments serve as a useful mechanism for simulating those models and observing outcomes. The experiments allow control of the elements of the institutional environment, and to some degree the characteristics of the individual, thus facilitating analysis of the relationship between the structure of that environment and the resultant outcomes. Some of the experiments developed and described in Ostrom et al. [43] form the basis for the development of the simulations described here.

In the baseline version of these experiments, eight subjects were presented with a situation in which they chose to invest tokens in two alternatives, or markets. Market 1, a safe alternative, provides a constant rate of return on investments. Market 2 provides a return that varies in relation to the total group investment and the investment of the individual. Market 2 is the CPR. The marginal return per token invested in the CPR increases to a point, and then begins to decrease as the resource is overharvested. Through an unknown number of rounds, each subject was endowed with ten tokens per round and asked to divide their tokens between the two markets. The subjects knew that each subject had a ten-token endowment and the returns that each market would provide at differing levels of investment. They also knew the number of tokens that they individually had invested and that the group had invested in each market, plus their earnings and the group's earnings after each decision round. Data was collected on the round-by-round investments of each participant, but this information was not made available to the subjects during the experiment.

The primary purpose of the experiments was to test theoretical predictions. Using a model of perfect rationality, an individual will invest in the CPR (market 2), so as to maximize his or her return given what the other subjects do. This is known as Nash equilibrium—it is the best outcome that a perfectly rational individual can achieve in this setting. The Nash equilibrium level of investment in market 2 was 64 tokens, or an investment level of 8 tokens by each participant. The Nash equilibrium, however, is deficient in that it does not maximize the group return. The group return from the CPR is maximized when 36 tokens are invested in it, or an investment level of between 4 and 5 tokens per individual. In order for perfectly rational individuals to make themselves better off by achieving the optimal group return, each individual would have to substantially cut back his or her investment. Yet, once this occurs individuals face temptations to cheat. An individual can improve his individual return if he invests 8 tokens and all other individuals invest 5 tokens. Once individuals succumb to temptation, they find themselves back at the Nash equilibrium. All are investing 8 tokens each. Just how deficient is the Nash equilibrium? The performance of the group is measured as rent as a percentage of optimum, or the return the group receives from market 2, minus the opportunity costs of investing in market 1, compared to the optimal level of investment. With this measure, the Nash equilibrium level of investment would return rent as a percentage of optimum at 39%.

2 COMMON-POOL-RESOURCE SIMULATIONS

This is the setting that human agents and intelligent agents acted within. Before we describe the outcomes of laboratory experiments and of simulations and compare them with the theoretical predictions that are based on perfect rationality, we must first briefly describe the simulations. The system utilized for these simulations was Swarm, a multiagent simulation platform developed at the Santa Fe Institute [37]. This platform consists of a collection of object-oriented libraries of reusable components for building models and analyzing, displaying, and controlling experiments on those models. The user writes a collection of classes to represent the unique components of the model itself, as well as the objects that display the output of the simulation and create instances of the model classes. Swarm adopts a modeling formalism that consists of a collection of autonomous agents, interacting via a time-stepped series of discrete events. The basic unit of a Swarm simulation, an agent, is an entity that generates events that can affect itself and other agents [37]. In a CPR model, individual agents can be created, with their own set of unique characteristics, to represent individual participants in an institution and the natural resource itself.

The initial simulation was developed to explore the baseline laboratory experiments. The base model for these simulations includes a collection of eight agents, who utilize a set of simple heuristics, referred to here as strategies, to make decisions about the investments they make in each of the two available markets. These agents display adaptive characteristics in the sense that they select, from amongst a collection of 16 available strategies, the one which best achieves their goal. In this case their goal is to maximize their return given information about the returns they received in previous rounds, and the actions of the other members of the group. In the initial simulations, no communication occurred between agents. Later modifications to the simulation explored the effects of alternative communication routines on the outcomes. See the Appendix for a detailed description of the simulation.

Let's explore the three models of individual decision making that are being used here. In the theoretical model, perfect rationality is assumed. Individuals possess a complete preference ordering over all possible outcomes in the CPR experiment. Individuals possess complete information about the setting. The only information that they do not have are the actual choices of each individual after each decision round; however, they do know the group outcome. They are capable of processing the information flawlessly, and are able to determine their best possible strategy, given the best possible strategies of the other individuals. They will choose the strategy that maximizes their payoffs. Thus, they will select the strategy that yields the Nash equilibrium. Finally, they possess resources sufficient to allow them to pursue such strategies.

In the laboratory experiment it is much more difficult to precisely specify the values of each of the characteristics of the model of the individual. Experimenters can control the values of some of the characteristics, but the human subjects through their own life experiences bring the values of other characteristics to the experiments. For instance, experimenters can control the information that subjects receive; however, they cannot control the ability of individuals to process the information and to identify and select strategies that most nearly achieve their preferences. Nor can the experimenters control the norms and values that individuals bring to the decision-making process. But experimenters can attempt to induce individuals to engage in maximizing behavior by paying individuals according to the outcomes that they achieve. The better the outcome, the more they are paid. Finally, in these experiments, subjects were endowed with sufficient resources to pursue their best strategies.

In the simulation, each of the characteristics of the individual can be carefully specified, much like the theoretical model. In the simulation, the intelligent agents are boundedly rational and possess the following characteristics. They possess a complete and stable set of preferences over the outcomes that can be achieved in the CPR setting. They possess complete information; however, their information-processing capabilities are limited. Instead of computing the strategy that will make them best off, given the actions of all other agents, they select a strategy, from a pool of 16 possible strategies. The agents employ a form of adaptation by credit assignment in which they utilize a current strategy but keep track of the performance of the current strategy and all the alternates. The agent has the opportunity to switch an alternate with the current strategy if the alternate is producing a higher performance. In this case, performance is measured as the total return that the strategy earns for the agent in each round. Because the agents employed a feedback mechanism that evaluated the relative strength of each strategy based on the return that it earned for the agent, the performance of any particular strategy depended upon the actions of the other members of the group in each round. Therefore, a strategy that worked well for one agent at a particular point in time may result in a considerably poorer performance later in the simulation.

Six of the 16 strategies are based on attempting to maximize the return received in each round by determining any trends in the relationship between the number of tokens invested in market 2 and the return received by the agent. The increment or decrement of bids to market 2 from one round to the next is varied in each of the strategies. Another six strategies are based on the comparison of average returns from market 1 and market 2. The strategies differ in the amount that market 2 bids are incremented when the average return from market 2 exceeds the average return from market 1. This increment varies from one token to dumping all tokens into market 2 when the right conditions exist. This strategy is one that some participants in laboratory experiments reported to have used in exit interviews [43]. The final four strategies are based on a comparison of an individual agent's bid with group average bid, and then submitting a bid at or above this average level. The 16 strategies are summarized in table 1 in the appendix.

2.1 BASELINE SIMULATIONS—NO COMMUNICATION

Note the three different models of individual decision making. The theoretical model includes a perfectly rational decisionmaker. The laboratory experiment involves boundedly rational individuals, although the experimenters set the values of the characteristics of the individual over which they had control to those of perfect rationality, they could not control the information-processing capabilities of the individuals who participated in the experiments. The simulations involve boundedly rational individuals, in the sense that the intelligent agents possess limited information-processing capabilities as exhibited through the strategies that they could select from. None of the strategies contain any information on the equations used to determine the return from market 2. In addition, none of the strategies contain any information on the behavioral characteristics of the other agents in the simulation. Instead, the strategies attempt to discover information about their environment based on the bids they make to the two markets and the returns they receive.

The theoretical model predicts that individuals will adopt strategies that produce a Nash equilibrium outcome. In other words, individuals will invest eight tokens in market 2, the CPR, and they will invest their remaining tokens in market 1. As a group, they will earn 39% of the optimal amount that they

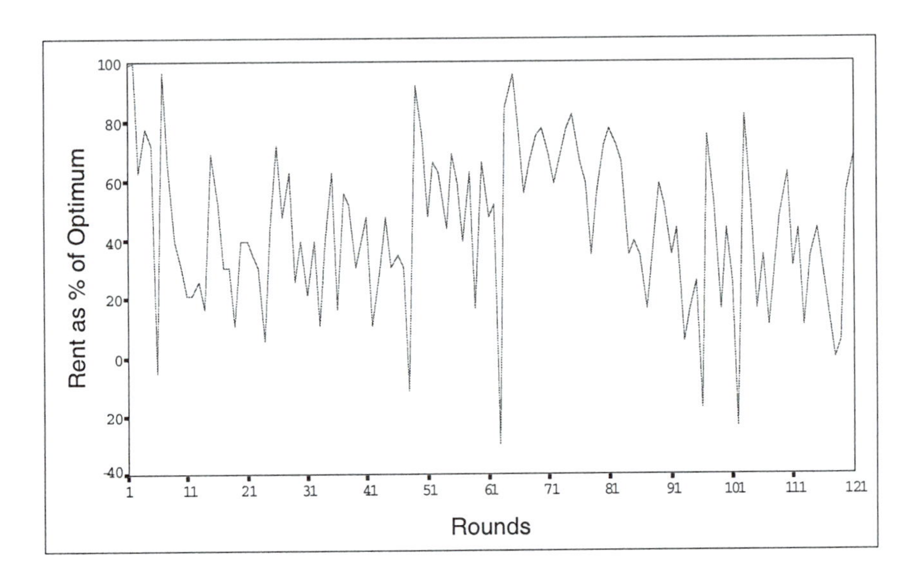

FIGURE 1 Group-level performance for one run of the simulation employing eight agents with ten tokens each and no communication.

could have earned. How do the subjects in the laboratory experiments and the intelligent agents in the simulations perform? In examining the group outcomes achieved, the model of the individual appears to make little difference. Group outcomes come very close to the Nash equilibrium. In the laboratory experiments, the eight subjects endowed with ten tokens each, group rent as a percentage of optimum averaged 37%—the Nash equilibrium. For the simulations, group rent as a percentage of optimum averaged just over 40%—very close to the Nash equilibrium (see fig. 1).

The model of the individual appears not to matter. The boundedly rational individual performs as well as the perfectly rational individual. Economists' assertion that models should be evaluated on the basis of the quality of their predictions and not on the realism of their assumptions appears born out. Perfect rationality, based on mathematically tractable and relatively clear and simple assumptions, should carry the day.

A closer inspection of the results, however, reveals two anomalies, which suggest that models of the individual matter. While group outcomes are very similar, the processes by which those outcomes are achieved are different. The theoretical model predicts that individuals will identify their best course of action, given the choices of all other individuals, and will adopt and remain with that course of action. Individuals play their Nash strategy. In the laboratory experiments and the simulations, that rarely occurred. Laboratory subjects and intelligent agents varied their investment in the CPR across decision periods. It is difficult to make sense of the investment decisions of an individual subject or agent. There appears to be no pattern or plan. However, at the group level, a pattern emerges. Unlike the theoretical model, the Nash equilibrium emerged from the varied and changing actions of subjects and agents. The subjects in the experiments and the simulations achieved average levels of performance that were very close to Nash, while seldom reproducing a Nash outcome in any one round. The Nash equilibrium was an emergent property of both the laboratory experiments and the noncommunication simulations.

Second, according to the model of perfect rationality, as long as an individual possesses sufficient resources to carry out the strategy that maximizes her welfare, that is the strategy she will choose, regardless if she possesses resources well in excess of that level. In other words, resources in excess of some minimum are irrelevant for outcomes. The laboratory experiments and the simulations were also run with subjects and agents endowed with 25 tokens. According to the theoretical model, based on perfect rationality, this larger endowment should not affect strategies pursued. Contrary to theoretical predictions, however, in this more resource-rich setting in which subjects and agents possessed substantially more resources to invest in the CPR, they did so. Group investments were initially far below the Nash equilibrium, but tended to approach it over time. In the laboratory experiments group rent as a percentage of optimum averaged −3%. That is, groups so over invested in the CPR that they received negative returns on their investments. Intelligent agents too performed poorly. At a 25-token endowment, group performance fluctuated around a mean of −10% (see fig. 2). The agents did not tend to approach the Nash equilibrium over time.

The difference between the behavior observed in the baseline laboratory experiments, the simulations, and the predictions of the theoretical model raises some interesting issues. It would appear that, in low endowment experiments and simulations, the theoretical model of neoclassical economics is capable of predicting the average behavior of the system, even though it does not capture the dynamic nature of the interactions between agents that led to that group-level outcome. The theoretical prediction that individuals are capable of calculating, and subsequently playing a stable strategy leading to a Nash equilibrium outcome is not supported by the experiments. In laboratory experiments, we could assume that only a limited number of students in any group would consider, or be capable of, figuring out the Nash level of investment at the beginning of the experiment. Even if they could, the information would have little value because they could not expect the others in the group to consistently play a Nash strategy. In fact, a member of the group who fully understood the nature of the experiment, and was interested in maximizing utility, might expect that there would be opportunities to take advantage of underinvestment in the CPR by the rest of the group. Individuals attempt a number of alternate investment strategies in an effort to determine the unknown behavioral characteristics of other participants. Such trial and error

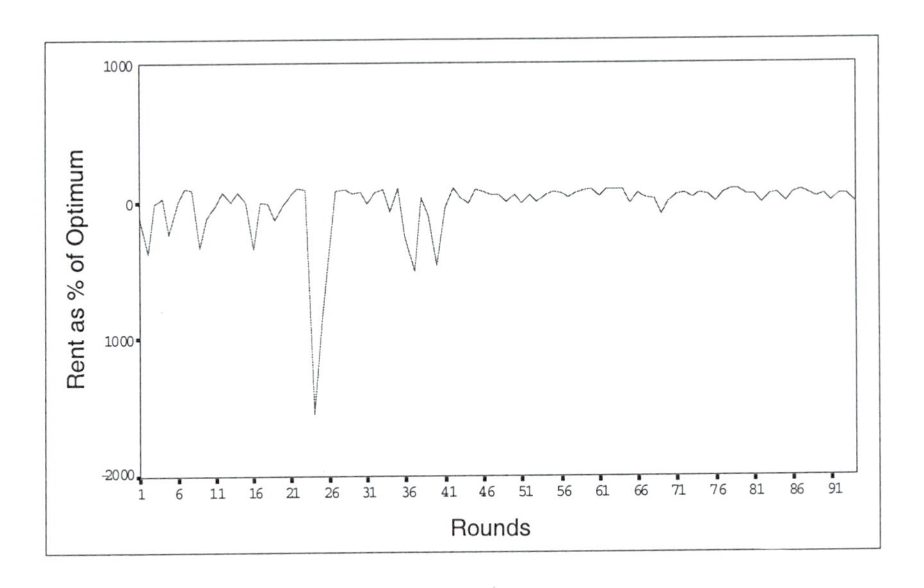

FIGURE 2 Group-level performance for one run of the simulation employing eight agents with 25 tokens each and no communication.

results in the observed fluctuations in group performance, and outcomes for low endowment experiments that are close to Nash.

In the simulations, agents have no knowledge of the Nash group level of investment, or the strategy required to produce that outcome. But they do utilize and evaluate a selection of alternate strategies in an attempt to maximize utility. The outcomes of the noncommunication simulations are close to those of the experiments. By utilizing a theoretical approach based on a model of limited rationality in which agents follow heuristics, it appears that the simulations have captured both the fluctuating nature of the individual bids and the overall group performance.

2.2 COMMUNICATION BETWEEN AGENTS

One of the strengths of agent-based simulation is the ease with which communication routines can be established between individual objects. Such simulations provide us with a tool to explore the effects of alternate communication routines on the behavior of groups in a commons dilemma. At its most basic level, communication provides an individual with additional information about her environment. In a CPR experiment this additional information may include the intended strategies of other individual participants, or the nature of the CPR return function and the optimal group level of investment.

The neoclassical model of rationality assumes that full information is available to the individual, so that any information acquired from others would be meaningless if it did not affect payoffs. Further, full rationality argues that promises from other participants to follow anything but the Nash strategy are meaningless if not enforceable. Therefore, fully rational agents in the experiments or simulations would not alter their behavior with the introduction of communication. But, as we will explore in this section, communication does affect the behavior of individuals in both the experiments and the simulations in ways that are not predicted by the neoclassical model of rational behavior.

Ostrom et al. [43] explored two communication routines in lab experiments that are relevant to the simulations discussed here. In the first version, subjects participated in ten decision rounds in which they could not communicate. After the tenth round, they were allowed to communicate face-to-face for ten minutes. During these communication rounds, subjects were allowed to openly discuss the decision problem facing them. No restrictions were placed on these discussions, other than: they were not allowed to discuss side payments, they were not allowed to make physical threats, and they were not allowed to see the private information on each others computer monitors [43]. Following the period of communication, they then participated in another series of decision rounds in which no communication was allowed. In the second version, subjects participated in ten decision rounds during which they could not communicate. In subsequent decision rounds, they were allowed a few minutes of communication after each round.

The first communication routine produced improved group levels of performance. In the first five decision rounds after communication, group's earnings averaged 74% of the optimal outcome [43, p. 152]. From that point on, earnings declined. One-shot face-to-face communication was found to promote cooperation, but the groups could not sustain it. The second communication routine produced clear improvements in group-level outcomes. Individuals identified the optimal group investment strategy, which was universally adopted. Repeated communication allowed the groups to sustain cooperation. These groups earned between 97% and 100% of the optimal group outcome [43, p. 154]. Thus, with both of these routines, communication was found to increase the frequency with which individuals chose joint income-maximizing strategies, even in situations where individual incentives conflict with the cooperative strategy [45].

Two alternate communication routines were explored in the CPR simulations. Each routine explores a different dynamic in the way information is exchanged and evaluated by the individual agents. For both communication routines, during specific rounds agents exchanged information as to the bid in the previous rounds that yielded the highest return. This information was developed by each agent during noncommunication rounds of the simulation as the agent kept track of the market 2 bid that yielded the highest return. Following the submission of this information, the information was evaluated in an effort to determine the bid that would yield the highest return. The suggestion yielding the highest apparent return was then selected for incorporation by the agents as an additional strategy. Therefore, agents utilizing a pool of sixteen strategies in the noncommunication rounds prior to the communication round adopted the selected suggestion as a seventeenth strategy for subsequent rounds. Initially each agent adopted this extra strategy as the current strategy. But in later rounds the agents evaluated the performance of the strategy acquired by communication against the alternates. As with the noncommunication simulations, the agents were able to switch away from the strategy acquired by communication if one of the alternates appeared to be capable of providing a higher return.

The two communication routines explored here differ in their approach to selecting the best of the suggested market 2 bids. In the first routine, the individual agents themselves evaluated all the market 2 bid suggestions and selected the one that appeared to give the best return relative to their current knowledge of the simulation environment. They then selected the suggestion that appeared to yield the highest return and adopted that as the additional strategy. In the second routine, a central authority, in this case the CPR itself, evaluated all the suggestions made by the individual agents and determined which one would provide the highest return if it was submitted uniformly by each member of the group. The central authority selected the submission that resulted in the best group performance and sent that bid to each agent to be adopted as its additional strategy. Simulations were run using each of the two communication and evaluation routines outlined above. Compared to the relatively wide-open communication of the laboratory experiments, the agents in these simulations employ a more restricted form of information exchange. Both communication routines are discussed in detail below.

2.2.1 Individual Evaluation of Best Group Bid. The first set of simulations explored a communication routine in which the agents individually evaluated the suggestions submitted by the others in the group. The agents selected the suggestion that would have given them the highest return, had they submitted that bid in the previous round. This communication routine does not allow the individual agents to consider the effects of their individual actions on collective outcomes. The agents are simply trying to maximize individual returns with the benefit of the additional information that they acquire by communication. Despite the lack of a global perspective in this communication routine, it was still possible for the agents in these simulations to adopt and retain a uniform group-level investment pattern. These simulations were run in part to determine if, and at what bid, the members of the group might lock in to a uniform, groupwide market 2 investment level.

The most important observation from the simulations utilizing this communication routine is that the agents did eventually lock in to a uniform market 2 bid. All agents eventually came to a tacit agreement on a bid that resulted in the best return, given the events that had occurred previously in that particular simulation run. However, this groupwide uniform market 2 bid, although always near or above the Nash equilibrium level, was frequently suboptimal. The amount of time required for the agents to lock into a uniform

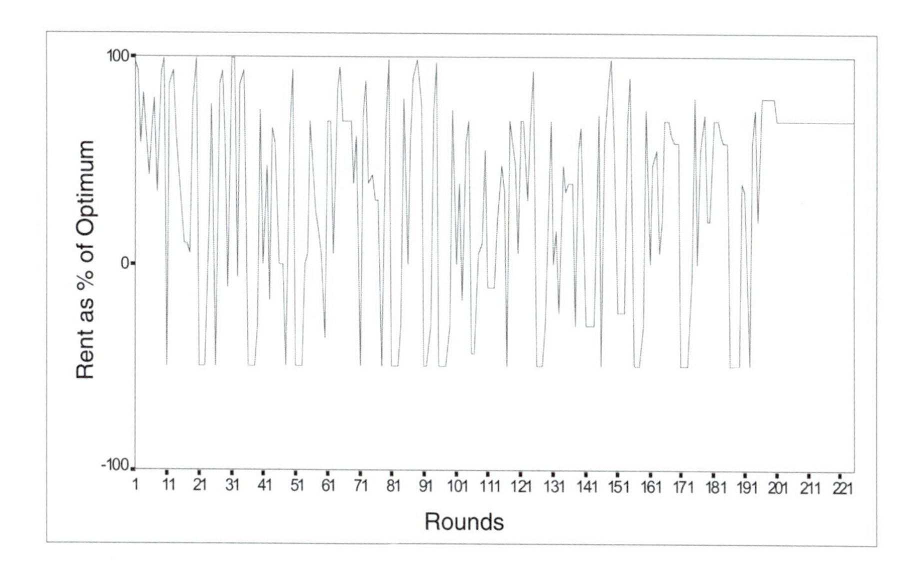

FIGURE 3 Group performance for one run of the simulation using eight agents with communication and individual evaluation of best bid. The agents eventually lock in to a uniform bid of seven tokens each.

level of investment frequently exceeded 100 rounds, and occasionally exceeded 200 rounds.

The length of time required to achieve tacit collusion is a product of the limited rationality of the agents and the mechanism they used to select from amongst the different strategies. Near the beginning of these simulations the agents suggested a wide variety of bids during the communication round. Frequently these bids suggested higher levels of investment in market 2 than bids that would result in the group optimum level of investment. However, when all members of the group implemented this bid, group performance dropped and the bid was discarded. Over time, the agents continued to implement a variety of strategies, evaluating their performance as they went along. Eventually, a bid was suggested and adopted by some members of the group that provided a return that was higher than the score of any alternate strategy. At this point the agent continued to submit that bid indefinitely. Over several communication rounds, more agents adopted this bid until all agents found that it performed better than any alternate strategy. From this point on, all agents continue to submit the same uniform bids in each round of the simulation (see fig. 3).

The communication mechanism changes the simulation significantly from the previous noncommunication simulations. It creates a simple self-reinforcing mechanism as discussed by Arthur [5]. According to Arthur, researchers have discovered that systems in many different fields of study, from theoretical biology to physics, tend to possess a multiplicity of asymptotic states, or "emergent structures." The initial configuration of the system, and some early, often random, events tend to push these dynamic systems into the domain of one of these asymptotic states, or attractors, and thus select a state that the system eventually "locks into" [2].

Arthur points out that such states exist in economic systems as well, citing examples from international trade theory, spatial economics, and industrial organization. The evolution of Silicon Valley in California is one such example from spatial economics. According to Arthur [3], these systems display four properties; multiple equilibria, possible inefficiencies, path dependence, and lock-in. Each of these properties has been reproduced by these simple simulations. Multiple equilibria are seen in these simulations as the groupwide level of investment that the agents eventually agree on changes in successive runs of the simulation. Inefficiencies are demonstrated as the agents settle on uniform group bids that are frequently well below optimum. Path dependence is ensured as the strength of alternate strategies is influenced by past performance. Events early in the simulation may cause certain strategies to be permanently discarded, thus influencing future decisions by individuals and overall group behavior. Finally, lock-in is clearly demonstrated as the agents settle on a uniform groupwide level of investment.

The agents in this simulation reach a form of tacit cooperation without direct interaction. They only go along with the final groupwide equilibrium bid because none of their alternate strategies appears to perform any better. This tacit agreement can take several hundred rounds to evolve as contrasted to the rapid agreement that can be achieved by subjects in the lab. The level of investment that the agents lock into is frequently suboptimal because the agents take an individual focus that attempts to maximize individual returns without considering the performance of the group as a whole.

2.2.2 Centralized Evaluation of Best Group Bid. A second communication routine was explored in which the CPR evaluated each submitted suggestion to determine the group performance if that suggestion had been played uniformly by all members. The central authority then instructed the agents to adopt the best performing of the submitted bids as the additional strategy. Although each agent adopts this bid as an additional strategy, they were allowed to switch to an alternate strategy later in the simulation if the alternate appeared to provide a higher return.

This approach to communication allows the individual agents the perspective of a centralized view of the relationship between individual actions and group-level performance. In this sense, the routine is closer to the communication employed in laboratory experiments. In the experiments, individuals calculated and exchanged information as to the optimum group investment in the CPR, and the pattern of individual bids that were required to achieve that optimum. Individuals in lab experiments frequently adopted this strategy and used it to govern their investments in subsequent rounds. Communication

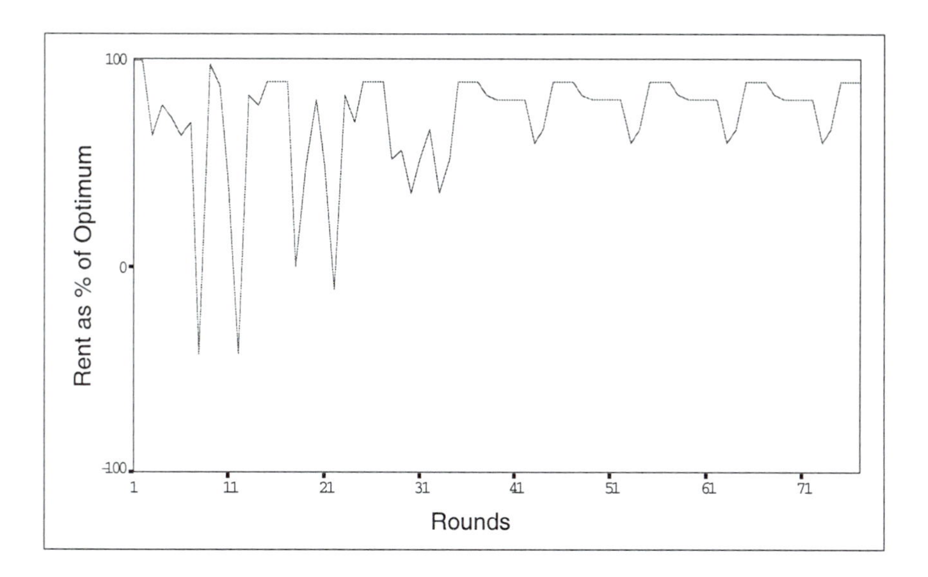

FIGURE 4 Group performance for one run of the simulation using eight agents with communication and centralized evaluation of the best market 2 bid.

facilitated increased group levels of performance in the short term following a communication round. As in the experiments, the agents now have access to information on the bid that will result in the best group performance.

They also adopt and implement this bid in the rounds immediately following communication.

The behavior of individual agents, and the performance of the group, produced by this communication routine differed from the first routine in two important ways. First, the group locked in to a uniform groupwide bid immediately following the first communication rounds. This groupwide bid was frequently at the near optimal level of 4 or 5 tokens each. Second, one or more agents in the simulation switched away from the group strategy as soon as possible. This behavior resulted in the establishment of a fluctuating pattern of group performance in which the agents adopted a single groupwide market 2 investment following communication, followed by a drop in performance as one or more members of the group switched to an alternate strategy (see fig. 4). In the simulation depicted in figure 4, the members of the group adopt the near-optimum investment level of five tokens each after each communication round. However, shortly thereafter four members of the group switched to an alternate strategy that appeared to provide a higher return, thereby causing an overinvestment in market 2 that lowered group performance. The fluctuating pattern we see is the result of this cycle of group-induced compliance at a communication round, followed by subsequent strategy changes by one or more members of the group. The level of token endowment does not appear to be correlated with the grouwide investment level, or the number of agents that will subsequently change strategies.

This form of communication does have some characteristics in common with the CPR experiments in which communication is allowed. As in the lab experiments, communication in these simulations does frequently result in the discovery of the optimal level of investment. However, there is an important difference in the subsequent behavior of human subjects and agents, following communication. Whereas humans in the lab are able to draw upon social norms favoring cooperation and verbal sanctions in subsequent communication rounds to ensure compliance, the agents in the simulations possess no mechanism to represent a social norm favoring cooperation. Therefore, they are not encouraged to cooperate with the imposed group bid by any mechanism other than an objective evaluation of the potential payoffs that may be earned by their alternate internal strategies. If it appears that an alternate strategy will yield a higher return in the next round, they switch away from the group strategy without any consideration of the impact of that action on, or the potential reactions of, the other agents.

A number of authors have explored the importance of norms in governing individual behavior during interactions with other individuals. Authors, such as Ostrom and Walker [45], Orbel et al. [41], and Palfrey and Rosenthal [47], have put forward a number of hypotheses as to why communication increases the selection and maintenance of cooperative strategies including: communication promotes generalized norms that favor cooperation, communication provides the opportunity to offer and extract offers of cooperation, and communication alters the perceptions of subjects about the likelihood of others contributing to the group good.

Other researchers have explored the effects of norms or cultural algorithms in agent-based social simulations. Axelrod [6] developed a simulation that explored how norms evolved in a collection of agents that made choices on whether or not to defect themselves, and whether or not to punish defectors that they observed. Reynolds [49] developed a series of simulations in which a cultural algorithm was embedded in agents representing herders in Peru. He found that the presence of a belief structure in the agents was essential to the evolution of an effective group-level cooperative strategy in a potentially hostile environment.

These recent developments present the opportunity to encode agents in the CPR simulations with an ability to utilize internal norms in their decision-making tasks. This approach would be grounded in the theoretical models of institutional analysis and would allow norms to be associated with the heuristics utilized by the agents described in these simulations. This approach draws on the grammar of institutions developed by Crawford and Ostrom [16]. The grammar of institutions is based on the idea that institutions are enduring regularities of human action in situations structured by rules, norms, and shared strategies. All three concepts are expressed in a series of institutional statements that advise, permit, or prescribe actions or outcomes for the actors involved [16].

The syntax of the grammar of institutions has five components: (1) attributes that describe to whom the institutional statement applies; (2) a deontic verb, "may," "must," or "must not"; (3) an AIM or description of the particular actions or outcomes to which the deontic applies; (4) condition variables that describe when, where, how, and to what extent an AIM is permitted, forbidden, or obligatory; and (5) an "or else" description of the sanctions to be followed for not following an institutional statement. The rules, norms, and shared strategies for any institution can be described utilizing these five components. According to this syntax, shared strategies contain attribute aims and conditions, which describe who does what under certain conditions. Norms add a deontic to these shared strategies. Adding the deontic to a statement implies that there may be a positive or negative payoff associated with the actions that an individual takes in relation to these statements. Such payoffs, which may be internally or externally derived, can be expressed in terms of a delta parameter.

The delta parameter expresses the total positive or negative payoff to an individual as the sum of internal and external payoffs associated with obeying or breaking a norm. For example, an internal payoff may come from obeying a norm if a person gains a certain feeling of satisfaction from that action. An external payoff may be associated with that same action if the individual's reputation is enhanced, or if there is a direct monetary reward associated with following that action. This approach to the definition of the delta parameter allows the researcher to incorporate normative aspects of the behavior of an individual directly into consideration when developing a model. The experimenter may set all of the delta parameters to specific values, or may leave some set to zero so as to observe the effects of others on outcomes. The experimenter may set all of the delta parameters equal to the same value to create a set of like individuals. Alternatively, the experimenter may adjust the values so as to create a set of individuals with different characteristics who behave differently in relation to a specific norm or rule. An adaptive mechanism could be added to the agent to allow for the adjustment of the delta parameter over time, based on the performance of the norms held by the agents.

The five criteria describing individual decisionmakers posited by Ostrom et al. [43] permitted a careful comparison between different models of the individual. The model of perfect rationality, as used by economists, did not perform as well as the model of bounded rationality used in the simulations. Unlike the model of bounded rationality, the model of perfect rationality cannot account for the emergence of outcomes from groups of interacting individuals, nor can it account for the effects of norms on individual choices.

The model of bounded rationality used in the simulations performed better than the model of perfect rationality in the noncommunication experiments, at least in terms of simulating the behavior of laboratory subjects.

Economic theory, based on the model of perfect rationality, predicted that individuals in the common-pool-resource setting would achieve the outcome described as a Nash equilibrium. The intelligent agents of the simulation consistently achieved such an outcome. However, perfectly rational individuals were predicted to individually pursue Nash equlibrium strategies. In other words, the group outcome would simply be a sum of the individuals' strategies. This did not occur among the intelligent agents of the simulations, who individually pursued a variety of strategies from which emerged a Nash equilibrium. The boundedly rational agents of the simulation behaved much more like the laboratory subjects.

The model of bounded rationality used in the simulations also performed better than the model of perfect rationality in the communication experiments. The behavior of perfectly rational individuals should not be affected by communication opportunities that do not affect payoffs. Decades of communication experiments have demonstrated the importance of communication for developing and sustaining cooperation. While the behavior of boundedly rational intelligent agents changed when allowed to communicate, the intelligent agents, lacking a means of developing norms of reciprocity, failed to sustain cooperation, unlike the laboratory subjects. Comparing the behavior of laboratory subjects who bring norms of behavior with them into the laboratory experiment with the behavior of intelligent agents who do not possess norms of behavior, clearly demonstrate the importance of such norms for developing and sustaining cooperation.

3 CONCLUSION

The simulations explored here have focused on modeling CPR laboratory experiments as a prelude to the development of other resource management or institutional models. Eventually the intention is to extend these models to link human systems and natural systems simulations with real-world resource management applications. Building effective simulations requires that the model of individual decision making validly represent the real system that is the subject of the simulation. Zeigler [60] outlines three degrees of validity for any modeling relationship: replicative, predictive, and structural. A model is replicatively valid if it is capable of reproducing the behavior of the system as determined from data already collected. A model is predictively valid if it capable of matching data from the real system before it is collected. A model is structurally valid if it both reproduces real system behavior and represents the way in which the real system operates to generate that behavior. Structural validity is the most difficult measure to achieve. In the field of social simulation we still lack effective tools for evaluating the structural validity of the individual agents utilized in simulations.

Addressing issues of validity in agent-based simulations with a resource management focus requires an approach that integrates the collection of data from the real system with the requirements of the model. During the development of the model, both the decision-making characteristics of the individual agent, and that agent's interactions with its simulated environment must be explicitly defined. This development activity must be supported by data collected from the real system at multiple scales. Data is required from the real system at a scale that matches the output of the simulation. Such data will facilitate the measuring of the replicative or predictive validity of the overall simulation. For the laboratory CPR experiments, this includes such measures as group rent as a percentage of optimum. But data must also be collected from the real system at a scale that matches the individual entities that are represented by agents in the simulation. For the CPR lab experiments, this means data on the investment decisions of individuals and the reported strategies utilized by individuals. At this scale of measurement, the framework outlined here can serve as a useful tool to guide both the design of intelligent agents and the collection of supporting data. Understanding the preferences, norms/values, information-processing capabilities, selection criteria, and resources of the real-world individuals who are to be represented by agents is an important early step in model design. Addressing these characteristics during simulation development can help ensure that the model of individual decision making is appropriate for the real system setting, thus achieving a higher degree of structural validity. Specifically, measures of structural validity could be developed for each of the five characteristics, and used to evaluate the design of the agents.

A number of such efforts, which integrate agent design and the collection of field data at the individual level, have already been documented (see the chapters in this volume on modeling rafting trips in the Grand Canyon). In one other experiment (see ch. 12) agent-based simulations are being developed to explore patterns of land use in the Amazon basin, based on the demographic characteristics of individual households. Such simulations seek to capture the decision-making processes of individuals (rafting trips or households) within the context of specific settings. Such efforts represent a step toward the development of effective and valid simulations.

As the field of agent-based social simulation continues to develop, modeling efforts can be expected to continue to branch out from their theoretical roots to explore a variety of real-world case studies. Such diverse efforts could be greatly aided by a framework that guides the development of individual agents. Such a framework will allow the researcher to tie models of individual decision making to real-world settings and established theory. Further, the framework will facilitate the comparison of alternate models of the individual explored in different simulations. As the field of social simulation evolves, such a device should prove to be an effective tool for developing, evaluating, and comparing alternative simulations of social phenomena.

4 APPENDIX 1: DETAILED DESCRIPTION OF THE SIMULATION

4.1 COMPONENTS OF THE MODEL

The individual components of the simulations described here are designed to represent the individual appropriators in the experiments, the CPR itself, and the other components necessary to run the simulation. These components are specified in the simulation as individual classes and described below. At the lowest hierarchical level lie the classes which represent the components of the CPR experiment itself, the participants in the experiment and the CPR. Above these, the simulation contains an instance of a class, called CprModelSwarm, which creates the CPR, the eight appropriator agents, and Strategies class. The CprModelSwarm is also where the schedule of agent actions is specified. This schedule is executed once in every round of the simulation to instruct specific agents in the simulation to perform certain methods in a certain order.

One level above the CprModelSwarm class lies an instance of a class called CprObserverSwarm. The CprObserverSwarm object creates the CprModelSwarm, controls the overall simulation, and creates objects and schedules that display output data. Throughout the different simulations described below, the CprModelSwarm and CprObserverSwarm classes remain basically unchanged.

4.2 THE COMMON-POOL-RESOURCE CLASS

The methods written for the CPR agent class specify the state of the CPR in relation to actions of the appropriators. Only one instance of this class is created in these simulations. The quadratic production function for market 2, as utilized by Ostrom et al. [43], is embedded in the code of the CPR agent and specified as follows:

$$F\left(\sum x_i\right) = a\left(\sum x_i\right) - b\left(\sum x_i\right)^2, \tag{1}$$

where $\sum x_i$ is the sum of all the market 2 bids submitted by the agents. By manipulating the **a** and **b** parameters, the shape and magnitude of the quadratic production function can be controlled.

The CPR agent receives from the CprModelSwarm object, and keeps track of: parameters used in the quadratic production function (a and b), the parameter w which is used to calculate the return from market 1, and the number of agents in the simulation. For all the simulations explored in this work, the a, b, and w parameters of the production function and market 1 fixed return were set to 23, 0.25, and 0.05, respectively. During the initialization phase of the simulation, the CPR object calculates the optimum group bid as:

$$\sum x_i \frac{(a-w)}{2b}. \tag{2}$$

The CPR object then calculates the subsequent group return from market 2 yielded by the optimum group investment. During each round of the simulation, the CPR agent collects the token bids for markets 1 and 2 from the individual agents. The CPR agent then calculates the total return from market 2, group rent as a percentage of optimum for market 2, and the return to each individual appropriator for that round. In addition, during each round of the experiment the CPR object outputs the above information to a data file for later analysis.

4.3 THE AGENT CLASS

The individual appropriator agents explored in this chapter are described as a class called agents. As described above, the overall goal of the agents is to maximize returns. They do this by employing the strategies described in table 1. At the beginning of each simulation, the agent is randomly assigned a specific number, either 4, 8, or all 16, of these 16 strategies which it may access from the strategies class. One of these strategies becomes the current strategy, the rest become alternates. The agents begin by playing the current strategy. But they also keep track of how the alternate strategies would have performed in each round had the agent used them. Every third round, the agents have the opportunity to switch their current strategy with best performing of the alternates. The performance of each alternate strategy is measured as the average return that the agent would have received in each round had it utilized that strategy. In this way, the agent simulates one of the mechanisms of an inductive process whereby it selects from alternative rules, based upon those rules relative strengths, in an effort to find the one with the best performance.

4.4 THE STRATEGIES CLASS

Building a complete list of strategies into each of the agents when the agent normally only has access to a portion of those strategies is an unnecessary duplication of code across the agents. Therefore, in these simulations, a pool of possible strategies were developed and specified in a separate class called strategies. Only one instance of this class is ever created. The strategies developed for this class are derived from the strategies utilized in the previous simulations. Six of the 16 strategies are based on attempting to maximize the return received in each round. The increment or decrement of bids to market 2 is varied in each of the strategies. Another six strategies are based on the comparison of average returns from market 1 and market 2, a heuristic that subjects reported using during the experiments. The strategies differ in the amount that market 2 bids are incremented when the average return from market 2 exceeds the average return from market 1. This increment varies from one token to dumping all tokens into market 2 when the right conditions

TABLE 1 Strategy descriptions for the strategies class.

Strategy	Number Description
1	Total return maximizing strategy—increment and decrement market 2 bid by one token.
2	Total return maximizing strategy—increment and decrement market 2 bid by two tokens.
3	Total return maximizing strategy—increment and decrement market 2 bid by three tokens.
4	Total return maximizing strategy—increment and decrement market 2 bid by four tokens.
5	Total return maximizing strategy—increment market 2 bid by all available tokens, decrement market 2 bid by three tokens.
6	Total return maximizing strategy—increment market 2 bid by all available tokens, decrement market 2 bid by five tokens.
7	Unit return maximizing strategy—increment and decrement market 2 bid by one token.
8	Unit return maximizing strategy—increment and decrement market 2 bid by two tokens.
9	Unit return maximizing strategy—increment and decrement market 2 bid by three tokens.
10	Unit return maximizing strategy—increment and decrement market 2 bid by four tokens.
11	Unit return maximizing strategy—increment market 2 bid by all available tokens, decrement market 2 bid by three tokens.
12	Unit return maximizing strategy—increment market 2 bid bid by all available tokens, decrement market 2 bid by five tokens.
13	Submit market 2 bid equal to group average bid in previous round.
14	Submit market 2 bid equal to group average bid in previous round plus one token.
15	Submit market 2 bid equal to group average bid in previous round plus two tokens.
16	Submit market 2 bid equal to group average bid in previous round plus three tokens.

exist. The final four strategies are based on a comparison of an individual agent's bid with group average bid, and then submitting a bid at or above this average level.

During a simulation run, agents repeatedly make calls to the strategies object, supplying it with the necessary information from their own variables and receiving information in return. In every round of the simulation. The agent first calls the numbered strategy that corresponds to the number of its current strategy from the strategies object. The agent supplies the strategies' object with the information necessary to enact the selected strategy, and receives in return the market 2 bid as calculated by the strategy. Following the submission of its market 1 and market 2 bids to the CPR object, each agent evaluates it alternate strategies by obtaining the appropriate bid from the strategies object and submitting it to the CPR. The CPR informs the agent of the return it would have received, had it submitted the alternate bid. The return received by the current and alternate bids are stored, and later compared in rounds when the agent has the opportunity to switch its current strategy.

REFERENCES

[1] Arthur, W. B. "Inductive Reasoning and Bounded Rationality." *AEA Papers and Proc.* **84(2)** (1994): 406–411.

[2] Arthur, W. B., ed. "Path Dependence, Self-Reinforcement, and Human Learning." In *Increasing Returns and Path Dependence in the Economy,* 135–158. Ann Arbor: University of Michigan Press, 1994.

[3] Arthur,W. B., ed. *Increasing Returns and Path Dependence in the Economy,* ch. 8, 135–158. Ann Arbor, MI: University of Michigan Press, 1994.

[4] Arthur, W. B. "On Designing Economic Agents that Behave like Human Agents." *J. Evol. Econ.* **3** (1993): 1–22.

[5] Arthur, W. B. "Self-Reinforcing Mechanisms in Economics." In *The Economy as an Evolving Complex System,* edited by P. W. Anderson, K. J. Arrow, and D. Pines, 9–32. Santa Fe Institute Studies in the Sciences of Complexity, Proc. Vol. V. Reading, MA: Addison-Wesley, 1988.

[6] Axelrod, R. "An Evolutionary Approach to Norms." *Amer. Pol. Sci. Rev.* **80(4)** (1986): 1095–1111.

[7] Bates, R., A. Greif, J.-L. Rosenthal, M. Levi, and B. Weingast. *Analytic Narratives.* Princeton, NJ: Princeton University Press, 1998.

[8] Berkes, F. "Success and Failure in Marine Coastal Fisheries of Turkey." In *Making the Commons Work: Theory, Practice, and Policy,* edited by D. Bromley, 247–264. San Francisco: Institute for Contemporary Studies, 1992.

[9] Bhargava, H. K., and W. C. Branley. "Simulating Belief Systems of Autonomous Agents." *Dec. Supp. Sys.* **14** (1995): 329–348.

[10] Caldas, J. C., and H. Coelho. "Strategic Interactions in Oligopolistics Markets—Experimenting with Real and Artificial Agents." In *Artificial Society Systems*, edited by C. Castelfranchi and E. Werner. Berlin: Springer-Verlag, 1994.

[11] Camerer, C. "Individual Decision Making." In *The Handbook of Experimental Economics*, edited by J. Kagel and A. Roth, 587–703. Princeton, NJ: Princeton University Press, 1995.

[12] Chattoe, E. "Just How (Un)realistic are Evolutionary Algorithms as Representations of Social Processes?" *J. Art. Soc. & Soc. Sim.* **1(3)** (1998). ⟨http://www.soc.surrey.ac.uk/JASSS/1/3/2.html⟩.

[13] Converse, P. "Public Opinion and Voting Behavior." In *Nongovernmental Politics, Volume 4, Handbook of Political Science*, edited by F. Greenstein and N. Polsby, 75–170. Reading, MA: Addison-Wesley, 1975.

[14] Conte, R., R. Hegselmann, and P. Terno, eds. *Simulating Social Phenomena.* Berlin: Springer-Verlag, 1997.

[15] Conte, R., and N. Gilbert, eds. "Computer Simulation for Social Theory." In *Artificial Societies: The Computer Simulation of Social Life*, 1–15. London: UCL Press, 1995.

[16] Crawford, S. E. S., and E. Ostrom. "A Grammar of Institutions." *Am. Pol. Sci. Rev.* **89(3)** (1995): 582–600.

[17] Deadman, P., and H. R. Gimblett. "A Role for Goal-Oriented Autonomous Agents in Modelling People-Environment Interactions in Forest Recreation." *Math. & Comp. Model.* **20(8)** (1994): 121–133.

[18] Doran, J., and N. Gilbert, eds. "Simulating Societies: An Introduction." In *Simulating Societies: The Computer Simulation of Social Phenomena*, edited by N. Gilbert and J. Doran, 1–18. London: UCL Press. London, 1995.

[19] Drogoul, A., and J. Ferber. "Multi-Agent Simulation as a Tool for Studying Emergent Processes in Societies." In *Simulating Societies: The Computer Simulation of Social Phenomena*, edited by N. Gilbert and J. Doran, 127–142. London: UCL Press. 1995.

[20] Epstein, J. M., and R. L. Axtell. *Growing Artificial Societies: Social Science from the Bottom Up.* Cambridge, MA: MIT Press, 1996.

[21] Findler, N. V., and R. M. Malyankar. "Emergent Behaviour in Societies of Heterogeneous, Interacting Agents; Alliances and Norms." In *Artificial Societies: The Computer Simulation of Social Life*, edited by N. Gilbert and R. Conte, 212–236. London: UCL Press, 1995.

[22] Freedman, M. *Essays in Positive Economics.* Chicago, IL: University of Chicago Press, 1953.

[23] Gilbert, N., and J. Doran, eds. *Simulating Societies: The Computer Simulation of Social Phenomena.* London: UCL Press, 1994.

[24] Goldberg, D. E. *Genetic Algorithms in Search, Optimization, and Machine Learning.* Reading MA: Addison-Wesley, 1989.

[25] Hardin, G. "The Tragedy of the Commons." *Science* **162** (1968): 1243–1248.

[26] Hobbes, Thomas [1651]. *Leviathan or the Matter, Forme, and Power of a Commonwealth Ecclesiastical and Civil*, edited by Michael Oakeshott. Oxford: Basil Blackwell, 1960.

[27] Hoffman, E., J. R. Marsden, V. S. Jacob, and A. Whinston. "Artificial Intelligence in Economics—Expert System Modelling of Microeconomic Systems." In *Artificial Intelligence in Economics and Management*, edited by L. F. Pau, 1–9. North Holland: Elsevier Science, 1986.

[28] Holland, J. H. *Hidden Order: How Adaptation Builds Complexity.* New York: Addison-Wesley, 1995.

[29] Holland, J. H., and J. H. Miller. "Artificial Adaptive Agent in Economic Theory." *Amer. Econ. Rev.* **81(2)** (1991): 365–370.

[30] Holland, J. H., K. J. Holyoak, R. E. Nisbett, and P. R. Thagard. *Induction: Processes of Inference, Learning, and Discovery.* Cambridge, MA: MIT Press, 1986.

[31] Holland, J. H. *Adaptation in Natural and Artificial Systems: An Introductory Analysis with Applications to Biology, Control, and Artificial Intelligence.* Ann Arbor MI: University of Michigan Press, 1975.

[32] Kephard, J. O., T. Hogg, and B. A. Huberman. "Collective Behaviour of Predictive Agents." *Physica D* **42** (1990): 48–65.

[33] Knight, Jack. *Institutions and Social Conflict.* Cambridge, MA: Cambridge University Press, 1992.

[34] Machiavelli, N. *The Prince and the Discourses.* New York: Modern Library, 1940.

[35] Markus, H., and R. B. Zajonc. "The Cognitive Perspective in Social Psychology." In *The Handbook of Social Psychology*, edited by Gardner Lindzey and Elliot Aronson, vol. 1, 137–230. 3d ed. New York: Random House, 1985.

[36] McCay, B., and J. Acheson. *The Question of the Commons: The Culture and Ecology of Communal Resources.* Tucson, University of Arizona Press, 1987.

[37] Minar, N., R. Burkhard, C. Langton, and M. Askenazi. "The Swarm Simulation System: A Toolkit for Building Multi-agent Simulations." Overview paper. Santa Fe Institute, Santa Fe, NM, 1996.

[38] Moss, S., and J. Rae. *Artificial Intelligence and Economic Analysis: Prospects and Problems.* Northampton, MA: Edward Elgar Publishing, 1992.

[39] Openshaw, S. "Computational Human Geography: Towards a Research Agenda." *Envir. & Plan. A* **26** (1994): 499–508.

[40] Openshaw, S. "Human Systems Modelling as a New Grand Challenge Area in Science: What Has Happened to the Science in Social Science?" *Envir. & Plan. A* **27** (1995): 159–164.

[41] Orbel, J. M., A. J. C. van de Kragt, and R. M. Dawes. "Explaining Discussion-Induced Cooperation." *J. Personal. & Soc. Psychol.* **54(5)** (1988): 811–819.

[42] Ostrom, E. "A Behavioral Approach to the Rational Choice Theory of Collective Action." *Amer. Pol. Sci. Rev.* **92(1)** (1998): 1–22.

[43] Ostrom, E., R. Gardner, and J. Walker. *Rules, Games, and Common Pool Resources.* Ann Arbor, MI: The University of Michigan Press, 1994.

[44] Ostrom, E. *Governing the Commons: The Evolution of Institutions for Collective Action.* New York: Cambridge University Press, 1990.

[45] Ostrom, E., and J. Walker. "Communication in the Commons: Cooperation without External Enforcement." In *Laboratory Research in Political Economy,* edited by T. R. Palfrey. Ann Arbor, MI: University of Michigan Press, 1991.

[46] Ostrom, Vincent. *The Political Theory of a Compound Republic: Designing the American Experiment,* 2d ed. San Francisco, CA: Institute for Contemporary Studies, 1987.

[47] Palfrey, T. R., and H. Rosenthal. "Testing for Effects of Cheap Talk in a Public Goods Game with Private Information." *Games & Econ. Behav.* **3** (1991): 183–220.

[48] Prietula, M. J., K. M. Carley, and L. Gasser, eds. *Simulating Organizations: Computational Models of Institutions and Groups.* Cambridge MA: MIT Press, 1998.

[49] Reynolds, R. G. "Learning to Co-operate using Cultural Algorithms." In *Simulating Societies: The Computer Simulation of Social Phenomena,* edited by N. Gilbert and J. Doran, 223–244. London: UCL Press, 1994.

[50] Sabatier, P., and H. Jenkins-Smith, eds. "The Dynamics of Policy Oriented Learning." In *Policy Change and Learning: An Advocacy Coalitions Approach,* 41–56. Boulder, CO: Westview Press, 1993.

[51] Sanglier, M., M. Romain, and F. Flament. "A Behavioral Approach to the Dynamics of Financial Markets." *Dec. Supp. Sys.* **12** (1994): 405–413.

[52] Scharpf, F. *Games Real Actors Play.* Boulder, CO: Westview Press, 1997.

[53] Simon, H. *Reason in Human Affairs.* Stanford, CA: Stanford University Press, 1983.

[54] Simon, H. *Administrative Behaviour.* New York: Free Press, 1954.

[55] Stender, J., K. Tout, and P. Stender. "Using Genetic Algorithms in Economic Modelling: The Many-Agents Approach." In *Artificial Neural Nets And Genetic Algorithms,* edited by In R. F. Albrecht, C. R. Reeves and N. C. Steele. Springer-Verlag, 1993.

[56] Taylor, Michael. *The Possibility of Cooperation.* Cambridge, MA: Cambridge University Press, 1987.

[57] Tocqueville, Alexis de. *Democracy in American,* edited by Philips Bradley. 2 vols. New York: Alfred A. Knopf, 1945.

[58] Tversky, A. "Contrasting Rational and Psychological Principles of Choice." In *Wise Choices: Decisions, Games, and Negotiations,* edited by Richard Zeckhauser, Ralph Keeney, and James Sebenius. Boston, MA: Harvard Business School Press, 1996.

[59] Williamson, O. *The Economic Institutions of Capitalism.* New York: Free Press, 1985.

[60] Zeigler, B. P. *Theory of Modelling and Simulation*. New York: John Wiley, 1976.

[61] Zeigler, B. P. *Object-Oriented Simulation with Hierarchical, Modular Models: Intelligent Agents and Endormorphic Systems*. New York: Academic Press, 1990.

An Agent-Based Approach to Environmental and Urban Systems within Geographic Information Systems

Bin Jiang
H. Randy Gimblett

1 INTRODUCTION

Both environment and urban systems are complex systems that are intrinsically spatially and temporally organized. Geographic information systems (GIS) provide a platform to deal with such complex systems, both from modeling and visualization points of view. For a long time, cell-based GIS has been widely used for modeling urban and environment system from various perspectives such as digital terrain representation, overlay, distance mapping, etc. Recently temporal GIS (TGIS) has been challenged to model dynamic aspects of urban and environment system (e.g., Langran [25], Clifford and Tuzhilin [9], Egenhofer and Golledge [14]), in pursuit of better understanding and perception of both spatial and temporal aspects of these systems.

In regional and urban sciences, cellular automata (CA) provide useful methods and tools for studying how regional and urban systems evolve. Because of its conceptual resemblance to cell-based GIS, CA have been extensively used to integrate GIS as potentially useful qualitative forecasting models. This approach intends to look at urban and environment systems as self-organized processes; i.e., how coherent global patterns emerge from local interaction. Thus this approach differentiates it from TGIS in that there is no database support for space-time dynamics.

An agent-based approach was initially developed from distributed artificial intelligence (DAI). The basic idea of agent-based approaches is that programs exhibit behaviors entirely described by their internal mechanisms. By linking an individual to a program, it is possible to simulate an artificial world inhabited by interacting processes. Thus it is possible to implement simulation by transposing the population of a real system to its artificial counterpart. Each member of population is represented as an agent who has built-in behaviors. Agent-based approaches provide a platform for modeling situations in which there are large numbers of individuals that can create complex behaviors. It is likely to be of particular interest for modeling space-time dynamics in environmental and urban systems, because it allows researchers to explore relationships between microlevel individual actions and the emergent macrolevel phenomena.

An agent-based approach has great potential for modeling environmental and urban systems within GIS. Previous work has focused on modeling people-environment interaction [13], virtual ecosystems [21], and integration of agent-based approach and GIS [21]. Rodrigues and Raper [31] have employed spatial agents to distinguish those agents for geographic information processing. They have defined spatial agents as agents that make spatial concepts computable for the purpose of spatial simulation, spatial decision making, and construction of interface agents for GIS. Ferrand has applied agent technology to both complex diffusion processes and cartographic generalization.

This chapter explores this possibility with some practical application examples from urban and environmental systems. The remainder of the chapter is organized as follows. Section 2 briefly reviews current approaches of cell-based GIS and CA, as both have certain conceptual resemblances to agent-based approaches. Section 3 introduces the autonomous agent systems, fundamentals, and software platforms of multiagent simulation (MAS). Sections 4, 5, and 6 present a set of examples of urban and environmental modeling using MAS and, finally, section 7 draws the conclusions.

2 CELL-BASED GEOGRAPHIC INFORMATION SYSTEMS AND CELLULAR AUTOMATA MODELING

Because of its spatial structure, cell-based GIS—still considered to be a very important type of GIS—is very suitable for spatial analysis. In particular, its data structure is very similar to a satellite image; cell-based GIS is considered to be very important for the integration of satellite data in GIS. From the analytical point of view, the cell-based data format is often considered to be an intermediate process for vector GIS. To date, cell-based GIS has been widely used in the following areas [19]:

- map algebra,
- distance mapping,

- topographic feature extraction and surface description, and
- surface interpolation.

Map algebra probably is one of most conceptual framework for spatial modeling in cell-based GIS. It is a set of formal languages for spatial analysis and modeling. The idea was developed from the notion of a map, so it is often referred to as cartographic modeling [34]. A map is a commonly accepted metaphor for spatial representation. Indeed, it has been used in the map algebra for spatial representation and analysis. A map is a model of space in reduced scale which represents multiple characteristics. A map layer (or simple layer) is much like a conventional map, but each layer has one single characteristic, such as a street layer, a land use layer, etc. Each layer consists of numerous locations (or cells). A set of locations at a specified cartographic distance and/or directions from a particular location is defined as neighborhood. A set of locations with the same category is referred to as zone. Thus a layer and its components (locations, neighborhoods, and zones) constitutes the basic notions of map algebra, on which a range of operations is defined.

Based on the above-introduced notions, a range of operations have been defined for the purpose of spatial modeling and analysis. These operations, in terms of their scopes of imposed operations, can be categorized as five types:

- per-cell (local)
- per-neighborhood (incremental)
- per-neighborhood (focal)
- per-zone (zonal)
- per-layer (global)

Local operators are functions of a specific cell on one or more layers, i.e., to compute a new value for every location as a function of one or more existing values associated with that location. Per-neighborhood operations can be classified according to the nature of the spatial relationship between each neighborhood and its focus: incremental operations and focal operations. Incremental operators are functions of specific locations and the geometric condition represented at those locations. Focal operations are those that compute each location's new value as a function of the existing values, distance, and/or directions of neighboring (but not necessarily adjacent) locations on a specified map layer. Zonal operators are functions of irregular neighborhoods on one or more layers. Global operations are used for the generation of Euclidean distance and weighted cost distance maps, shortest path maps, nearest-neighbor allocation maps, for the grouping of zones into connected regions, for geometric transformations, for raster-vector interconversion, and for interpolation.

Distance mapping involves calculation of Euclidean distance, isotropic cost distance, and directional path distance, from or to a set of source locations. It can help to generate a buffer zone of geographic objects such as

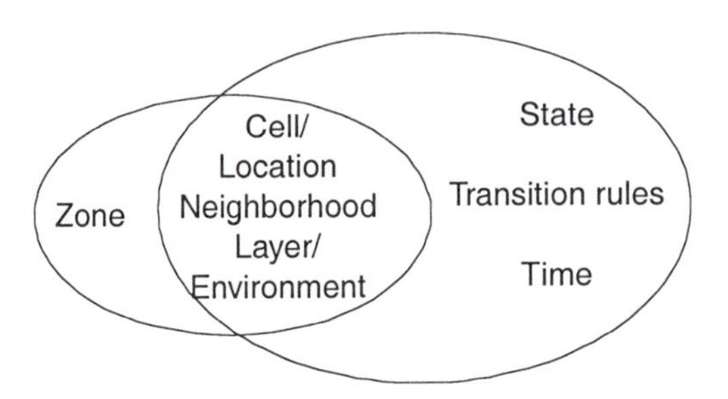

FIGURE 1 Basic notions of CA with overlaps to those of cell-based GIS.

rivers and roads. Topographic feature extraction and surface description have wide application in hydrological modeling such as the derivation of watershed, stream networks, flow accumulation, and flow length. Cell-based GIS is often used for surface interpolation. All these functionalities involve expensive computation.

Cell-based GIS provides a valuable tool for spatial modeling and analysis, which has wide application in urban and environmental systems. However, it inherently lacks the ability to deal with the temporal dimension. As stated by Takeyama and Couclelis [33], map algebra does not deal explicitly with spatial relations and interaction among locations. In contrast, CA, initially developed from computer science, are essentially designed for spatial interaction and dynamic phenomena. It shares with cell-based GIS a number of concepts, such as cell, neighborhood, and environment. Additionally CA provide three more notions to handle dynamics, i.e., transition rules, states, and time (fig. 1).

States or cell states represent the states of each cell, e.g., dead or alive. Transition rules are the heart of CA since they represent the process as time goes by. A typical example of transition rule (Life) reads as "if a cell is off, it turns on if exactly three of its neighbors are on. If a cell is on, it stays on if exactly two or three neighbors are on; otherwise it turns off." Thus interaction not only occurs at the space level but also on the time scale.

Because of its ability to deal with both space and time, CA have been widely used for space-time dynamics modeling in the context of GIS. In this connection, the most influential application field is urban dynamics. However, cities are even more complex and are beyond the capability of standard CA. Thus, in order to deal with more complex situations, standard CA have been extended in various ways, by considering, for instance, more states, a larger neighborhood, and more complex transition rules. Various efforts have been made to use CA for space-time dynamics, including among others:

- *CA for urban and regional dynamics.* Urban and regional systems are essentially dynamics and complex processes and CA have been intensively used in the context of simulating and predicting these dynamic processes (see White [36] for an overview). Both standard CA and extended CA have used and have been proposed for very complex urban and regional dynamics. These efforts provide deeper insights into urban and regional systems from both the microlevel and the macroscale.

- *CA for environmental modeling.* Environmental modeling is probably one of promising application areas of GIS, particularly in modeling space-time phenomena such as wildfire propagation [8] and ecosystem [6]. Burrough [5] has recently done a comprehensive overview of dynamic modeling as a set of tool kits for geocomputation.
- *Integration of GIS and CA.* Full integration of GIS and CA has also been considered. For instance, Wagner [35] has examined the similarities of CA and raster GIS, and the potential to implement one to another is demonstrated. Of particularly interest is the effort made by Couclelis and Takeyama [11] who proposed a general mathematical framework for the integration of GIS and CA based on the notion of proximal space [11].

However, many geographic phenomena, essentially involved with space-time dynamics, can be thought to have emerged from individual interactions. Usually there is more than one type of agents involved, which is beyond the capacity of CA modeling. In this respect, autonomous agents seem to have a high possibility for extension to modeling space-time dynamics.

2.1 AGENT-BASED MODELING

What is an agent or autonomous agent? It has been a very controversial topic these days. Based on a comprehensive survey on the existing definitions of autonomous agent, Franklin and Graesser [18, p. 25] have formalized an autonomous agent as "a system situated within and a part of an environment that senses that environment and acts on it, over time, in pursuit of its own agenda and so as to affect what it senses in the future." Thus an autonomous agent could be humans, other animals, autonomous mobile robots, artificial life creatures, and software agents. A few things are quite important for an agent.

First of all, an agent is only adapted to its own environments; if an agent leaves the environment, it may no longer be an agent. We know that a certain kind of animal lives in a certain kind of natural environment. Change of the environment will dramatically change their adaptation (which is limited). In other words, different agents have different environments. Real-world agents live in the real world; software agents "live" in computer operating systems, databases, networks, etc.; artificial life agents "live" in artificial environments such as on a computer screen or in its memory [18]; and spatial agents live in geographic space. Generally, two kinds of environments are identified for different modeling situations. A distributed environment is a CA-like space

TABLE 1 Properties of agents (after Franklin and Graesser [18]).

Property	Meaning
reactive	responds in a timely fashion to changes in the environment
autonomous	exercises control over its own actions
goal-oriented/ proactive/ purposeful	does not simply act in response to the environment
temporally continuous	is a continuously running process
communicative/ socially able	communicates with other agents, perhaps including people
learning/ adaptive	changes its behavior based on its previous experience
mobile	able to transport itself from one machine to another
flexible	actions are not scripted
character	believable "personality" and emotional state

which consists of a set of cells, whereas a centralized environment has a unique structure [16].

Secondly, sense and action are two important properties of an agent, which determine how they behave in their environment. Agents can be named as reactive agents and cognitive (or deliberative) agents, which are respectively the low and high end of being agents, according to the range and sensitivity of their sense, and the range and effectiveness of their actions. In response to what is sensed, agents take action autonomously. The differences between reactive agents and cognitive agents can be further characterized as follows. Humans, when they navigate in a complex urban system, can be treated as the high end of being agent, in that they not only interact with each other as reactive agents, but also remember what they have sensed, and they can also do some global planning by the use of maps, relevant sources of information, and even previous experience. Agents do things with their own agenda and, in an agent system, none of them acts as a sort of leader or coordinator.

An agent is treated in the above definition as a system. To describe an autonomous agent, it is necessary to describe its environment, sensing capabilities, and actions. On the other hand, an agent can also be treated as a part of an environment, which has a variety of properties such as reactive, autonomous, goal-oriented, temporally, continuous, communication, learning, mobile, flexible, and character (table 1, Franklin and Graesser [18]). The range of properties is ordered in the sequence of intelligibility, from low to high end of being autonomous agents. So agents are not just objects; they are those objects with spatial communication mechanisms that allow them to interact each other.

There is a special kind of agent called a "real-life agent" which aims to simulate the real-world counterpart by means of intuitive visualization. SimCity system is a very good example in this respect. It is a computer game for children of all ages. Agents in SimCity could be various vehicles, pedestrians, and other objects with senses which can act on city environments. So real-life agents are directly visible to users. This property provides the possibility for scientists to construct an exploratory simulation of real life, and to use the computer as a laboratory for studying the informational structure of complex systems.

3 MULTIAGENT SIMULATIONS

Multiagent simulation (MAS) is an agent system with multiple agents. By using multiagent simulation rather than multiagent systems [16], we intend to stress the SimCity-like agent systems which combine the capacities of visualization and modeling together. In contrast to SimCity, MAS usually can be customized in an exploratory way, which means end users can set a range of parameters for exploratory purposes.

Such simulations can be summarized as a set of the following elements: agents, objects, environments, and communications. These are described by the quadruplet:

$$\langle \text{agents, objects, environments, communications} \rangle$$

where *agents* are the set of all the simulated individuals; *objects* are the set of all represented passive entities that do not react to stimuli (e.g., buildings, street furniture in urban environments); *environments* are the topological space where agents and objects are located, where they can move and act, and where signals (sounds, smell, etc.) propagate; and *communications* are the set of all communication categories, such as voice, written materials, signs, etc. Behaviors are generated by the ways in which agents interact or communicate with other objects and their environment(s), and thus can be seen as properties of any of these although they are usually considered to be properties of agents. Thus an MAS can be thought of as the combination of CA and autonomous agents (fig. 2).

A MAS provides a platform for space-time dynamics. We are developing agent-based dynamic models in a number of different contexts. The MAS treats a population of interacting objects in a decentralized or distributed manner, each agent having an independent behavior but with the ability to communicate (with each other). Two types of agent can be identified: reactive agents whose behavior depends entirely upon how they react to their environment; and cognitive agents with plans or protocols who usually interact with one another, are influenced by their environment, but whose behavior is largely self-driven. In the examples sketched here, we will deal exclusively with

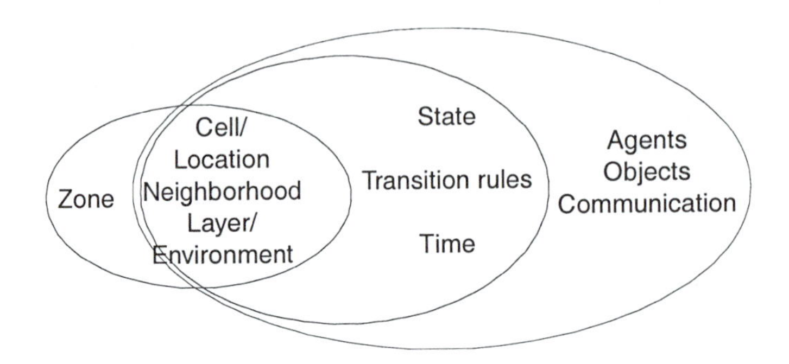

FIGURE 2 Notions of MAS and those of cell-based GIS and CA.

reactive agent-based models where the environment is extremely important to the simulation of behavior.

Over the past years, much effort has been made in order to provide a easily used software platform for scientists to undertake studies on complex systems. Among others, the Swarm project [26] is one of the ambitious projects, which intends to construct an MAS platform for exploring various complex systems. It is designed to serve as a generic platform for modeling and simulating the complex behavior of space-time dynamics. It provides a set of classes for defining agents' behavior, properties, etc. using the computer language Objective-C. Based on Swarm system, various projects have been undertaken, e.g., Transims [32] and Sugarscape [15]. However, Swarm does not, in contrast to what it promised, provide an easily used platform for MAS for noncomputer experts. Attempts have been made to provide a more easily used platform for average users based on Swarm engine [22].

StarLogo [30], a MAS platform with exploratory capability, provides an experimental counterpart of real-world complex systems. It was developed from Logo, a programming language for children [28]. Now the new developed StarLogo has dramatically expanded the simulation of complex systems; various applications have been developed for simulating real-life phenomena such as bird flocks, traffic jams, ant colonies, and market economies (for a set of extendible models, see homepage on ⟨http://www.ccl.tufts.edu/cm/models/⟩). StarLogo consists of three characters: turtles, patches, and observer. Turtles are actually autonomous agents living in CA-like space, each cell of which is called a patch; interaction can occur between turtles, or between turtles and patches through visual and chemical senses. In response to what is sensed, turtles can move around with behaviors such as speed up/down, and heading differently. It should be noted that the observer is not the leader or coordinator, but simply responsible for creating agents in the virtual world. In other words, global patterns created by agents are not due to the coordinated work of the observer. The architecture of the system pictured in figure 3 indicates

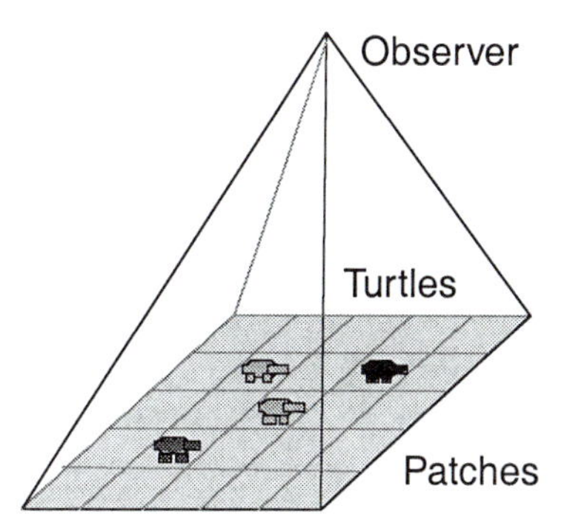

FIGURE 3 The patch-turtle-observer model.

in spatial terms that turtles react to patches, but both turtles and patches are subject to the controls posed by the observer.

We have applied this model to many different kinds of problems, all of which involve some kind of motion or movement. Here we will describe three different processes that depend on moving agents in response to the attributes of their environment which are coded in patches. We will present traditional models of motion in a spatial environment based on traffic which we have developed for pedestrian movement, models of flow dynamics which relate to how watersheds and rivers are formed, and models of how agents can be used to explore the geometry of local environments in buildings through ideas pertaining to what can be seen relative to the view (see Itami [23] for intervisibility analysis).

Many space-time dynamics can be characterized as the interaction at both space and time dimensions. For human or vehicle movement in an urban system, each individual interacts with each other within the neighborhood and the environment to make a movement decision. For example, a driving behavior can be simply defined as, "a car speeds up if there is no cars ahead, otherwise slowdown or overtakes." These behaviors lead to traffic congestion [30]. With an MAS, many space-time dynamics can be modeled.

4 PEDESTRIAN MOVEMENT IN URBAN SYSTEMS

The patterns of people's movement in urban systems characterize one of the very important research areas in urban studies. It has been a big concern in a range of disciplines such as traffic engineering, urban design, and planning. Attraction and spatial configuration are traditionally considered to be very important in characterizing the complex phenomena. It is considered to be very important that spatial configuration is the basic driver for people's movement in urban systems. Often it is found that streets which are well connected attract more people. In other words, integrated streets tends to have more people movement and, on the other hand, segregated streets tend to have less people movement.

Within an urban system, there are two elements that have some direct effect on people movement. From the ecological psychology [20] point of view, human behavior in urban systems can be thought of as a stimulus-reaction model; i.e., what is perceived determines how to act. Thus people's movement can be considered in some sense to be self-organized phenomena through interaction between each other and their environment.

In the model following, we intend to set up a counterpart of pedestrian movement in an urban system, in order to explore this complex phenomena. The aim is to investigate how people's movement is affected by urban morphological structure. Here the structure is described by space syntax, with an integration value describing the properties of urban structure. In the simulation system, the virtual pedestrians have no sense of global structure. They just explore the open space locally and learn themselves from what they have explored. At every moment, we collect pedestrian flows in each street segment for the analytical purposes. The procedure can be described as follows:

Step 1: create a number of pedestrians in the center of an urban system.
Step 2: let all pedestrian move around without encountering obstacles; count the pedestrian rates in each street segment.
Step 3: visually check if all pedestrians have distributed all around; if yes, output pedestrian rates for analysis; if no, go to **step 2**.

As an example, consider figure 4, which shows a small urban system with a relatively regular grid structure of an urban system. Let us first use space syntax to analyze the structure. The analysis result is shown in (a), where the structure parameter of local integration is colored by a spectrum legend; i.e., red represents highest value and blue represent lowest value. Figure 4(b) is a snapshot of the simulation process. The detailed scatter plot is shown in figure 5, where the r-square tends to be 0.7.

Now we slightly change the simulation, and let only one live pedestrian in the system, to see how the shortest paths emerge from locations. In the same urban system, pedestrians can be modeled as working out the route from a starting point or origin called (A) to a preset destination (B), as we show in figure 6. As each pedestrian reacts locally to what is in the surrounding neighborhood, we can compute the crow-fly distance from the point reached in the path so far to the ultimate destination, and then move the agent toward this point in terms of the local geometry which usually poses many obstacles to moving in a straight line. At each stage the crow-fly distance is recomputed, and the agent adapts locally. This process is a crude simulation of the dynamic programming algorithm used to compute shortest paths, first suggested by Bellman and Dijkstra. A typical path is shown in figure 6.

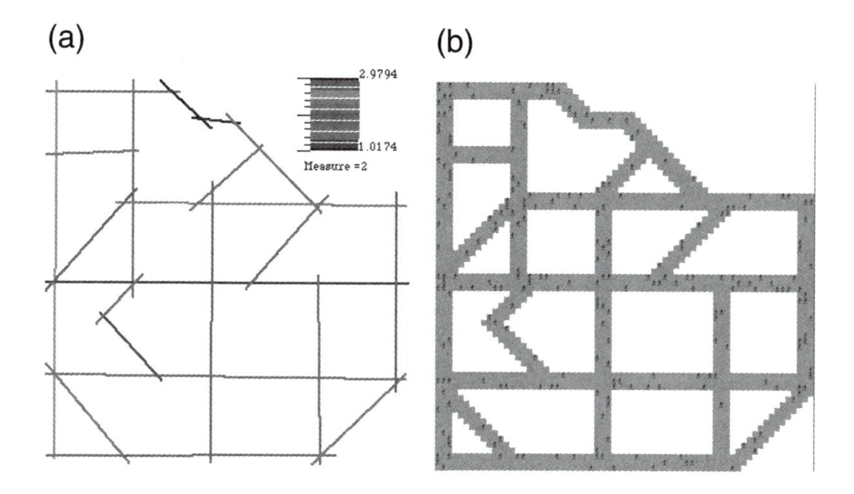

FIGURE 4 A small urban system with a relatively regular grid structure of an urban system.

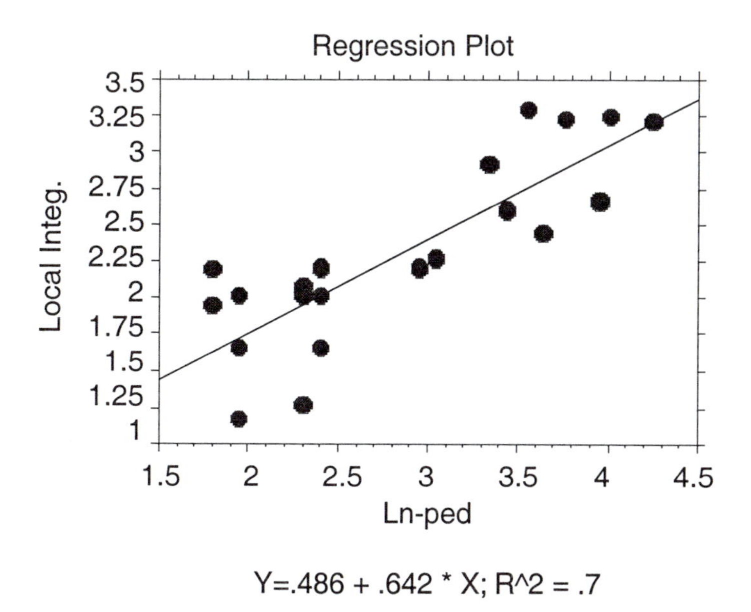

FIGURE 5 The regression plot between local integration and pedestrian rates.

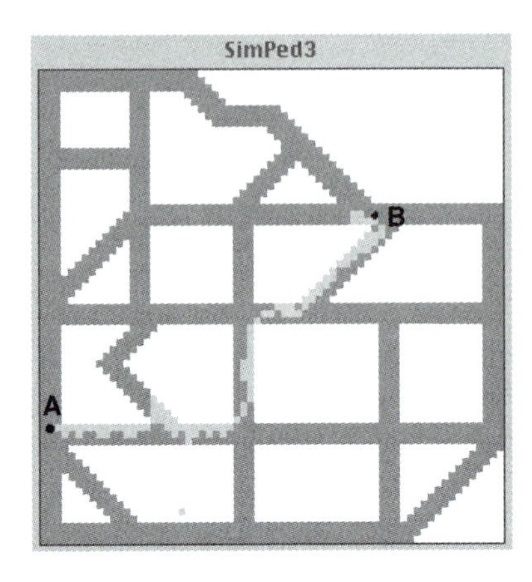

FIGURE 6 A shortest path emerging from local interactions.

As an exploratory platform, the program allows us to experiment repeatedly with the behaviors of agents. Surprisingly, each time the agent works out the route, the routines differ slightly, but in principle are the shortest paths. In the simulation, we only set the destination, and then let the agent explore the space in terms of local interaction. It should be noted that the shortest path does not emerge every time. For instance, if we set starting point and destination at opposite side of a street block, most likely the agents will be confused, and have problems working out a routine. Incorporating some global knowledge in the agent solves this problem. One option would be to apply the following visibility parameter to patches; i.e., the agent continuously interacts with patches and obtains visibility properties to guide its navigation.

5 VISUAL FIELDS IN URBAN SYSTEMS

A view shed, or what can be seen from a certain location, has been a very important issue in environmental modeling. View-shed analysis using the digital terrain model (DTM) has long been one of standard functions of the GIS. However, visual field in urban systems has not received as much attention in the GIS community. In contrast, researchers in architecture and urban studies have paid much attention to how people perceive space. Visual fields are determined by local geometry, moving agents systematically through space in contrast to computing geometric lines of sight [2]. It has been a very important to understand how people perceive and understand and move around their environment. For instance, Hillier's space syntax covered in the above section is based on the notion of visibility, and Peponis et al. [29] have taken this a step further by expanding the partition space into an infinite number of convex spaces in terms of visual perception.

In the fields of robotics and computer vision, visual fields have been very important for robot navigation—i.e., path planning [7, 27]. For instance, before beginning a forward movement, a robot has to check whether the movement is safe. The same check is performed repeatedly during the movement to prevent collisions. In the above example about pedestrian movement, we have used a conflict avoidance procedure, but that was a local check, which means that the pedestrian only can see one step away.

Computation of visual field is a very intensive task [12], because it involves computing the line of sight or possible occlusions between obstacles. Instead it appears that agent-based approaches provide a fascinating platform to achieve the task. The idea is that we fill the space with agents and that we get each agent to explore their environment as far as they can before they come to an obstacle that impedes their path. They make this exploration in every direction or rather in enough directions to cover the entire space which is represented as a raster. In essence the technique depends upon setting as many agents as there are raster cells in the open space between obstacles—rooms, buildings etc.—and then exhaustively computing all areas which they can visit from their particular starting point. Later on with what we have explored, we can explore the space dynamically to show the visual fields from each location. The algorithm is described as follows,

Step 1: fill all open space with agents.
Step 2: alpha = 0.
Step 3: let all agents move to this direction of alpha; and accumulate distance until hit the spatial obstacle.
Step 4: alpha = alpha + increment.
Step 5: check if alpha = 360 if yes, stop; if no go back to **step 3**.

As an example, figure 7 illustrates part of an urban system where the blocks are supposed to be spatial obstacles such as buildings. After the computation preprocess shown above, each location of the space has a parameter which shows how much one can see from the standing point of view. As the mouse moves around, a series of visual fields will be seen dynamically on screen. One typical series is shown in figure 7 (b).

The above model has been applied to some real urban systems such as Wolverhampton town center and London Tate Gallery. With the model, we have constructed spatial properties such as most visible space (see Batty and Jiang [1] for details). Various models have demonstrated that agent-based approaches to the computation of visual fields have been very efficient and effective. However, there exist some problems. For instance, because of the cellular space adaptation, there is an unavoidable gap in the displayed visual field, regardless of the fineness of the adapted grid.

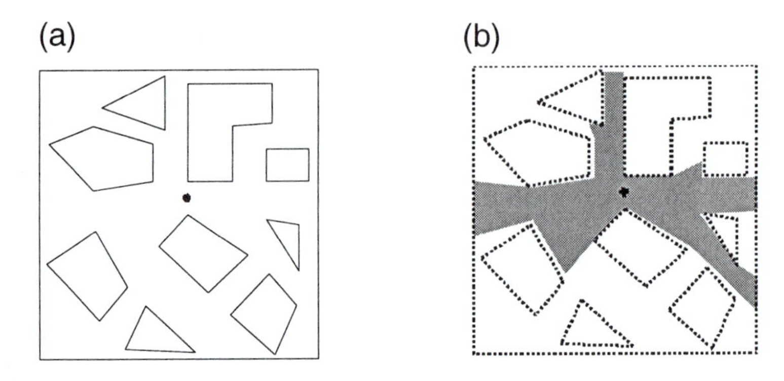

FIGURE 7 The visual fields dynamically shown using agent-based approaches.

In the visual perception of environment, Gibson [20] proposed the idea of an optic array which is considered to be critical in human behavior. However, it is indeed difficult to visualize an optic array. Benedikt and Burham [3] have investigated how Gibson's optical arrays can be objectively simplified as isovists. It is also found that isovists, indeed, do affect human perception in urban environments. Since isovists can be considered in a global sense, it appears to be possible to extend an agent's sense from local to global through interaction with the environment. In the mean time, the visual fields are stored as a patch variable with which agents can interact, and this could lead to more deliberate or cognitive agents.

6 WATERSHED DYNAMICS AND WILDFIRE DIFFUSION IN ENVIRONMENTAL SYSTEMS

There are many phenomena in environment systems that can be characterized as a kind of interaction between agents and environment. For example, spatial diffusion such as air pollution, heat diffusion, and water floods can be considered as interaction between those objects, such as pollutants, heat, and water, with their respective environments. In the following discussion, we discuss two examples from environmental systems.

The first example is about watershed dynamics. We first generate a terrain, and smooth the surface and apply color to make it appear more realistic. Then we create a group of water droplets and let them randomly distribute over the surface. We use greedy search strategies to find the lowest location within a neighborhood. The key step is to use a greedy search to find local maximum (minimum). The procedure can be described as follows:

Step 1: create a random terrain surface and color it.

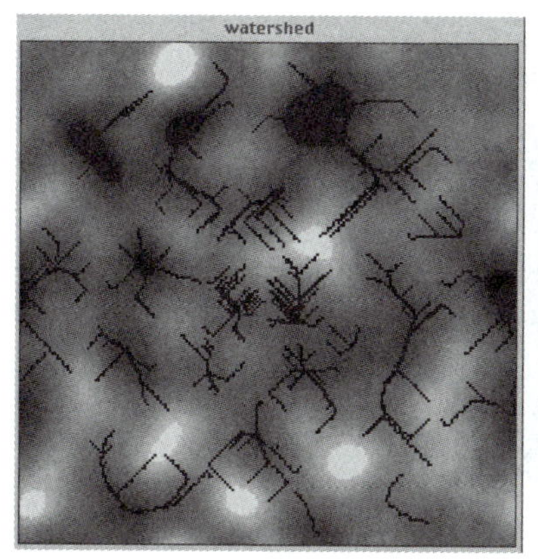

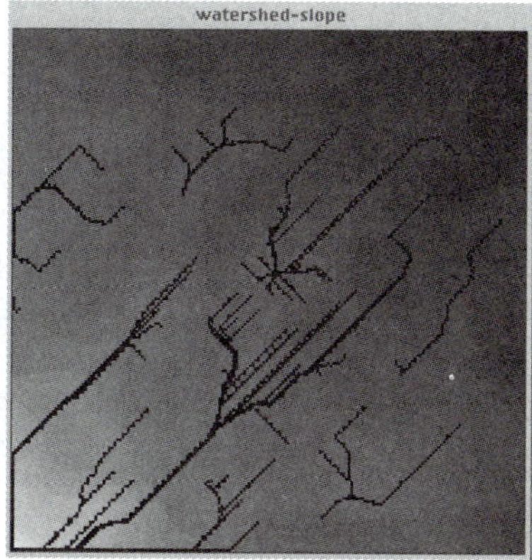

FIGURE 8 Stream networks created from the local interaction of agents with their environment.

Step 2: create a number of water droplets and randomly locate them on the terrain surface.

Step 3: use a greedy search to find the highest elevation and move to the local minimum.

Step 4: check if all water droplets are in local minimum; if yes, stop; if no, go to **step 3**.

As an example, we assume an idealized terrain in which the patches code the height of a surface without any vegetation or forest cover. Stream networks are created by dropping water randomly onto the terrain, and then treating each droplet as an agent whose heading (direction of flow) is computed as a function of elevation in its neighborhood. Figure 8 shows two different terrain models and stream networks that were generated. In the experiment, there are a number of things we can explore. For instance, we can change the number of water droplets, we can change terrain surface, and we can change radius of the greedy search.

The second example is about wildfire diffusion that has been discussed elsewhere by Resnick [30]. Since it is a very typical example for environmental systems in GIS, we briefly discuss the example as evidence to support our position. How does wildfire spread over the forest? It depends on many factors. One critical factor is the density of trees; i.e., if the forest is dense enough, the fire is likely to spread all over; otherwise, it is likely to extinguish. So the exploration of fire diffusion with different forest density could be very interesting in investigating spatial diffusion phenomena. Using a different adaptation than Resnick, we assume that the fire is the agents and the forest is the en-

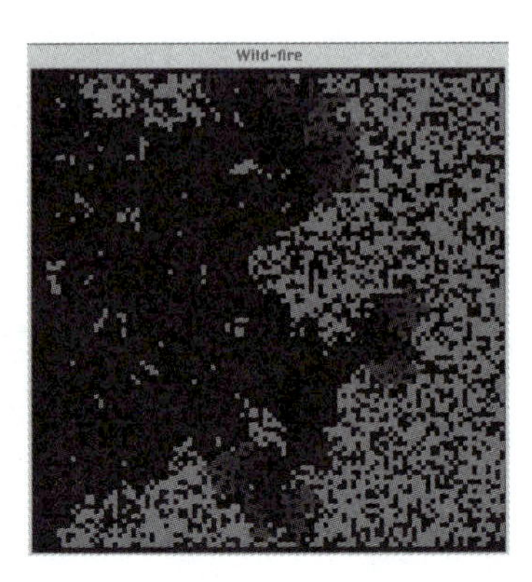

FIGURE 9 A snapshot of fire diffusion in forest with density of 61%.

vironment. Thus the fire diffusion can be thought of as spatial interaction between fire and trees. A simple fire behavior can be described as

```
IF a cell is surrounded by more than one tree
THEN [fire diffuse]
ELSE [fire is extinguished]
```

As an example, figure 9 shows a fire diffusion pattern assuming a density of 61%.

From both the watershed and fire examples, we have reached a conclusion, as in the above example, how the local action gives rise to a global pattern. Obviously this example is far from complete, since factors such as gravity or speed of flow have not yet been considered. In addition, the fire diffusion example could consider wind factor or different types of trees in order to generate a more robust model.

7 CONCLUSIONS

Agent-based approaches have been shown to have many advantages over existing approaches for modeling environmentation and urban systems problem. MAS provides an exploratory platform for users to test hypotheses behind the space-time dynamics. It provides a platform for researchers to experiment and play what-if games with complex spatiotemporal processes, i.e., using a computer as a laboratory for the study of complex, adaptive systems. In this chapter, we have explored and illustrated the potential of the MAS as useful tools for space-time dynamics. We have shown its advantages over existing approaches such as cell-based GIS and CA, not only for space-time dynamics but also for some tasks which need expensive computation. Future work should attempt to go beyond the limitation of reactive agents, to have cognitive or deliberative agents incorporated in the MAS. Future work also implies a fully integration of the MAS and the GIS.

REFERENCES

[1] Batty, M., B. Jiang, and M. Thurstain-Goodwin. "Local Movement: Agent-Based Models of Pedestrian Flows." Working Paper Series, CASA, 1998.

[2] Benedikt, M. L. "To Take Hold of Space: Isovists and Isovist Fields." *Envir. & Plan. B* **6** (1979): 47–65.

[3] Benedikt, M. L., and C. A. Burnham. "Perceiving Architectural Space: From Optical Arrays to Isovists." In *Persistence and Change*, edited by W. H. Warren, R. E. Shaw, and N. H. Hillsdale, 103–114. Hillside, NJ: Lawrence Erlbaum, 1984.

[4] Berry, J. K. "Cartographic Modeling: The Analytical Capabilities of GIS." In *Environmental Modeling with GIS*, edited by M. F. Goodchild, B. O. Parks, and L. T. Steyaert, 58–74. New York: Oxford University Press, 1993.

[5] Burrough, P. A. "Dynamic Modeling and Geocomputation." In *Geocomputation: A Primer*, edited by P. A. Longley, S. M. Brookes, J. McDonnell, and B. MacMillan, 165–191. Chichester, UK: Wiley, 1998.

[6] Camara, A. S., R. Ferreira, and P. Castro. "Spatial Simulation Modeling." In *Spatial Analytical Perspectives on GIS*, edited by M. Fisher, H. J. Scholten, and D. Unwin. London: Taylor & Francis, 1996.

[7] Cameron, S., and P. Probert. *Advanced Guided Vehicles: Aspects of the Oxford AGV Project.* London: World Scientific, 1994.

[8] Clarke, K. C., and G. Olsen. "Refining a Cellular Automaton Model of Wildfire Propagation and Extinction." In *GIS and Environmental Modeling: Progress and Research Issues*, edited by M. F. Goodchild, L. T. Steyaert, B. O. Parks, C. Johnstone, D. Maidment, M. Crane, and H. Glendinning, 333–338. Fort Collins, CO: GIS World, 1993.

[9] Clifford, J., and A. Tuzhilin, eds. *Recent Advances in Temporal Databases.* Berlin, Springer-Verlag, 1995.

[10] Couclelis, H. "Cellular Worlds: A Framework for Modeling Micro-Macro Dynamics." *Envir. & Plan. A* **16** (1985): 141–154.

[11] Couclelis, H. "From Cellular Automata to Urban Models: New Principles for Model Development and Implementation." *Envir. & Plan. B: Plan. & Design* **24** (1997): 165–174.

[12] Davis, L. S., and M. L. Benedikt. "Computational Models of Space: Isovists and Isovist Fields." *Comp. Graphics & Img. Proc.* **11** (1979): 49–72.

[13] Deadman, P., and R. Gimblett. "The Role of Goal-Oriented Autonomous Agents in Modeling People-Environment Interactions in Forest Recreation." *Math. & Comp. Model.* **20(8)** (1994): 121–133.

[14] Egenhofer, M. J., and R. G. Golledge, eds. *Spatial and Temporal Reasoning in GIS.* New York: Oxford University Press, 1998.

[15] Epstein, J., and R. Axtell. *Growing Artificial Societies: Social Science from the Bottom-Up.* Princeton, NJ: Princeton University Press, 1996.

[16] Ferber, J. *Multi-agent Systems: An Introduction to Distributed Artificial Intelligence.* Reading, MA: Addison-Wesley, 1999.

[17] Flake, G. W. *The Computational Beauty of Nature: Computer Explorations of Fractals, Chaos, Complex Systems, and Adaptation.* Cambridge, MA: MIT Press, 1998.

[18] Franklin, S., and A. Graesser. "Is It an Agent, or Just a Program?: A Taxonomy for Autonomous Agents." In *Intelligent Agents III: Agent Theories, Architectures, and Languages*, edited by J. P. Muller, M. J. Wooldridge, and N. R. Jennings, 21–35. Berlin: Springer, 1997.

[19] Gao, P., C. Zhan, and S. Menon. "An Overview of Cell-Based Modeling with GIS." In *GIS and Environmental Modeling: Progress and Research Issues*, edited by M. F. Goodchild, L. T. Steyaert, B. O. Parks, C. Johnstone, D. Maidment, M. Crane, and H. Glendinning, 325–331. Fort Collins, CO: GIS World, 1993.

[20] Gibson, J. J. *The Ecological Approach to Visual Perception.* Boston: Houghton Mifflin, 1979.

[21] Gimblett, R. "Virtual Ecosystems." *AI Appl. Natl. Res., Agric. & Envir. Sci.* **8(1)** (1994): 77–81.

[22] Gulyás, L., T. Kozsik, P. Czabala, and J. B. Corliss. "Telemodeling—Overview of a System. Short paper presented at "Teleteaching '98," part of IFIP World Computer Congress, held September 1998, in Vienna-Budapest. 1998. Available at ⟨http://www.syslab.ceu.hu/telemodeling/index.html⟩.

[23] Itami, R. "Mobile Agents with Spatial Intelligence." This volume.

[24] Jiang, B. "Multi-agent Simulations for Pedestrian Crowds." In *Simulation Technology: Science and Art*, edited by A. Bargiela and E. Kerckhoffs, 383–387. Also in *Proceedings of the 10th European Simulation Symposium and Exhibition*, held October 26–28, 1998, in Nottingham.

[25] Langran, G. *Time in Geographic Information Systems.* London: Taylor & Francis, 1992.

[26] Langton, C., M. Nelson, and R. Burkhart. "The Swarm Simulation Systems: A Tool for Studying Complex Systems." Santa Fe Institute, Santa Fe, New Mexico, USA. 1995. ⟨http://www.santafe.edu/projects/swarm/⟩.

[27] Moutarlier, P., and R. Chatila. "Incremental Free-Space Modeling from Uncertain Data by an Autonomous Mobile Robot." In *Geometric Reasoning for Perception and Action*, edited by C. Laugier, 200–213. Berlin: Springer-Verlag, 1991.

[28] Papert, S. *Mindstorms: Children, Computers and Powerful Ideas.* New York: Basic Books, 1980.

[29] Peponis, J., J. Wineman, M. Rashid, S. Bafna, and S. H. Kim. "Describes Plan Configuration According to the Covisibility of Surfaces." *Envir. & Plan. B: Plan. & Des.* **25** (1998): 693–708.

[30] Resnick, M. *Turtles, Termites, and Traffic Jams: Explorations in Massively Parallel Microworlds.* Cambridge, MA: MIT Press, 1997.

[31] Rodrigues, A., C. Grueau, J. Raper, and N. Neves. "Environmental Planning using Spatial Agents." In *Innovations in GIS 5*, edited by S. Carver, 108–118. London: Taylor & Francis, 1996.

[32] Smith, A. P., M. Dodge, and S. Doyle. "Visual Communication in Urban Planning and Urban Design." CASA Working Paper 2. University College, London, 1–19, Torrington Place, Gower Street, London, 1998.

[33] Takeyama, M., and H. Couclelis. "Map Dynamics: Integrating Cellular Automata and GIS through Geo-Algebra." *Intl. J. Geograph. Infor. Sci.* **11(1)** (1997): 73–91.

[34] Tomlin, C. D. "Cartographic Modeling." In *The Accuracy of Spatial Databases*, edited by D. J. Maquire, M. Goodchild, and S. Gopal, 115–122. London: Taylor & Francis, 1991.

[35] Wagner, D. F. "Cellular Automata and Geographic Information Systems." *Envir. & Plan. B: Plan. & Design* **24** (1997): 219–234.

[36] White, R. "Cities and Cellular Automata." *Discrete Dynamics in Nature and Society* **2** (1998): 111–125.

Mobile Agents with Spatial Intelligence

Robert M. Itami

1 RECREATION BEHAVIOR SIMULATION IN GEOGRAPHIC SPACE

Recreation behavior simulation (RBSim) is a computer program that simulates the behavior of human recreators in high-use natural environments. Specifically RBSim uses concepts from recreation research and artificial intelligence (AI) and combines them with geographic information systems (GIS) to produce an integrated system for exploring the interactions between different recreation user groups within geographic space. RBSim joins two computer technologies:

- Geographic information systems to represent the environment, and
- Autonomous agents to simulate human behavior within geographic space.

RBSim demonstrates the potential of combining the two technologies to explore the complex interactions between humans and the environment [1, 2, 3, 4]. The implications of this technology should also be applicable to the study of wildlife populations and other systems where there are complex interactions in the environment.

RBSim uses autonomous agents to simulate recreator behavior. An autonomous agent is a computer simulation that is based on concepts from artificial life research. Agent simulations are built using object-oriented programming technology. The agents are autonomous because, once they are programmed, they can move about the landscape like software robots. The agents can gather data from their environment, make decisions from this information, and change their behavior according to the situation in which they find themselves. Each individual agent has its own physical mobility, sensory, and cognitive capabilities. This results in actions that echo the behavior of real animals (in this case, humans) in the environment.

The process of building an agent is iterative and combines knowledge derived from empirical data with the intuition of the programmer. By continuing to program knowledge and rules into the agent, watching the behavior resulting from these rules, and comparing it to what is known about actual behavior, a rich and complex set of behaviors emerge. What is compelling about this type of simulation is that it is impossible to predict the behavior of any single agent in the simulation and, by observing the interactions between agents, it is possible to draw conclusions that are impossible using any other analytical process.

RBSim is important because, until now, there have been no tools for recreation managers and researchers to systematically investigate different recreation management options. Much of the recreation research is based on interviews or surveys, but this information fails to inform the manager/researcher how different management options might affect the overall experience of the user. For example, if a new trail is introduced, we might expect that conflicts might be reduced, but to what extent? If we go to a system of scheduling use, what is the impact on the number and frequency of users? More importantly when you have different, conflicting recreation uses, how do different management options increase or decrease the potential conflicts?

None of these questions can be answered using conventional tools. These questions all pivot around issues such as time and space as well as more complex issues such as intervisibility between two locations. By combining human agent simulations with geographic information systems, it is possible to study all these issues simultaneously and with relative simplicity.

2 RECREATION-BEHAVIOR-SIMULATION OBJECT MODEL

RBSim is developed using object-oriented programming technology. Figure 1 shows a diagram of the principle components of the simulation program. RBSim is comprised of five major components:

1. A graphical user interface (GUI) for *model parametrization*. This is comprised of a set of forms for setting values for the remaining components described below.

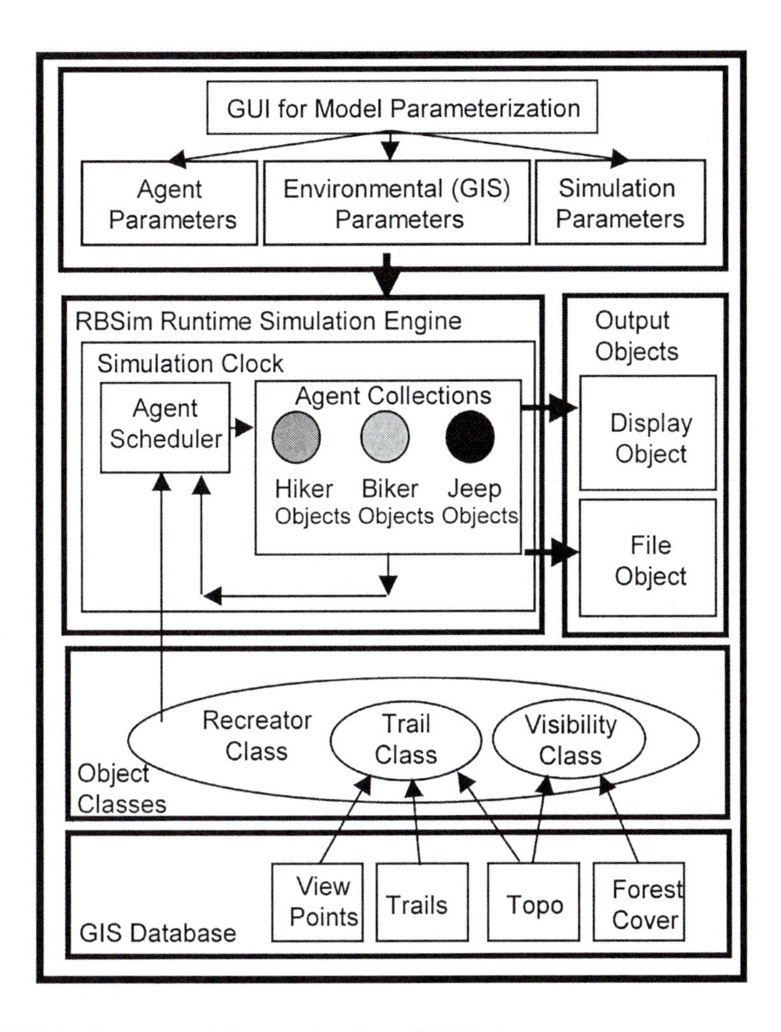

FIGURE 1 Recreation behavior simulator (RBSim).

2. *Output classes* including:
   - the *video display* showing a shaded relief map as a backdrop to the agent type and location displayed as graphic objects during the simulation run, and
   - a *file object* for saving simulation statistics.
3. *Object classes*, including:
   - *the recreator* class, the base class for the hiker, biker, and jeep recreation agents,
   - *the visibility class* which provides the visual system for the recreator class, and

- a *trail class* which represents the trail as a list which contains the location, elevation, and viewpoints at each trail location.

4. *The GIS database* which is used to parametrize the trail and visibility classes.

3 MODEL PARAMETRIZATION

RBSim allows the user to specify the following parameters for the recreation agents (see fig. 2):

1. The total number of agents in each class (landscape hikers, landscape bikers, social hikers, social bikers, and jeeps).
2. The age distribution of recreators in the hiker and biker classes.
3. The frequency within which each recreator begins a journey through the trail system.
4. The files containing data for trail configurations to be simulated.
5. The duration of the simulation run.
6. Parameters for setting up visibility for the agents (to be discussed in more detail later).

4 THE VISIBILITY CLASS

Crowding is one of the major factors influencing satisfaction levels during a recreation experience. Since much of the perception of crowding is based on visual contact as well as physical contact with other recreators, a vision system is designed for the agents. The visibility class is a modification of standard GIS line of sight or intervisibility analysis (fig. 3). To reduce the computations required to check for visibility between two points, the visibility class checks for intervisibility only between points occupied by other recreators referenced in the trail object. The line of sight is calculated taking into account the screening effects of intervening topography and vegetation from the GIS databases for elevation and forest cover.

5 THE RECREATOR CLASS

The recreator class is the most complex object class. It is comprised of a set of properties for age and personality type (landscape or social agent). These properties determine the behavioral rules that the agent will follow and the mobility and energetics of the agent. Behavioral rules (fig. 4) relate to how the agent responds to the views, and the number and type of other recreation agents. These rules may result in the agent changing speed to overtake or

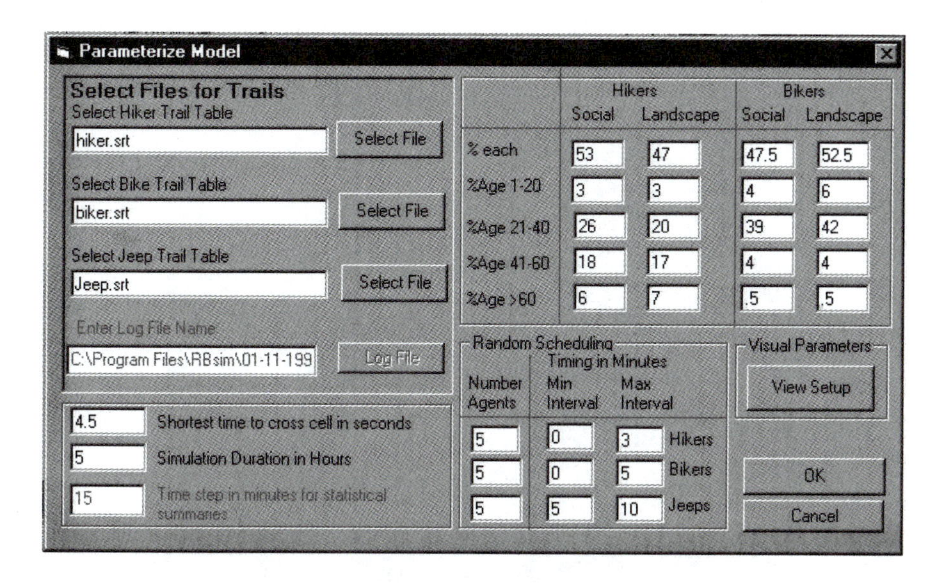

FIGURE 2 Form for parametrizing simulation model. Top left allows user to select files for trails. Alternative trails configurations can be loaded to test impact of trail design on crowding. The top right allows the user to set the age and personality profiles for hikers and bikers. Default values are from research by Gimblett [3, 4]. Lower left sets the duration and speed of the simulation. The Random Scheduling section in lower center sets the total number of agents of each type and the frequency which they start their journey. Finally, the visual parameters button on the center right brings up the form in figure 3.

catch up with other agents, or slowing down and stopping to rest or spend time at important landscape features or viewpoints. In the RBSim program, three recreator types are generated: hikers, bikers, and jeeps.

Hiker and biker agents also have a system of energetics programmed. Energy levels and speed of travel are related to the age of the agent. Very young and older agents will move more slowly than agents in other age groups will. In addition, as energy is expended during the simulation, these agents will also need to rest to rebuild energy levels. The length of resting time is determined by the estimated time it takes agents of different age groups to recover. Energy expended is calculated incrementally as the recreator moves along the trail. Uphill travel expends more energy than downhill travel. Resting times are randomized between preset time thresholds to represent variability between real human recreators.

Each agent object has a single method called "Move" which the simulation engine calls when that particular agent's turn in the execution queue comes up. The move method then calls a private "DecideBehavior" method where all the internal rules for the agent are executed.

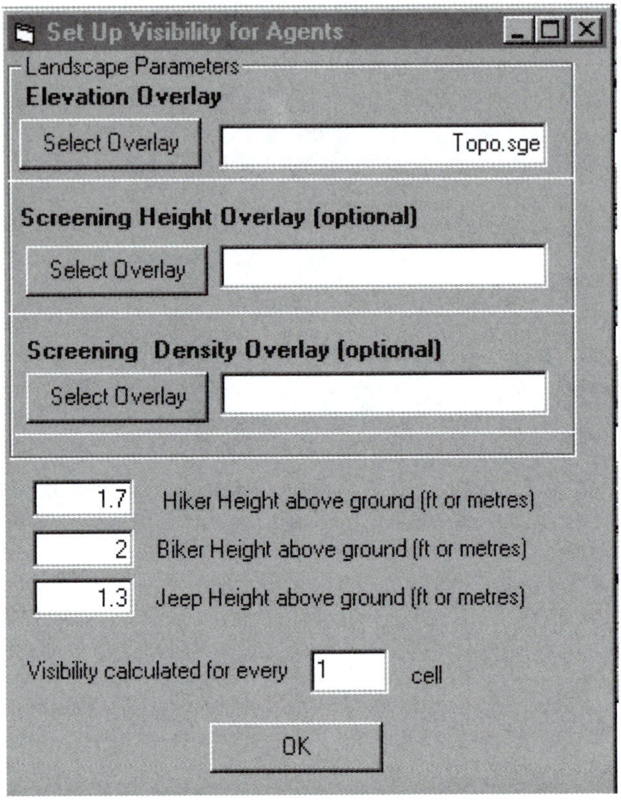

FIGURE 3 This form allows the user to set parameters for the visibility system for agents. The landscape parameters are GIS overlays for elevation, screening heights (e.g., vegetation height), and screening density (e.g., vegetation density). The user can also set the height of the viewer's eye above the ground and how frequently each agent checks visibility to other agents.

6 THE TRAIL CLASS

Trails are specified for each agent, for each run. The trails are stored in the trail object, which is constructed as a linear list of cells derived from a grid-based GIS. For each trail cell the distance from the trail head, the elevation, and landscape features associated with the cell is stored. During the simulation run the trail object also stores the number of recreators in each cell. This data structure is designed to minimize the computing time for agent navigation through the trail system. All agents of the same class (hiker, biker, or jeep) share the same trail object. The trail object therefore acts as a "collective memory" for the agents in that respective class. Each agent can reference

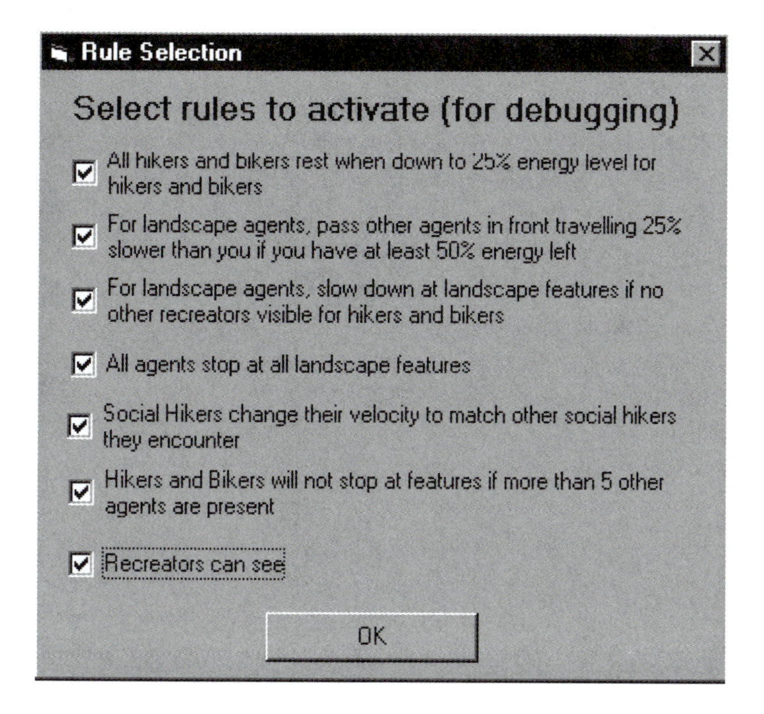

FIGURE 4 Rules for agents can be turned on and off with this form.

the trail object to determine the location of other agents on the trail and to determine the trail conditions. As the agent moves from one cell to the next, it de-references its location from the last cell (by subtracting one from the recreator count field for that cell) and references its location in the next cell. Since the hikers, bikers, and jeeps follow different trails, a unique trail object is created for each recreation type. To test management alternatives for new trails, the user may specify different trail files for each simulation run.

7 THE RUN-TIME SIMULATION ENGINE

The RBSim run-time simulation engine runs in discrete time steps. At each time step in the simulation, each recreator class (hikers, bikers, and jeeps) is evaluated to determine if a new instance (agent object) of that class should be created. For each class of recreator a timer is set which begins incrementing from the start of the simulation run and is reset to zero each time a new recreator agent is generated. In the model parametrization, the minimum and maximum times between agents is specified. A random time is generated between the minimum and maximum time each time a new agent is generated.

A new agent of the respective class will be generated once the timer reaches the randomly generated time.

The new agent object is generated as an instance of the generic recreator class. When the agent is created, properties are set for age, personality, and agent type. These properties are set based on a randomly generated number (between 0 and 1) which sets the probability for each property. For instance, if 25 percent of the biker agents are of the landscape personality type and 75 percent are of the social personality type, then, if the random number is between 0 and .25, the simulation engine will generate a landscape bike agent. If the number is greater than .25 and less than or equal to one, the simulation engine will generate a social bike agent. This same strategy applies to the age distribution as well.

Recreator agents of the hiker, biker, and jeep types are placed in collections for each type. The simulation engine then tracks each agent in each collection. Since the simulation engine is running on a synchronous clock, the order in which the agents are executed will affect consequences such as crowding and visibility. In order to avoid order effects from executing agent movement in a set sequence, the sequence is randomized within each collection for every iteration of the simulation. Each agent has a single method called "Move" which triggers the execution of the internal rules, energetics, and mobility for that agent. Once the agent has completed execution of all its behaviors for that time step, the run-time simulation engine then executes the move method for the next agent in the randomized list for that iteration. This process continues in a loop until either all agents have completed their journey or the maximum time set for the simulation run is reached.

At the conclusion of the simulation, each trail object writes its contents to the output file object. RBSim then returns control to the user.

The remainder of this chapter will focus on particular aspects of the object model and will look at how each component can be generalized or expanded to provide a generic framework for other applications of mobile agents in geographic space.

7.1 GENERALIZING RECREATION BEHAVIOR SIMULATION

The above discussion has described the first version of RBSim. This version represents a "proof of concept" to show how agent models can be integrated with GIS for simulating recreation behavior. In version 1 of RBSim a key innovation is the adaptation of intervisibility operators in GIS for use as the vision system for agents. With minor reprogramming, the complex issue of seeing across terrain and vegetation cover in a simulation model is easily implemented. The spatial processing algorithms in GIS provide a potentially rich set of capabilities that can be incorporated into agents for a variety of purposes. These will be explored in more detail in the following sections.

7.2 INTELLIGENT AGENTS IN GEOGRAPHIC SPACE

A fundamental innovation in RBSim is the integration of GIS data into an agent-based simulation. The map data from the GIS provides an accurate representation of real-world landscape features. The GIS represents a geographic region as a set of map overlays that can be conceptualized as a series of transparent sheets upon which certain map features are drawn. On one sheet one could find water features such as rivers, lakes, springs, and wetlands. On another layer one could find vegetation types. On the next, transportation networks such as roads, railways, and trails. Other layers might include elevation, climate data, geology, location of hospitals, or other features commonly found on topographic maps. Because common features are grouped on separate overlays, it is efficient to retrieve objects within a given class. Furthermore, only those layers that are pertinent to the problem at hand need to be loaded. Because these features are accurately located in a given coordinate system, and are in digital form, GIS data sets provide ready-made environments for agent simulations.

GIS systems are complex systems, with their own command language, graphic display systems, user interfaces, analytical capabilities, and data formats. This complexity makes it difficult to integrate with other systems. There are essentially two ways of integrating GIS with agent simulation models:

1. Have the simulation program work independently of the GIS, passing parameters to the GIS to run conventional GIS operations, and then returning results in the form of maps or variables from the GIS back to the simulation package.
2. Accessing the GIS map data directly from the simulation package and encapsulate only those GIS functions that are needed in the simulation.

The first option has the advantage of requiring little additional programming to gain the full functionality of the GIS. The disadvantages revolve around the appropriateness of standard functions in the GIS to the agent simulation model and the slow speed at which some of these functions operate. If a GIS function takes more than a few seconds to run (which is common), this makes the simulation run unacceptably slow. If the GIS will not perform a function that is needed in the simulation model, then one must build a custom program to produce the desired result anyway.

The second option has the disadvantage of increasing the size and complexity of the simulation program by requiring the need to support input/output functions to read and write to GIS map databases as well as the need to write the analytical routines to process the geographic data sets. The advantages are that custom functions can be written and the code can be optimized for speed to ensure better performance from the simulation run. RBSim takes this approach.

7.2.1 Geographic Information Systems Map Layers. RBSim uses a raster or grid-cell GIS database for representing environments. A grid-cell GIS represents a map as an evenly spaced set of grid cells. A grid-cell GIS has many advantages for agent simulations. First, grid cells simplify the problem of representing continuous data such as elevation. Second, the need for performance in simulation models is facilitated by grid-cell representation since it is easy to subset spatial data for local processing.

The detail or resolution of the map is determined by the cell size. A 10-meter cell represents an area 10 meters by 10 meters (100 square meters). A 20-meter cell represents an area 400 square meters but stores only a quarter of the detail of a 10-meter cell. Features are represented by numeric values assigned to each corresponding cell in the map. For instance, in a map of trails, a value of 1 might be assigned to walking trails; a value of 2 represents biking trails; and a value of 3 might represent jeep trials.

In the case of simulating recreation behavior, it is necessary to have a fine-scale database to adequately capture the characteristics of the landscape that are important to the recreation experience. Factors impacting the selection of cell size relate to the scale of the original map data, the complexity and density of site features captured in the map data, and at what level of accuracy movement of agents through the database are to be tracked. The first environment simulated in RBSim was a forest recreation area called Broken Arrow Canyon near Sedona, Arizona. The cell size selected for this simulation model was 10 meters. This captured the complexity of the trail layout and the major site features along the trails. In a simulation for the Grand Canyon a cell size of 90 meters has been selected because of the length of the canyon, and the relatively coarse level of information along the course of the river (see Gimblett et al. [5]).

Currently the cell size for trails, digital elevation model, and vegetation cover must be the same. However, it would be desirable to support variable cell resolutions for each map layer depending on the accuracy of the data and the type of analysis required.

7.2.2 Geographic Information Systems Analysis and Performance Problems. The nature of spatial analysis in a GIS presumes that any analytical procedure will be applied to the entire spatial extent of the region of interest. This "global" processing is inefficient in the context of an environmental simulation where thousands of iterations may be required. This problem can be resolved by changing the frequency in which map-based analysis is repeated. For example, if the environment is changing slowly in relation to the number of iterations of the simulation, it would be better to update the map data every tenth or fiftieth iteration of the model depending on how rapid the environmental change is.

Another method can be employed in a situation where only local regions of the study area are changing in the simulation. In this case, analytical models could be run only on small portions of the dataset, speeding up the simulation.

A good example of this type of application is where an agent may wish to summarize the local geographic area according to maximum local relief. This operation is fast for a single point in the database for a local region of cells, taking only a fraction of a second to execute. This approach is proposed to the spatial reasoning for mobile agents discussed in the next section.

7.3 MOBILE AGENTS WITH SPATIAL REASONING.

7.3.1 Agent Spatial Reasoning—Putting GIS Analytics Inside the Agent. As agents move in the GIS landscape, it is desirable to be able to acquire knowledge about the local landscape in terms of elevation, slope in the direction of travel, ruggedness, the type of vegetation, percent vegetation cover, the diversity of vegetation cover, and many other factors. Some of these attributes of the landscape can be determined by simple query of the GIS database. For example, elevation can be determined by simply querying the elevation map at the cell location of the agent. However, many factors, such as slope in the direction of travel, cannot be determined using conventional GIS techniques for calculating slope since "direction of travel" is not a variable in calculating slope in a GIS. Similarly a factor such as diversity of vegetation cover cannot be determined by simple query of the GIS vegetation map. It must be calculated by first defining the geographic region of interest and then counting the number of different vegetation types within that region.

Grid GIS systems have a set of analysis classes which have been defined by Tomlin [6] in his Map Algebra language. The five classes of spatial analysis that Tomlin defines are:

1. *Local functions*—computation on one cell location by simple interrogation of the cell value on a single map or by computation on a single cell by operations with one or more other maps for the same cell location. A common example of this type of analysis is where one may have an equation in which cell values on a map are used as variables in the equation. The result is a map where each cell value represents the result of running the equation on every location on the map.
2. *Focal functions on immediate neighborhoods*—characterizing a cell according to a computation on neighboring cell values on the same or different map, diversity, averages, and other statistical summaries of neighborhoods.
3. *Focal functions on extended neighborhoods*—measuring or calculating values across a map. These include analyses such as intervisibility analysis, distance measure, travel time, travel cost, hydrological analysis, and others.
4. *Zonal functions*—characterizing a region of cells based on a computation on cell values within the region for the same or a different map. Examples of this type of analysis include calculation of area and perimeter of a region; calculating average elevation; minimum, maximum, standard deviation; and other statistical measures of cells in a region or zone.

5. *Incremental functions*—examples of this type of analysis include calculation of density, slope, solar aspect, accumulated runoff, slope direction, surface volume, and others.

We may add another class, which is simply a modification of the above classes, which is customized analyses that are specific to functional requirements of an agent's reasoning. An example of this is "slope in the direction of travel." As was noted earlier, slope is a standard function of a GIS, but slope in the direction of travel can only by computed in the context of an agent's location and direction of travel.

The above spatial analytic classes provide a "tool box" of analytical capabilities that can be programmed into the agent. Again, rather than applying these analytical techniques to an entire map, the agent implementation would only compute these values for a localized region around the agent's position. This decreases computation time, and provides the agent with an enormous pool of spatial reasoning abilities. Which of these analytical tools to apply depends on the decisions that the agent needs to make about the environment. This suggests that the reasoning system of the agent must be defined in relation to personality profiles.

7.4 A FRAMEWORK FOR ENVIRONMENTAL REASONING

An example of the application of environmental reasoning might be at a fork in the trail where a decision should be made as to which route to take. Such a decision might be based on the following criteria:

- Destination—if one trail only leads to the desired destination than the choice is simple. However, if both trails ultimately lead to the same destination, then the following factors may come into play.
- Distance—the recreator may wish to take the shortest distance, or on the other hand, if a longer walk is intended, the longer distance may be preferable.
- Steepness—if one fork of the trail is steeper than the other, the agent may opt for the easier trail.
- Views—if one trail looks like it leads to better views, this trail may be selected over the other.
- Environmental diversity—the most interesting or diverse route may be more desirable.

How does the agent decide which trail to take if both trails ultimately lead to the same destination? The answer cannot be answered simply. The answer may differ from one agent to the next depending on the goals, intentions, and physical condition of the agent, as well as energy levels, time to complete the journey, and other factors. In other words, the decision is based on a complex hierarchy of factors that may be changing as the agent continues its journey.

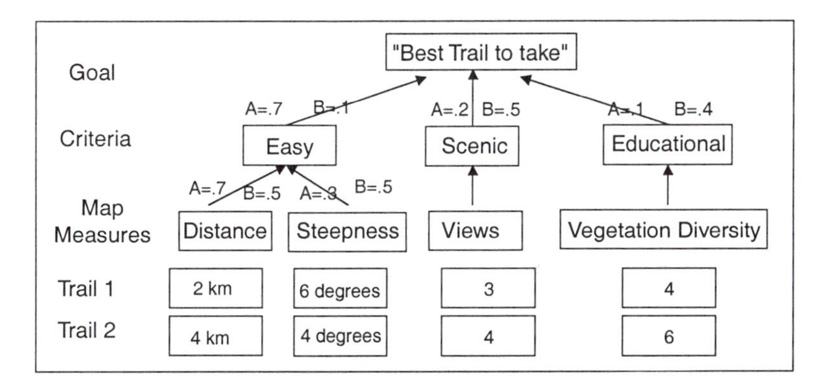

FIGURE 5 An example of a decision hierarchy that could be constructed to integrate environmental measures from maps into criteria for trail selection. In this context, the structure of the decision hierarchy would be identical for each class of recreation agent; however, the weightings on the criteria could be different between agents resulting in different choices.

Remember that most judgments are not of a life and death nature. They are guesses as to which choice may produce the best outcome given a mix of criteria. Also remember that the human in a real-world environment is neither omniscient in that he or she does not have comprehensive knowledge about environmental conditions for each route, nor does he or she expect to make optimal decisions. In reality most visitors will be making decisions on partial knowledge of the environment, and most often localized knowledge in terms of geographic space.

This type of decision making falls nicely in the class of fuzzy decisions. Yet any given decision must marry a range of environmental factors against a set of decision criteria. An approach to dealing with this problem is to prioritize the criteria and then measure the performance of environmental factors that contribute to each of the criteria. One method which is simple to implement and will allow for this type of decision making is the analytical hierarchy process (AHP) [7].

In AHP, decision criteria are defined and weights are defined through a method of pairwise comparison. If the criteria can be measured from environmental factors (either local or global), it would then be possible to make decisions incrementally based on local environmental conditions or globally based on measurements of entire environments. Figure 5 shows an example of a decision hierarchy for the above factors and how the decision criteria for selecting a trail can be measured using GIS operators on a map database. Note that the structure of the decision hierarchy would be the same between agents, but the weightings of the criteria could be different, say, between a 25-year-old backpacker and a family walking with small children. The figure shows two

TABLE 1 Criteria weights and trail ratings. Criteria weights for two different recreation groups. Group A prefers easy walks, Group B prefers interesting walks with good views (weights for each group sum to 1). Normalized ratings for each trail are calculated by dividing each value by the sum of all values for each criteria. Distance and Steepness are subtracted from one since the criteria is easiness and the scale must be reversed to measure easiness in the positive direction. The normalized ratings for Trail 1 and Trail 2 sum to 1 for each criteria.

	Criteria weights for Group A	Criteria weights for Group B	Trial 1 normalized ratings	Trial 2 normalized ratings
Easy—Distance	.7 × .7 = .49	.1 × .5 = .05	1–2/6 = .67	1–4/6 = .33
Easy—Steepness	.3 × .7 = .21	.5 × .1 = .05	1–6/10 = .4	1–4/10 = .6
Views	.2	.5	5/12 = .43	7/12 = .57
Veg. Diversity	.1	.4	4/10 = .4	6/10 = .6

TABLE 2 Weighted scores for each criteria for Trails 1 and 2 for Groups A and B. These scores are derived by multiplying the group weights by the criteria ratings for each trail.

	Trial 1 scores for Group A	Trial 2 scores for Group A	Trial 1 scores for Group B	Trial 2 scores for Group B
Distance	0.33	0.16	0.03	0.02
Steepness	0.08	0.13	0.02	0.03
Views	0.09	0.11	0.21	0.29
Veg. Diversity	0.04	0.06	0.16	0.24

trails with the measurements for each of the map measures. Weights for the criteria can be generated for two different groups: (A) a family with young children and (B) nature-loving young hikers. Note that the weights vary for the two groups. The choice of trail to take can be calculated as in tables 1 and 2.

The difference between local and global decisions is important in recreation management, since signs are often placed at trail intersections. Often, the only information provided at these points is linear distances to destinations from that point. However, often estimated travel time is also provided along with an indication of trail conditions such as difficulty ratings or warnings about hazardous conditions. When this "global" knowledge is provided, it will generally influence the decision of a recreator. He or she still must weigh this criteria against other criteria; however, it reduces the amount of uncertainty in key factors and from a computing point of view can reduce the number of calculations that need to be made at run time.

TABLE 3 Trail preference scores for each group for Trail 1 and Trail 2. Family Group A will choose Trail 1 (.54) over Trail 2 (.46) because of the shorter length. For Group B (young, nature-loving hikers) easiness is not important so higher preference is placed on views and vegetation diversity. Trail 2 (.57) is selected over Trail 1 (.46).

	Trial 1	Trial 2
Preference scores for A	0.54	0.46
Preference scores for B	0.43	0.57

7.4.1 Improving the Agent Mobility Model. The original RBSim model used a simple approach to move agents through the trail system. In that model, trails were one-way with a single origin and destination. Although distance and elevation were assigned to each cell, there were no attributes relating to trail width, surface characteristics of the trail, or other factors that would impact mobility. A generalized approach to landscape mobility would have to account for two different types of travel: travel on a network or travel "cross country." Immediate improvements can be made to travel on networks simply by supporting a true network topology with one-way and two-way travel. In addition, nodes in the network where the trail branches are natural decision points where agents must decide on which branch of the trail to take.

Trail networks work well for most recreation simulations. However, there are instances where agents are "free range"; that is, they can traverse the landscape with or without reference to a formal trail system. The mobility system for free-roaming agents is more complex, but not insurmountable. A reasoning system using AHP as described earlier can be used to make incremental decisions as the agent traverses across the landscape.

The trail and road networks can be stored in a database that contains the attributes such as width and surface characters for each section of the network. The agent model requires that the spatial reasoning capabilities described in the earlier section are implemented. For off-trail mobility, GIS map layers supply the required data for terrain and land cover.

7.4.2 Agent Visual System. A fundamental characteristic of human or animal behavior in the landscape is the ability to see. Decisions relating to crowding, attractiveness of a view, and the perceived difficulty of access, are all based on judgements from information collected through the visual system.

The screening effects of terrain, vegetation cover, buildings, and other obstructions influence vision in the landscape. As one moves through the landscape the view is constantly changing. Calculating visibility is already a standard function of GIS systems. "Line of Sight" calculations also called "intervisibility" or "view shed" analysis produces a map showing all areas visible from one or more points. More sophisticated intervisibility calculations take into account the screening effects of terrain, vegetation cover, and curva-

TABLE 4 The following table shows the characteristics of generalized trail and agent models as they relate to mobility.

Trail/Road Model	
Trail topology	Support network topology with one-way and two-way links. Standard network data structures can be implemented to store either raster or vector networks.
Trail characteristics	Support width, surface characteristics, and elevation.
Agent Model	
Agent goal hierarchy	Agents should have primary and secondary goals in terms of destinations. These are necessary to make choices at trail intersections. Goals need to have a geographic location relative to the network so the agent can determine if destination goals have been achieved or not and whether they are ahead or behind of them relative to the network.
Agent mobility	Agents should be able to move off trails in situations where this is desirable. They should be able to travel cross-country using mobility rules that relate to terrain characteristics, destination, and obstacles.
Agent energetics	Agent mobility should be modeled to be influenced by fitness, air temperature, energy expended/duration of travel, difficulty of travel (trail conditions), and recovery rates from fatigue.

ture of the earth. Figure 6 illustrates how intervisibility is determined using line of sight calculations.

8 POINT-TO-POINT VISIBILITY

In the first version of RBSim, intervisibility was only calculated between recreation agents. This requires that a vector or single line of sight be calculated between an agent and each other agent in the simulation. This calculation is fast since shortcuts can be taken to terminate the calculation. For example, because of the data structure for the simulation, we know the elevation of the agent from which we are calculating the line of sight and the elevation of the agent we are looking to. We also know the x, y location of both agents.

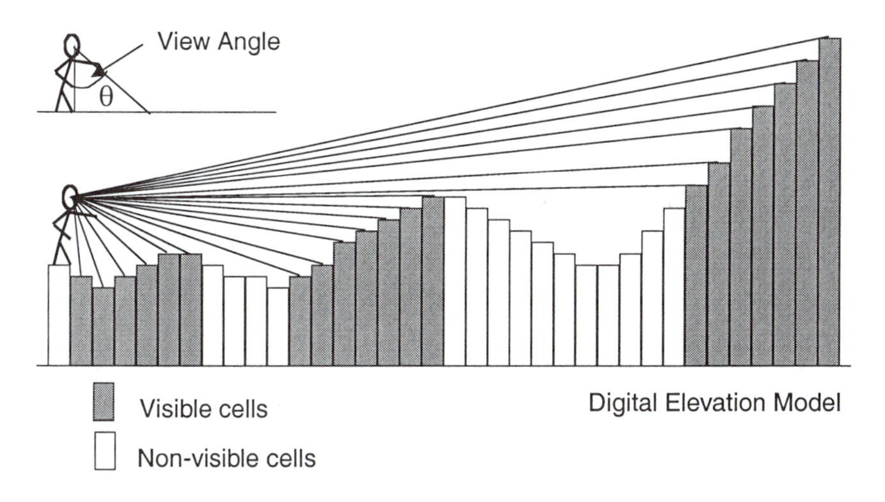

FIGURE 6 Intervisibility is calculated in geographic information systems by calculating the angle of view starting from the observer position. As the angle of view increases moving away from the viewer's position, the target is visible. If the angle of view decreases from the last maximum angle of view, the target is not visible.

From this information we can calculate the angle of the line of sight from thev viewing agent to the target agent (MAXTHETA). Next we begin calculating the angle of view starting from the cell closest to the agent and proceeding cell by cell in the direction of the target agent. As long as the angle is less than MAXTHETA, we know that the agent is still visible. If, at any point along the line of sight, the angle of view exceeds MAXTHETA, then we know the target agent is obscured from view, and we can terminate the line-of-sight function and return a value of zero (not visible). This type of time saving is not possible if the call were made to a GIS. In addition, there would still be a need to query the map generated by the GIS visibility function at the location of the target agent to see if it is visible. This generates a needless map, affords no opportunity to save computation time, and demands an extra step to query the map to find out if the agent is visible or not.

9 POINT-TO-AREA VISIBILITY

Point-to-point visibility as described above is useful in many different simulations. However, there are many, more complex visibility problems that also should be considered. For example, there is the point-to-area intervisibility problem. For example, we may wish to find out what percent of a landscape feature, such as a rock outcropping or lake, is visible from a point. This type of operation takes much more computation time than point-to-point visibility and could substantially slow down the simulation. One solution method is to precompute the visibility. This is especially appropriate where we have known points, such as scenic viewpoints, where we can predict where viewers will want to see significant views. This type of precomputation is probably best handled in the GIS. For example, we might have five important features for which we wish to compute visibility.

We would compute visibility for one feature at a time. Using a lake as an example, we would compute visibility from the lake and map out all areas that are visible from each grid cell in the lake. The GIS has an option to record frequency counts, so that every cell that has visibility to the lake will have the number of cells of the lake that are visible from that point. The percent of the lake visible from each point can then be computed by dividing this figure by the total number of cells in the lake and multiplying by 100. This can be done for an entire map in one calculation since the GIS does calculations on all cells in the map simultaneously.

By extracting all relevant cells (e.g., cells corresponding to trails) and storing the visibility values for each landscape feature, agents can then be programmed to respond to views of one or more features based on the type of feature, the amount visible, and association with other features. This can all be done without the computational overheads of the point-to-point approach to visibility.

9.1 THE SITE AS AN OBJECT

The previous discussion has concentrated on the landscape as represented by the GIS. There are, however, many details of specific sites that are important in an agent simulation that are not always captured within the GIS data structure.

An example of a typical site is a parking lot. A parking lot may appear on the GIS as simply a single dot. However, there are many attributes of a parking lot that are important in an agent simulation such as the number of parking spaces for buses and cars, whether or not there are interpretive signs and toilets, and during the simulation, how many parking spaces are available. CAD drawings often depict these features graphically, but may not have an associated file that quantifies these features. In a simulation there may be no advantage to showing these details graphically. Instead, sites that have many attributes that influence the behavior of agents can be represented as objects in much the same way that agents are. The advantage of this is that a parking lot can be moved by changing its location properties or parking spaces may be added or removed simply by changing the properties for the number of parking spaces for buses or cars. This simple structure can easily be expanded by adding additional properties such as the number of rows of parking, the number of parking bays in each row, etc. If a visual representation of the parking lot is desired, the object could be given a graphic interface which accurately shows the configuration of elements in the site object either as a bitmap or as a vector graphic.

Other sites which can be represented in this way are playgrounds, picnic areas, campsites, or buildings. The amount of detail and complexity is entirely up to the needs of the simulation.

10 CONCLUSIONS

This chapter has shown the great potential of incorporating spatial data from geographic information systems into agent simulations. It also shows the potential of recasting the spatial analytic algorithms in GIS into the spatial reasoning systems for agents as they traverse through environments represented by GIS maps. With these tools, it has been demonstrated how the intervisibility analysis capabilities of the GIS can be adapted to supply the visual system for an agent. By expanding this concept to include a more complete set of spatial analytic methods into the agent's reasoning system, a complex set of decisions can be made.

The reasoning system for an agent with spatial mobility cannot be based only on a set of methods for collecting and summarizing information about the environment. There must also be a mechanism for weighing this information against a set of goals, desires, and intentions. The analytical hierarchy process has been suggested as an approach to cope with the problem of prioritizing a set of choices based on environmental factors.

The framework here has been focussed on the problem of simulating recreation behavior; however, there is no reason why the framework cannot be applied in simulating other types of behavior such as movement of animals, transportation systems, or any other system where environmental factors influence behavior, communication between agents, and movement through a landscape.

Clearly there is strong motivation for integrating GIS with agent-based simulations. The combination of the two technologies adds richness of function and behavior that are impossible to realize within the confines of each individual technology.

REFERENCES

[1] Gimblett, H. R., B. Durnota, and R. M. Itami. "Spatially-Explicit Autonomous Agents for Modeling Recreation Use in Complex Wilderness Landscapes." *Complex. Intl. J.* **3** (1996).

[2] Gimblett, H. R., R. M. Itami, and D. Durnota. "Some Practical Issues in Designing and Calibrating Artificial Human Agents in GIS-Based Simulated Worlds." *Complex. Intl. J.* **3** (1996).

[3] Gimblett, H. R. "Simulating Recreation Behavior in Complex Wilderness Landscapes using Spatially-Explicit Autonomous Agents." Unpublished Ph.D. diss., University of Melbourne, Parkville, Victoria, 3052 Australia, 1997.

[4] Gimblett, H. R., and R. M. Itami. "Modeling the Spatial Dynamics and Social Interaction of Human Recreators using GIS and Intelligent Agents." Paper presented at MODSIM 97—International Congress on Modeling and Simulation, held December 8–11, 1997, in Hobart, Tasmania.

[5] Gimblett, H. R., Catherine A. Roberts, Terry C. Daniel, Michael Ratliff, Michael J. Meitner, Susan Cherry, Doug Stallman, Rian Bogle, Robert Allred, Dana Kilbourne, and Joanne Bieri. "An Intelligent Agent-Based Model for Simulating and Evaluating River Trip Scenerios Along the Colorado River in Grand Canyon National Park." This volume.

[6] Tomlin, C. Dana. *Geographic Information Systems and Cartographic Modeling.* Englewood Cliffs, NJ: Prentice Hall, 1990.

[7] Saaty, Thomas L. *Decision Making for Leaders: The Analytical Hierarchy Process for Decisions in a Complex World.* Pittsburgh, PA: RWS Publications, 1995.

Simulating Wildland Recreation Use and Conflicting Spatial Interactions using Rule-Driven Intelligent Agents

H. Randy Gimblett
Merton T. Richards
Robert M. Itami

1 INTRODUCTION

Ecosystem management, in the ideal sense, gives appropriate consideration to the complex and interdependent ecological and social systems that comprise forestlands. One prominent and growing arena where ecological and social systems interact is in the recreational use of wildlands. Recreational uses of forestlands are among an extensive array of commodities and amenities that are increasingly demanded of forest managers. An in-depth understanding of the relationships between recreational and other important uses is essential to effective ecosystem management [46]. Within the human dimension of ecosystem management, recreation and amenity uses of forestlands and the associated benefits of those uses, constitute an important component of management decisions.

2 A COMPREHENSIVE CONCEPTUAL MODEL

Forestland recreation is a special form of leisure behavior not only because it takes place outdoors, but because it depends upon a "natural" setting. Particular environmental settings are crucial to the fulfillment of forest recreation goals, because the recreationist seeks meaningful and satisfying experiences rather than simply engagement in activities. Importantly, wildland recreation takes place in settings that result from management actions of one form or another, whether the management objective is recreation opportunity, wildlife habitat improvement, or timber production, among others [46].

The recreation opportunity spectrum [7, 18] (ROS) provides a conceptual framework for relating opportunities for particular behaviors and experiences to specific settings. The ROS argues that recreator's pursuits of certain activities in specific settings reveals their demand for experiences that are satisfying and that may give long-term benefits. The ROS framework describes a spectrum of recreation opportunity classes that relate a range of recreation experiences to an array of possible settings and activities. Setting structure is composed of three components: an ecological component, a social component, and a managerial component [56]. The ecological component comprises the physical-biological conditions of the setting. These are typically delineated by the relative remoteness of the setting, its size, and evidence of human impact (number and condition of trails, structures, or roads, alteration of vegetation, etc.). The social component is typically defined by the number of users at one time (density) in the setting, delineated by the number of encounters or sightings a recreation party has with others. The managerial component connotes the degree and obviousness of regulation, restriction, and regimentation imposed on the setting.

The ROS is usually expressed as a functional relationship between recreational experience opportunities and a variety of environmental settings and recreational activities: experience opportunities depend on a particular combination of ecological, social, and managerial settings, and recreational activities. An opportunity class is a subjectively defined segment of a continuum of recreational opportunities. Changes in activities or settings will likely result in changes in a recreation experience, and often will result in changes in a recreation opportunity class. Further, changes in the experience may alter, or negate, beneficial outcomes. Thus, a recreator in pursuit of solitude or a spiritual connection with nature can be negatively affected by too many encounters with others, or by visually incongruous elements in a setting. Conversely, someone whose primary beneficial objective is improved physical fitness may find the same setting quite suitable. The manager, by altering the components of the setting, can have a significant influence on the recreational opportunities and beneficial outcomes of wildland visitors.

The ROS framework, is commonly used by public land managers to guide the planning and management of wildlands for recreational purposes. Even so, these managers rely on their own informed or expert judgment to provide the best experience opportunities for their visitors. While their judgments are informed by experience and an array of research findings, this information is historical and static, by nature.

The recreation assessment of forestlands in ecosystem management requires a more elaborate modeling structure that incorporates a dynamic interplay of time, space, and human response. A more appropriate concept calls for the interaction of four models: a model of desired recreation settings; a model that expresses the outcomes of recreation behavior in those settings; a model of recreation behavior that predicts the number of users per unit of time in those settings from which personal, social, and economic value estimates can be made; and a model that minimizes conflicts within and between recreation groups [47].

A considerable body of scholarly work has been completed that provides needed components of the models described above, but key linkages among and between the models have not formally been developed for practical use. In particular, the spatial orientation and temporal nature of encounters, conflicts, psychological states, experience opportunities, and associated benefits between and within groups of recreationists is still not well understood. Various authors [14, 20, 21, 31, 36, 37, 38, 44, 51, 57] have all focused on the nature and extent of conflict between members of specific recreation groups, none to date have examined conflict from a spatial and temporal perspective. Work by Hull et al. [33, 34], Lee and Driver [42], Stewart [53], Hammitt [30], Williams et al. [59], Manfredo [43], and Tinsley and Tinsley [55] have employed a variety of techniques including on-site surveys, recording devices, and an experience sampling method [11, 12, 13, 32, 41] to quantify immediate psychological states and desires to get at this issue of dynamic, multimodal experience. Recent work by Hull et al. [33, 34] and Hammitt [30] are among the few to have successfully used these methodologies to analyze a recreationists dynamic experience patterns on-site and found that they varied predictably over the course of an outing and were strongly influenced by site characteristics and site management.

An application of computer simulation technology, presented in this chapter, is intended to demonstrate the potential for developing a practical model readily useable by wildland recreation managers. The application will show:

- How decisionmakers, such as natural resource managers, would benefit from inexpensive, simulation techniques that could be utilized to explore dynamic recreation behavior—develop thresholds of use and test ideas or theories—before expensive management plans are implemented;
- How resource managers can have confidence in the use and results of these simulations since the design of the behavioral systems which are utilized in the simulations are grounded in observations of and data captured from actual human behavior in the physical settings in which they naturally occur;
- How simulation technology can be used to refine and enhance the use of ROS and other management strategies, as well as promote greater public understanding of management decisions; and
- Reveal additional research needs that will improve the utility of the technology.

3 AN APPLICATION

The focus in this chapter is to demonstrate simulation techniques using artificial intelligent agents to explore the complex interactions between recreationists and the environment, and interactions among recreationists as a means to improve the utility of recreation theory and concepts.

Broken Arrow Canyon near Sedona, Arizona was used to capture visitor use data and demonstrate the prototype software to simulate conflicts between recreation groups over time. The Canyon is popular for day hikers, mountain bikers, and people on commercial jeep tours because of the unique spectacular desert scenery of eroded red sandstone. The popularity of this canyon is a problem common to many popular wildland recreation destinations. People are "loving the place to death" by overuse. This overuse not only has negative impacts on the landscape but also in the quality of the experience people have when they visit. Crowding, conflicts between hikers, mountain bike enthusiasts, and jeep tours can create negative experiences in what should be a spectacular and memorable landscape setting, but very little is known about where, why, and how these impacts are occurring.

The U.S. Forest Service, who manages the resource as part of the Coconino National Forest, seeks guidance on what actions to take to protect the environment and provide the best possible recreation experience for an increasing diversity of visitors. While conventional survey techniques and public meetings have assisted in acquiring a better understanding of use, the spatial and temporal nature of the recreation experience still remain grossly misunderstood. There are no tools currently available for natural resource managers to study and quantify the complex spatial dynamic interactions and resulting impacts of recreational use over time.

The recreation behavior simulator (RBSim), formally described by Robert Itami in chapter 9, was developed to address these complex issues by using computer simulation technology. By simulating human behavior in the context of geographic space, it is possible to study the number and type of interactions, especially recreation encounters, that visitors will have within each group and between groups. Interactive modeling techniques are used to instill humanlike behavior into artificial agents to explore recreation planning alternatives. If resource managers are to have confidence in the use and results of agent-based simulations, it is crucial that the design of the behavioral systems of these agents is grounded in observations of actual human behavior in the physical settings in which they naturally occur. The behavior of RBSim agents is guided by a set of parameters whose values are set by the user. These behavioral parameters determine how an agent reacts when encountering other agents, at what speed an agent travels through a landscape derived from a geographic information systems (GIS) database, how often, and for how long an agent must rest, the recreational goals of the agent for a given landscape, the route the agent will follow through the landscape, and for how long the simulation will run. In effect, the user is able to create different behavioral patterns

and personality types for classes of agents based on social and demographic data and examine the efficacy of the agents in a rapid series of simulations. A system of this kind can provide a better understanding of recreation conflict and provide a mechanism to test and assess assumptions and theories of recreation behavior in a realistic temporal and spatial environment. In addition, the results of the simulations yield spatially explicit, social setting data, that could ultimately be used to improve the overall predictability and mapability of ROS and the beneficial outcomes of experiences.

4 DEFINING PERSONALITY TRAITS AND INTERACTION RULES FOR ARTIFICIAL AGENTS

To simulate an individual's behavior within an environment requires an understanding of their personal characteristics (which include personal recreational goals, expectations, length of visit, age, etc.), how they interact with other individuals they encounter, and how they react to the physical world where they are engage in their recreational activity. This practical application consists of three phases; first, the collection and analysis of recreational use data for providing artificial agents with personalities and rules that closely reflect actual recreator behavior; second, the development of a GIS-based agent simulator for mimicking recreator behavior over time; and finally, testing the agents in their simulated world under typical use conditions, and after imposing alternative management strategies.

An on-site visitor use survey was employed over a nine-month period to capture data on recreational use, desired beneficial outcomes, and conflicting recreational uses in the canyon. Trip motives, expectations, use density, reported contacts, and place of encounters have been identified as key factors in a satisfactory recreational experience [50]. A two-phased measurement was used to solicit response on the type of benefits that were desired (trip motives and expectations) during their visit and to what degree they were able to obtain them. The focus was on recreation as essentially goal-directed behavior [17]. Expectations have been acknowledged as extremely important to goal-oriented approaches to recreation behavior. This measure coincided with Jacob and Schreyers's [51] goal-interference definition of conflict. Visitors were asked if a range of benefits were desirable (*goals and intentions*) and whether they could obtain those benefits over time (*goal interference*). The benefit types used in this study are well documented in Bruns et al. [8] and Lee and Driver [42] based on research undertaken on other public lands. Achieving desired benefits such as getting away from crowding, reducing stress, and increasing physical fitness can be strong indicators of recreational satisfaction. Crowding has been shown to be one of the major predictors of user dissatisfaction. In a recent study by Behan [3], hikers and mountain bikers were queried in the same setting (Broken Arrow Canyon) as to their ability to achieve their desired benefits (using the same benefit types as used here) when increasing numbers of other bikers or jeeps were visible to them. Viewing digitally manipulated photo images of the setting that showed increasing numbers of either jeeps or mountain bikes visible in the landscape, respondents indicated the degree to which such conditions affected their ability to attain their most desired benefits. The results of the study show that ability to attain benefits became significantly more negative at each level of increased jeep or bike density.

The survey for this study was used to identify anything that either made the setting an ideal place for achieving, or interfered with acquiring, the desired benefits. So negative detractors and the inability to obtain desired benefits together are used to measure goal interference and conflicts, and imply an inability to obtain desired recreational experiences leading to unsatisfactory outcomes [27, 28].

To derive meaningful recreator profiles of the visitors to Broken Arrow Canyon, cluster analysis was first run on the recreation activity respondent data to isolate visitors by activity groups and then later used to aggregate visitors within each group based on desired benefits (goals and intentions). K-means cluster analysis allows one to specify the number of clusters desired or in the case of this research to explore the number of significant recreator types that could be found within each activity group. In addition, cross tabulation was used to calculate the frequency with which respondents within the classes identified the significance of each benefit type. Similarly they were asked to indicate their ability to obtain each of the benefits. This measure provides an indication of how often the respondents loaded on the benefit types by cluster and what particular benefits could not be obtained. This analysis was subsequently used to determine the statistically relevant number of agent types within each activity class for subsequently programming artificial intelligent agents with these behavior traits. Since it is possible to specify hundreds of agent personality profiles, for purposes of demonstrating the method, this procedure was used to aggregate agent classes into a reasonable number for final implementation.

Of the ($n = 1041$) visitors sampled, three significant recreation use groups were identified; day-use hikers ($n = 337$), mountain bikers ($n = 393$), and commercial jeep passengers ($n = 319$). For more detailed demographic data see Gimblett [27]. While there could be many combinations of personality traits derived from the visitor data collected, to demonstrate the utility of the agent modeling system the recreator patterns resulting from the cluster analysis were aggregated into two unique types for both the hikers and mountain bikers. These two types are referred to as either a "*landscape*" or "social" recreator type. Each desires significantly different benefits from their recreation experience. Due to the nature and mode of travel, commercial jeep passengers were modeled as a jeep unit.

Figures 1 and 2 illustrates the differences in the two recreator types. A *landscape recreator* or *agent type* is one that seeks out landscapes that are *physically challenging*, and *avoids crowds* subsequently leading to a *reduction*

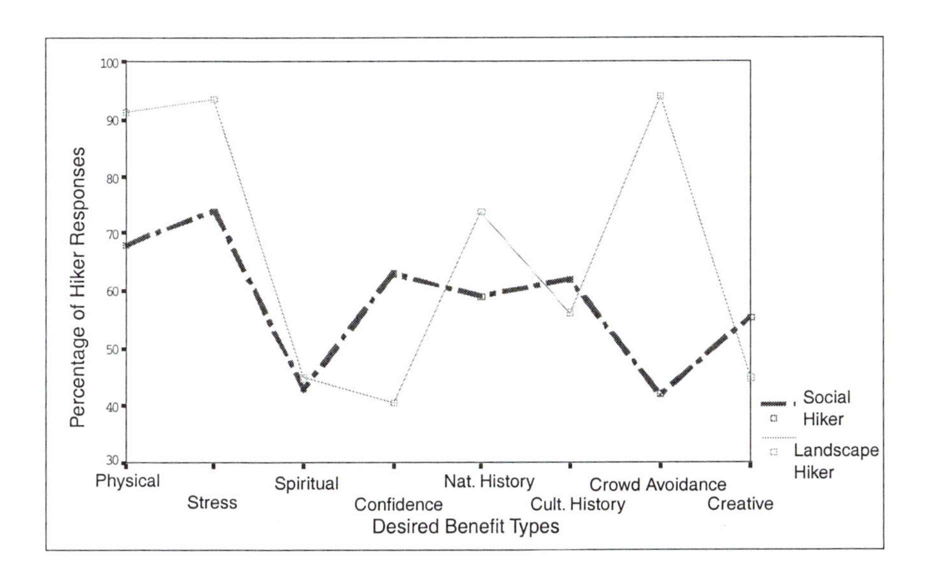

FIGURE 1 The frequency of responses of aggregated hikers to desired benefits.

in stress. This recreator type typically avoids others at all costs. In the exit interviews, recreationists that were representative of this agent class indicated that they would only stop in locations where there are no other recreationists and move as fast as possible along the trails. Physical exercise was a strong motivation in this recreation group and common to both hikers and mountain bikers. These recreator types fall within the personal well-being and health benefits class as identified in Bruns et al. [8].

A *social recreator* or *agent type* is more group oriented, one who seeks out those landscapes which *are not necessarily physically challenging* but tend to build *self-confidence*, and provide more *opportunity to learn more about the natural and cultural history* of the area and interact with others who share these goals. This is evident in figures 1 and 2 where their desires to obtain certain types of benefits are not as strong as those representative of landscape recreationists. Social agents did not mind encountering other social recreationists along the trail in the case of either hikers or mountain bikers. During the exit interviews, recreationists that represented this class indicated that they liked social interaction while engaging in their favorite recreational activity and would spend longer periods of time wandering through the landscape, sitting in special locations, and contemplating life. Tables 1 and 2 outline the aggregation of agent clusters and their associated age groups into agent classes for both hikers and mountain bikers. For more details on the statistical analysis see Gimblett [27].

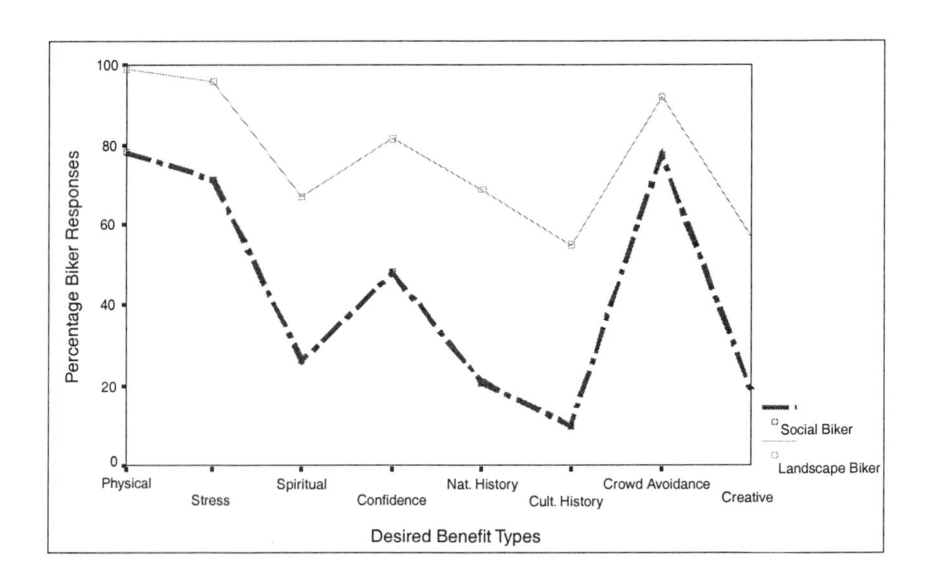

FIGURE 2 The frequency of responses of aggregated bikers to desired benefits.

TABLE 1 Aggregation of hiker agent types for simulations based on benefit preferences and age group.

	Hiker Social Agents Clusters 1, 2, & 3		Hiker Landscape Agents Clusters 4 & 5	
	Agents	Hikers Ratio	Agents	Hikers Ratio
Age Group 1–20	10	3%	13	3%
Age Group 21–40	101	26%	77	20%
Age Group 41–60	73	18%	64	17%
Age Group >60	24	6%	29	7%
Proportion of Agents		53%		47%

TABLE 2 Aggregation of biker agent types for simulations based on benefit preferences and age group.

	Biker Social Agents Clusters 1, 2, & 3		Biker Landscape Agents Clusters 4 & 5	
	Agents	Bikers Ratio	Agents	Bikers Ratio
Age Group 1–20	41	4%	60	6%
Age Group 21–40	393	39%	426	42%
Age Group 41–60	38	4%	46	4%
Age Group >60	4	.5%	1	.5%
Proportion of Agents		57.5%		52.5%

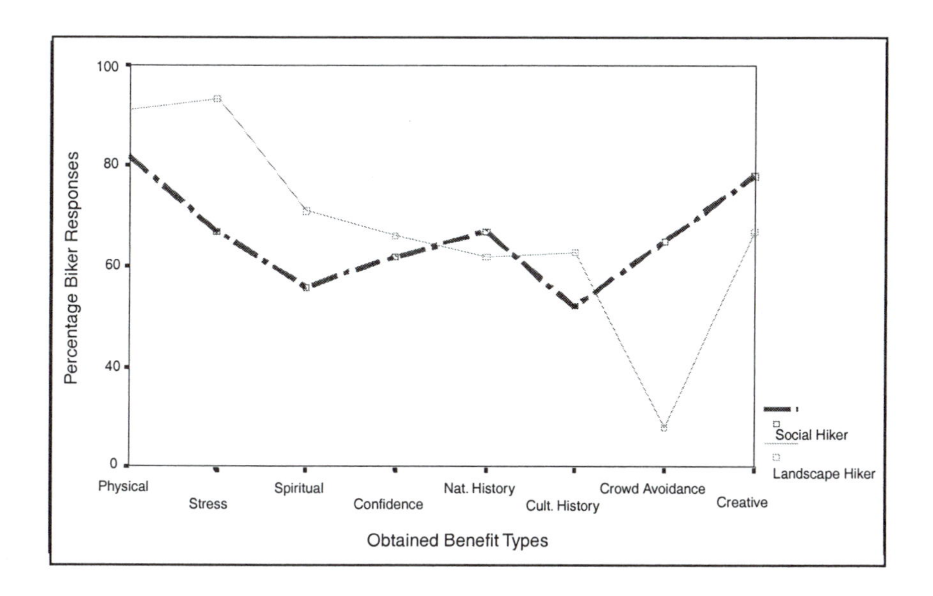

FIGURE 3 The frequency of responses of aggregated hikers ability to obtain benefits.

Figures 3 and 4 represent the same recreationist agent classes as presented in figures 1 and 2 but includes data for responses to the questions about their ability to obtain the type of benefits that they desired. Figure 3 illustrates there was a reasonable agreement by both groups of recreationists that they could satisfy their desired goals, except in the case of the landscape hikers who could not avoid crowds as opposed to the social agents who could or who felt it simply did not matter to them. Figure 4 reports on the mountain biker's ability to obtain their desired benefits. The landscape bikers reported an inability to obtain their desired physical benefits and strongly agreed that it was too crowded. Crowds could be the reason they were unable to achieve the physical benefit. The social bikers seemed to agree that most of their desired benefits could be obtained.

While these results are certainly not conclusive, they do provide a method for assessing the goals and intentions of the recreationists visiting Broken Arrow Canyon and also provide a measure of how well they were able to meet those goals. While none of the visitors indicated they were totally unsatisfied with their experience, many seemed frustrated with the numbers of encounters they had with other recreationists using the canyon.

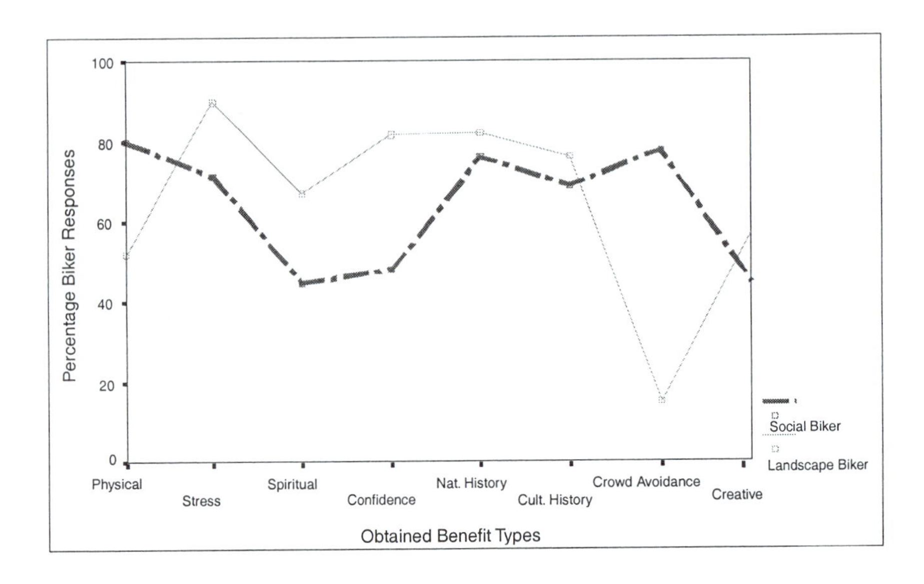

FIGURE 4 The frequency of responses of aggregated bikers ability to obtain benefits.

5 RULES FOR ARTIFICIAL AGENT BEHAVIOR

Rules for providing simulation agents with the social behaviors of human recreationists were derived from surveys of what respondents told us about their experiences, statistical analyses presented earlier, and through interviews following their outing. The respondents were all asked to explain the types of behaviors that they exhibited along the trails when encountering other recreationists. While the surveys clearly documented that visitors spent a minimum of two hours performing their activities, the benefit questions provided the goals and intentions for their visit. Survey maps were used to provide a clear indication of where they rested, their final destination, or where they stopped to view cultural and geologic features. Many of those recreationists that subsequently fell into the social agent class indicated that they stopped at all the locations regardless of the numbers of other hikers or bikers that were present and stayed primarily on the appropriate trail for their activity. Some of the mountain bikers and hikers who fell into the landscape agent classes indicated in both the survey and later in the exit interviews that they would stop at the cultural and geologic features only if there were a limited number of other recreationists present. They also indicated that they would go out of their way to pass others along the trails quickly and avoid them if possible.

For the testing of the prototype agent simulator, table 3 presents six rules which were developed that conformed to what was learned about the intensity

TABLE 3 Mobility rules for agents.

	Behavior rules for recreationist agents
Rule 1	Hiker and mountain biker agents rest when down to 25% energy level.
Rule 2	Landscape agents pass other agents in front traveling 25% slower than you if you have at least 50% energy left.
Rule 3	Hiker and mountain biker landscape agents slow down at landscape features if no other recreationists.
Rule 4	Hiker, mountain biker, and jeep agents stop at all landscape features.
Rule 5	Social hiker agents change their velocity to match other social hiker agents they encounter.
Rule 6	Hiker and mountain biker agents will not stop at features if more than five other agents are present.

of use and interactions of both social and landscape types of recreationists using Broken Arrow Canyon. To accommodate the solitude-seeking and crowd-avoidance desired benefits of the landscape recreationist, a rule was developed that prohibits a landscape agent from stopping if there are more than five other agents present at the cultural or geologic features. Landscape agents are programmed to avoid crowds at all costs. They will speed up, if they have 50% energy remaining, to pass other agents on the trails if they are within fifty meters and traveling slower. This rule conforms to what some of the hikers and mountain bikers told us about their trail experiences and adds to the physical challenge that they sought.

To accommodate the needs of the social hiker and biker visitors, the corresponding agents are programmed to hike or ride to areas in the simulated landscape to learn more about natural features and to socialize with other agents. The agents in these classes generally spend at least two hours performing that activity. They have lower desires for extremely challenging physical fitness, but will seek out areas where they can spend time, such as at the cultural or geologic sites. If a social agent encounters a small group or perceives the ability to catch up to another social agent, they will increase their speed to do so. They will remain with them throughout the duration of the simulation, unless they expend too much energy and are forced to slow down and rest. These rules conform to what the hikers and mountain bikers reported about the type of behaviors they exhibited on the trails. Social agents will stop at all cultural or historic features no matter how many other recreationists are present.

Since there are four different recreationist age groups being represented in the simulations, they all will move at different rates along the trail, some will run out of energy sooner (older ones) and will be forced to rest. The behavior rules for these agents are summarized in table 4.

TABLE 4 Rules that modify agent behavior.

	Behavior Rules	
	Landscape Agents	Social Agents
Hiker Agents	Rules 1, 2, 3, 6	Rules 1, 4, 5
Mountain Biker Agents	Rules 1, 2, 3, 6	Rules 1, 4, 5
Jeep Agents*	Rule 4	Rule 4

*The jeeps are assumed to move continuously throughout the landscape and stop at all features for interpretation.

The behavior of individuals who are involved in the jeep tours are not as robust as those hiking and mountain biking. Jeeps contain between four and six visitors on each trip. These visitors spend from two to three hours on the jeeps interacting with the driver and other tourists. Jeep agents are modeled as a group of passengers in one jeep which vary their jeep speed according to the topographic conditions. They speed up or slow down according to the degree of slope. Jeep agents conform to only one action rule that defines their behavior and that is rule 4 (table 3) to stop at all cultural or geologic features. The time they spend at these features is predetermined and conforms to what the jeep tour drivers typically spend at each location.

6 SIMULATING CONFLICTS

Solitude seeking is an important reported desire, goal, or expectation in this study. The degree of interference with that goal is related to the number of encounters one has with other recreationists. Table 5 illustrates that perceived negative encounters with certain types of recreationists have some impact on the experience. The degree of that impact is not yet known. However, from the comments on the survey such as "seeing too many people," "too many people," "too many jeep tours on trails," and "seeing jeeps along the trails," suggests that both visual and physical encounters are important measures of the degree to which a goal or desired benefit is interfered with. Based on the number of negative encounters that were reported, and on the fact that both the hikers and mountain bikers reported that they were not able to obtain a crowd-avoidance benefit, visual and physical encounters within and between agent types are used as a measure of goal interference or the inability to achieve a desired or perceived benefit. The hypothesis is that the higher the number of encounters, the less the ability of the agent to obtain some of the other desired benefits and the less the perceived quality of the overall recreational experience.

From a management perspective what is needed is to identify the spatial locations along the trails where there are significant visual encounters. To accomplish this task, each agent keeps track of the number of encounters it

TABLE 5 Recreation conflicts (negative encounters) between and within recreationist groups.

	NEGATIVE ENCOUNTERS						
	Other Hikers/Hikers		Other Bikers		Other Jeeps		Frequency of Responses*
	No. of Hikers	Freq.	No. of Bikers	Freq.	No. of Jeeps	Freq.	
Hikers	70	**20%**	75	**30%**	101	**41%**	246/338 (**72%**)
Bikers	80	**58%**	6	4%	53	**38%**	139/393 (**35%**)
Total Conflicts	150		81		154		

*The number of times respondents identified negative detractors (conflicts) per total number of respondents in that group.

TABLE 6 Frequency of visits to the canyon on a weekly basis and time of day.

	VISITOR FREQUENCY								
	Days of the Week							Time Period	
	Mon.	Tues.	Wed.	Thur.	Fri.	Sat.	Sun.	AM	PM
Hiker	4%	7%	5%	11%	15%	**40%**	18%	55%	45%
Mountain Bikers	6%	7%	4%	9%	16%	**41%**	17%	44%	56%
Jeeps	**23%**	24%	9%	4%	7%	14%	19%	40%	60%

has in each cell along the trail and also stores the type and number of visual encounters it has with other agents on other trails. These encounters are summarized, graphed, and mapped to examine areas where there are levels of encounters that interfere with the recreator or agent's goal to obtain a desired benefit.

7 SYNTHETIC GIS WORLD AND INHERENT SPATIAL ASSESSMENT CAPABILITIES

The synthetic world that the simulated recreationists utilize is a georeferenced, raster database consisting of 513 rows × 522 columns, each cell represents 10 square meters. The database consists of topography, vegetation, adjacent primary and secondary roads, existing and proposed trails, trailhead, jeep staging area, significant geologic features, and scenic stops. These geographic themes were deemed important for this work, but many more could be incorporated as the sophistication of the modeling increases.

The approach taken in this research was to provide each agent with spatial analytical capabilities that is imperative to them processing information necessary for functioning in the simulated worlds. Each agent is provided with the ability to calculate distance or proximity to other agents and significant features in the landscape. Each calculates the percentage of slope from the topographic map and whether it is going up or down hill and, in turn, speeds up or slows down accordingly. They utilize neighborhood functions to identify trail cells or the location of significant geologic features and scenic stops. Most importantly each agent has visual capabilities for detecting other agents, how far away they are and whether they can or cannot be seen. This algorithm uses forest cover and topography as constraints to detecting other agents.

8 SIMULATING TYPICAL USE DAYS AND MANAGEMENT ALTERNATIVES IN THE CANYON

In order to test some of the ideas and concepts presented in this research and to determine the efficiency of the simulation system in identifying conflicting recreation behavior, a number of experiments were constructed. During the interview and survey phase of this research, visitors to Broken Arrow Canyon were asked (in addition to the information already discussed) to record the month, day, and time that they entered the canyon.

Table 6 presents a statistical summary of the visits of those recreationists sampled over the duration of the study. As can be seen, peak times throughout the canyon are weekends. Over 40% of hikers and bikers frequent the canyon during these time periods, about equally distributed between morning and afternoon periods. Jeep tourists typically visit the canyon Saturday through Tuesday, with 40% of visits in the morning, and 60% in the afternoon.

In order to test RBSim, many simulation runs were undertaken mimicking various peak and off-use times to examine the dynamic interactions of recreationists and the resulting visual and physical conflicts. This study reports on two of those experiments, a midweek day (Wednesday) which typically is not a peak use period but contains a moderate number of hikers and bikers and relatively low jeep usage, and a weekend day (Sunday), an especially busy day for bikers and commercial jeep users.

In order to demonstrate a potential management action such as restricting biking use on a heavily used trail, two alternative bike trails and an alternative jeep trail were substituted for the originals and the simulations rerun to evaluate the differences in recreational use and resulting perceived conflicts from all recreationists' perspectives (see fig. 5).

9 INITIAL RESULTS OF SIMULATION RUNS

Table 7 is an example of mixed recreational use along the trails on a typical weekday. Figure 6 illustrates the intensity of hiker encounters with other agents from the hiking trails and illustrates a significant number of encounters with both other hikers and bikers. The number of encounters with bikers is high from the beginning of the simulation and peaks at Chicken Point and is chaotic until completing the journey.

Encounters with hikers, on the other hand, peak at Chicken Point and then remain consistently high thereafter. What is of interest is that where

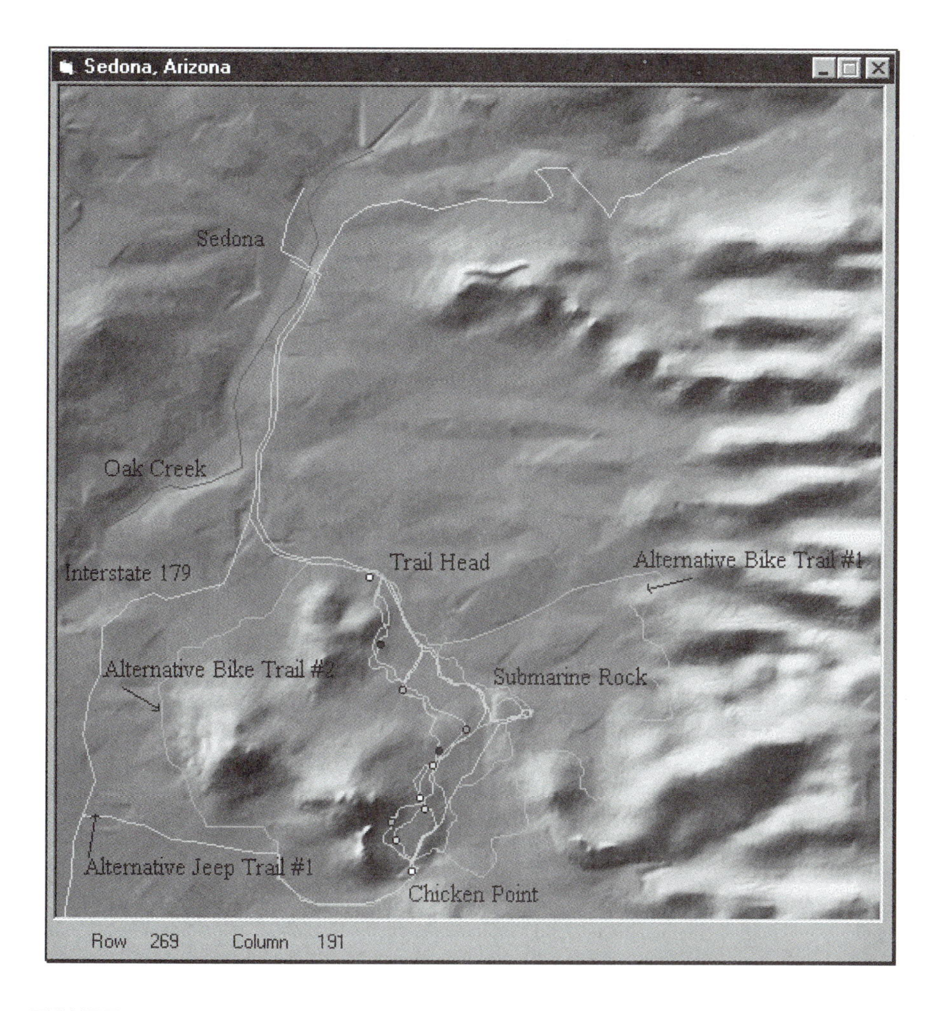

FIGURE 5 Original and alternative trail layouts for Broken Arrow study area.

the encounters with hikers peak, biker encounters drop off, and visa versa. Encounters with jeeps remain relatively low.

As summarized in table 5, over 40% of the negative encounters that occur to hikers are with jeeps and 30% with other hikers or bikers. It is interesting that even with the number of hikers, bikers, and jeeps included in this simulation, there are very few encounters with jeeps. On the other hand, the high amount of conflicts with bikers and other hikers may have a detrimental effect on the recreation experience. But because the place, time, and duration of encounters that occur between biker and hiker agents are not consistent, this may reduce the cumulative impact of the encounters on those hiking.

TABLE 7 A typical midweek entrance times by recreationists into Broken Arrow Canyon.

Wed.	Biker Data (AM)				Hiker Data (AM)				Jeep Data (AM)			
10/11	1–20	21–40	41–60	>60	1–20	21–40	41–60	>60	1–20	21–40	41–60	>60
9:00									1	1	0	1
9:00									2	0	0	1
9:00									0	2	0	1
9:00									2	0	0	1
9:00									1	2	0	1
10:30	1	2	0	0								
10:30	1	2	0	0								
11:30	0	1	0	0								
12:00	2	2	0	0								
12:00	1	2	0	0								
12:00	3	2	0	0								
1:30	0	2	0	0	0	0	6	7				
Percent	38%	62%	0%	0%	0%	0%	46%	54%				

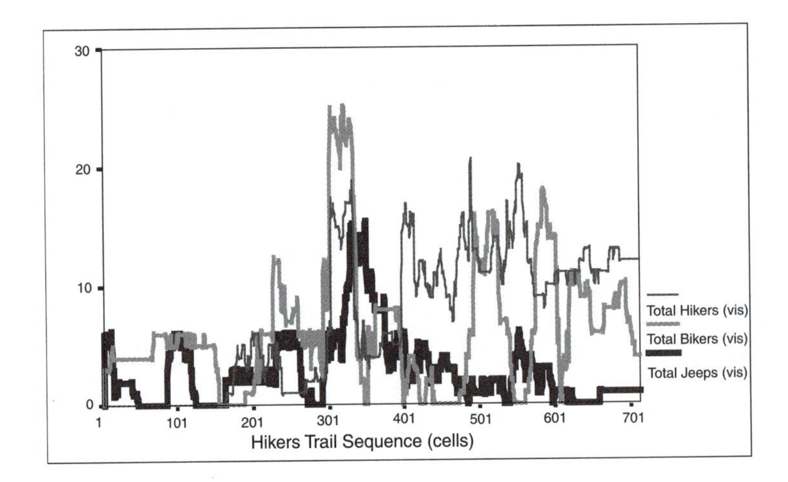

FIGURE 6 Graphed results of hiker encounters with other agents from along hiking trail.

Figure 7 illustrates biker encounters with hikers, jeeps, and other bikers from along the biking trail. The patterns are similar to those found in figure 6 except bike encounters steadily increase throughout the life of the simulation, but are not as high as in figure 6. Biker's encounters with hikers are more sporadic than is outlined in figure 6 dropping off at the end of the simulation. As with hikers, there are virtually no encounters with jeeps.

Figure 8 illustrates a high number of jeep encounters with hikers and bikers and only a minimal number of encounters with other jeeps. The encounters occurring with hikers and bikers are concentrated around Chicken Point and

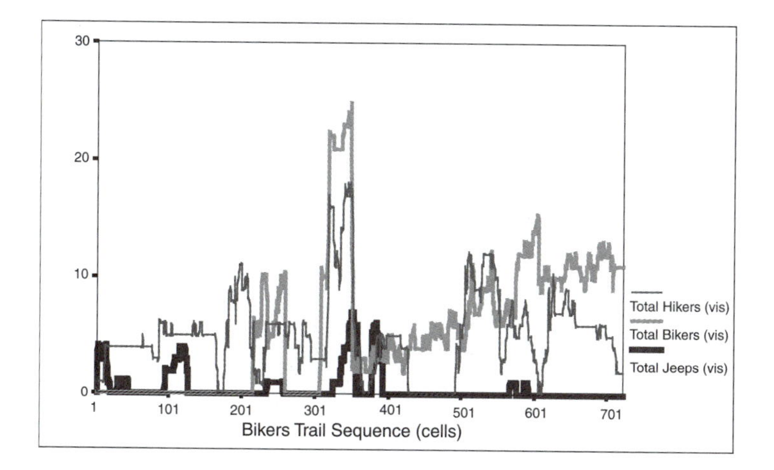

FIGURE 7 Graphed results of biker encounters with other agents from along hiking trail.

TABLE 8 Recreationists start times for experiment 2.

Sun.	Biker Data (AM)				Hiker Data (AM)				Jeep Data (AM)			
10/8	1–20	21–40	41–60	>60	1–20	21–40	41–60	>60	1–20	21–40	41–60	>60
12:30	1	1	0	0	0	1	1	0				
					0	0	2	0				
					0	0	2	0				
1:00	0	3	1	0	0	2	0	0	1	0	0	1
					0	1	0	0				
					0	2	0	0				
1:30					0	1	1	0				
2:00	1	0	1	0								
2:00	0	0	0	0								
3:00	0	2	0	0								
Percent	18%	64%	18%	0%	0%	54%	48%	0%				

Submarine Rock during the last half of the trip with virtually no encounters occurring for the first and last quarter of the simulation.

In the second experiment, the data shown in table 8 were used in the simulations. Although there were no reported hikers on the trail during the times the survey crew were collecting these data, the canyon is frequented excessively by mountain bikers and jeep tours. Figure 9 reveals the encounters along the bike trail between bikers and jeeps as well as other bikers. It is evident from both of these figures that from the bike trails, jeep encounters were not continuous and that jeeps could only be seen from specific points along the trail. However, what is more dramatic is the number of encounters that bikers have with other bikers.

It is evident from the simulation results that bikers tend to encounter more bikers at rest or scenic views along the way. This is evident around

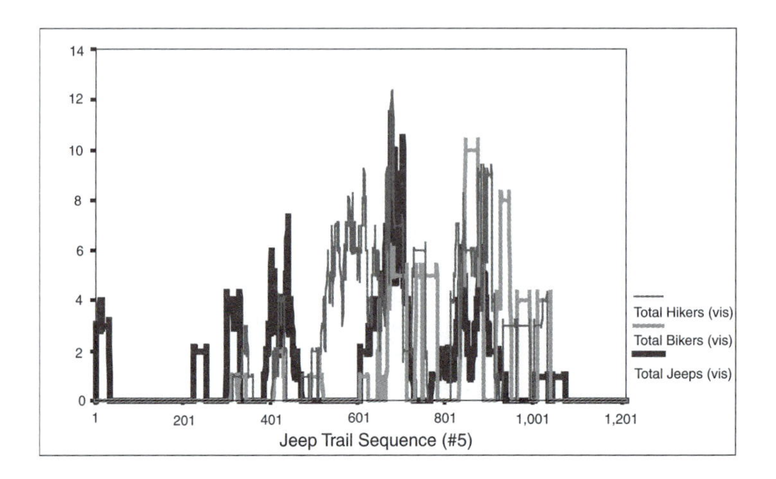

FIGURE 8 Graphed results of jeep encounters with other agents on along jeep trail.

Chicken Point lookout. From there on until the completion of the trip, there are significantly more encounters along the way. This would be a concern if higher numbers of negative encounters between bikers and other bikers was a problem, but since this not the case, the numbers of biker encounters is acceptable.

Figure 10 illustrates the number of jeep encounters with bikers and other jeeps. It is clear that encounters with bikers is random and is relatively low. There is only one particular point along the trail were congestion reaches a significant level and that is at Chicken Point, the farthest location that jeeps will travel to. Like jeep encounters, this is expected since all trails converge at this location and cause congestion. Jeep encounters with other jeeps are virtually nonexistent due to self-scheduling by jeep tour operators. The occasional encounter with other jeeps will have minimal effect on experience outcomes.

In summary it appears that with the increased number of hikers and bikers in the canyon that encounters with bikers are the most dominant impact. There are very few encounters with jeeps throughout all the simulations.

10 RESULTS OF SIMULATIONS USING ALTERNATIVE TRAIL LAYOUTS

One of the reasons for developing the simulator was to provide the land manager with a tool for assessing existing and proposed recreation trail layouts in terms of movement and distribution of recreationists along these trails and the resulting conflicts. Ultimately this tool has been developed to assist the manager to dynamically manage recreational use in the canyon over time. In

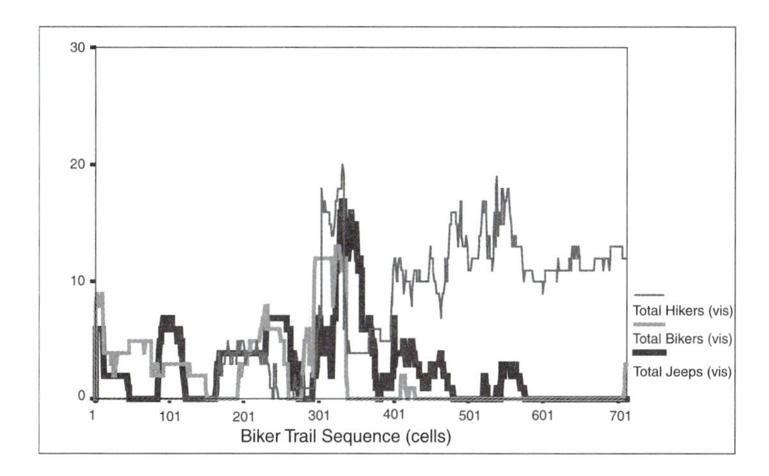

FIGURE 9 Graphed results of biker encounters with other agents from experiment 1 along biking trail.

order to demonstrate this concept and assess the effect of alternative trail use and conflicts within and between recreation groups, two alternative bike trails and one alternative jeep trail were used to demonstrate RBSim. The two alternative trail layouts were extracted from the mountain bike surveys. Those mountain bikers who were sampled tended not to use the conventional trails in the canyon, but outlined on the map in the survey where they preferred to ride.

The reason these alternative routes were popular amongst the bikers was that they were physically challenging, secluded, and provided extraordinary scenic views. While there were only a limited number that took the time to make these suggestions, it was thought important for demonstrating the use of the simulator to attempt to assess these trails to determine the number of visual encounters that would occur under the same conditions used in the initial experiments.

The alternative jeep trail came out of a proposal from the commercial jeep tour operator. This alternative was proposed as an overall plan to reduce conflicts in the canyon and increase numbers of jeeps that would be permitted.

11 SIMULATIONS USING ALTERNATIVE BIKE TRAIL #1

As illustrated in figure 11, selecting alternative bike trails can have a major impact on the number of encounters that occur along the trails. It can be seen that when alternative bike route 1 is used in the simulations, the number of biker encounters that the hikers will have decreases significantly to the point that it is negligible after Chicken Point. When compared to figure 6

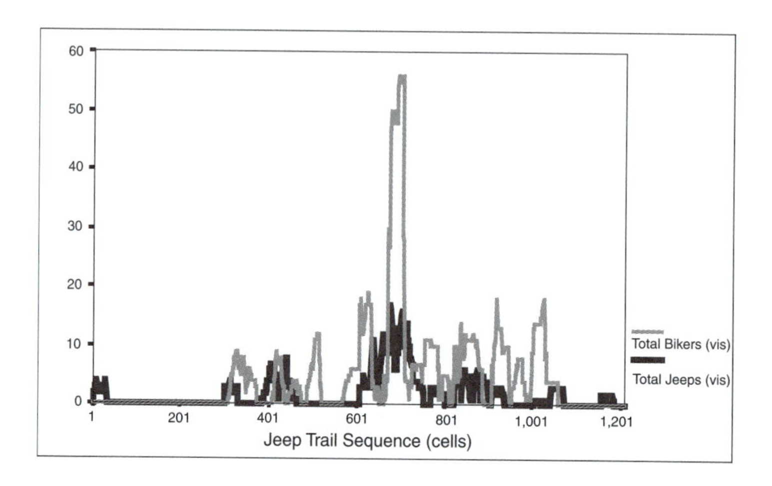

FIGURE 10 Graphed results of jeep encounters with other agents from experiment 1 along jeep trail.

and summarized in table 9, by altering the trail layout the mean number of encounters has dropped by two thirds and the maximum number of encounters by half.

In figure 12 the number of visual encounters with other recreationists that bikers will have when using the alternative bike route reveals a dramatic decline in both hikers and jeeps, but a steady increase in number of bikers. In fact, an evaluation of table 9 illustrates how the visual encounters with hikers declines to one fifth of those that occurred in figure 7, with the same number of hikers still using the trails. This strongly suggests that by using the alternative trail, the distribution of hikers and bikers within the canyon is more conducive to minimizing conflicts.

Figure 13 illustrates a significant number of encounters with jeeps from both bikers and hikers in the canyon. As in figures 7 and 8, encounters with other agents declined. Of significance are the encounters with hikers and jeeps. But interestingly enough, increasing the number of bikers from 11 to 27 has little effect on the mean number of encounters that occur, but does affect the maximum. In other words, while the number of encounters remains the same, the encounters are more evenly dispersed along the trail, rather than peaking at specific locations. From a management perspective, if the objective is to disperse the impacts of encounters over time and reduce high impact areas of conflict, then this alternative bike route would offer a solution to this problem.

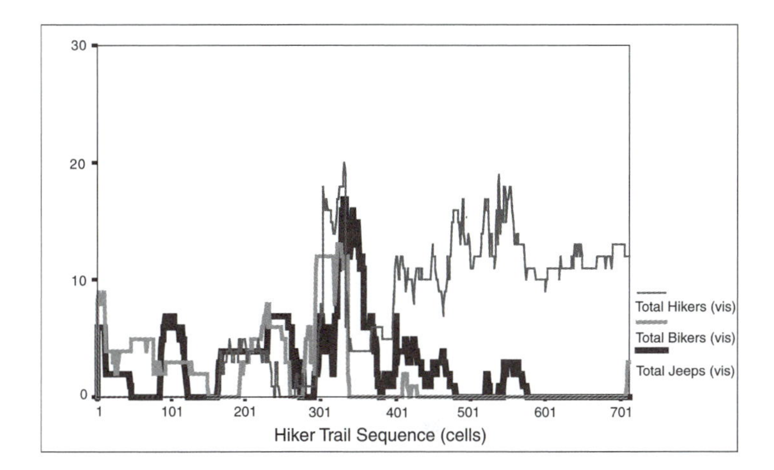

FIGURE 11 Graphed results of hiker versus other agent visual encounters using biking alternative trail #1.

TABLE 9 Comparison between existing and alternative bike routes for experiments 4, 7, and 9.

	Exp. 4		Exp. 7 (4a)		Exp. 9 (4b)	
	Mean	Max.	Mean	Max.	Mean	Max.
			HVIS			
Hike	7.0	20	6.9	20	6.9	22
Bike	6.4	25	2.1	13	2.6	10
Jeep	2.9	15	2.7	17	2.7	18
			BVIS			
Hike	5.1	18	1.4	17	1.9	18
Bike	5.5	24	4.5	12	2.0	12
Jeep	.51	7	.41	8	.67	7
			JVIS			
Hike	2.1	12	2.1	7	2.3	15
Bike	1.2	10	.4	18	.33	5
Jeep	1.3	11	5.1	19	4.5	18

12 SIMULATIONS USING ALTERNATIVE BIKE TRAIL # 2

It can be seen in figure 14 that when alternative bike route 2 is used in the simulations, the number of biker encounters that the hikers will have decreases significantly to the point that it is again negligible after Chicken Point. When compared to figure 6 and summarized in table 9, by altering the trail layout the mean number of encounters has dropped by two thirds and the maximum number of encounters by more than half.

Figure 14 illustrates that while hiker encounters with other hikers and jeep encounters are not significantly affected by this alternative bike trail,

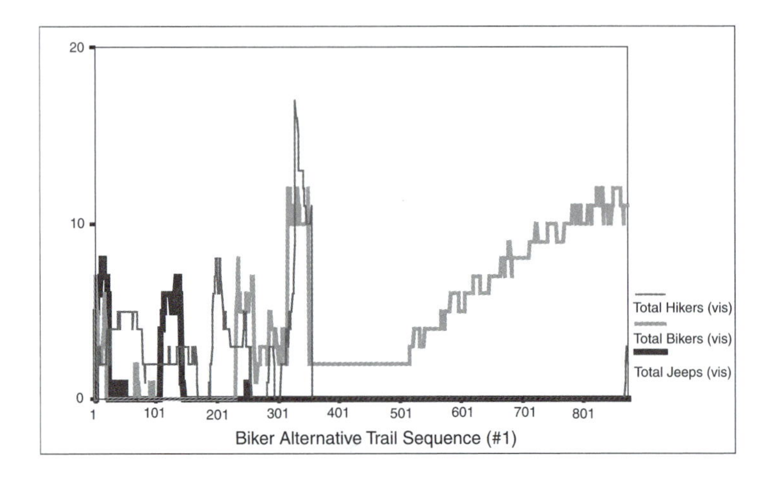

FIGURE 12 Graphed results of biker versus other agent visual encounters using biking trail alternative #1.

encounters with bikers are severely reduced. Reducing biker encounters would, at least from a hiker's perspective, reduce 30 percent of the conflict. Figure 15 strongly illustrates the effect of alternative bike route #2 on other agents. Encounters with all other agents are reduced by approximately two thirds as was the situation in earlier analyses.

However, what is of interest is, when comparing figure 6 using the original bike route with figure 15, encounters are extremely low and there are places in the landscape where there are no encounters at all. Between trail sections 401 and 801 there are no other encounters. Encounters with other biker agents tend to steadily increase over the remainder of the trail. It is only near the end, where encounters with other agent types occur again. Since biker encounters with other bikers has no significant negative impact on each other, this solution offers distinct advantages.

13 SIMULATIONS USING ALTERNATIVE JEEP TRAIL

Figure 16 illustrates the number of encounters bikers have with other bikers and jeeps using an alternative jeep route. Table 10 illustrates a decline in both jeep and other biker encounters when compared to the number of encounters on the original trails. When compared to figure 9, figure 16 shows a decline in total jeep encounters over time and includes locations where there are no jeep encounters at all.

Figure 17 illustrates the number of encounters that jeeps have with other agents using the alternative jeep route. When compared to figure 10, using the alternative trail tends to increase the number of jeeps while reducing the

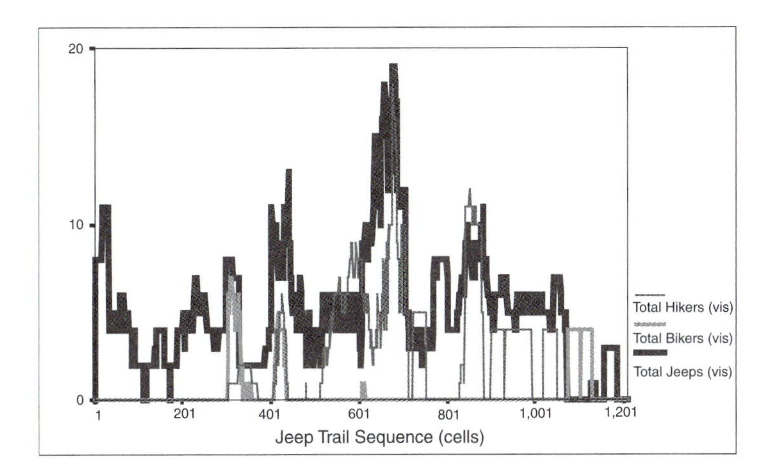

FIGURE 13 Graphed results of jeep versus other agent visual encounters using biking trail alternative #1 for experiment 7.

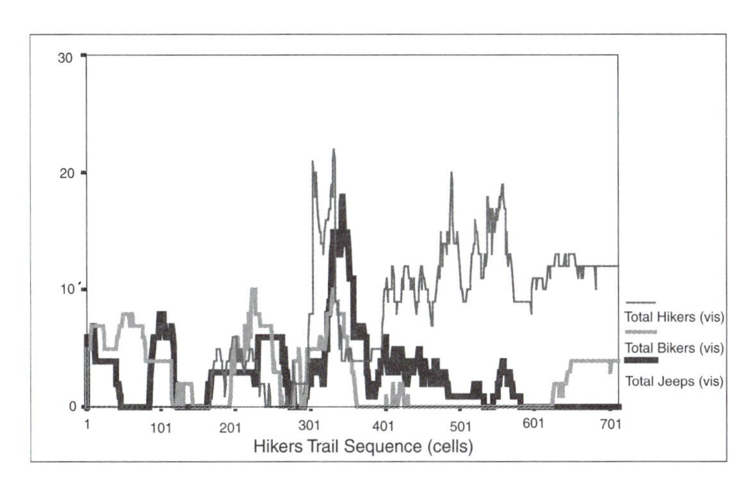

FIGURE 14 Graphed results of hiker versus other agent visual encounters using biking alternative trail #2.

TABLE 10 Comparison between existing and alternative jeep routes.

	Exp. 1		Exp. 11	
	Mean	Max.	Mean	Max.
		BVIS		
Hike				
Bike	43.1	120	30.9	108
Jeep	5.9	34	2.2	19
		JVIS		
Hike				
Bike	4.8	56	1.5	34
Jeep	1.8	17	2.7	24

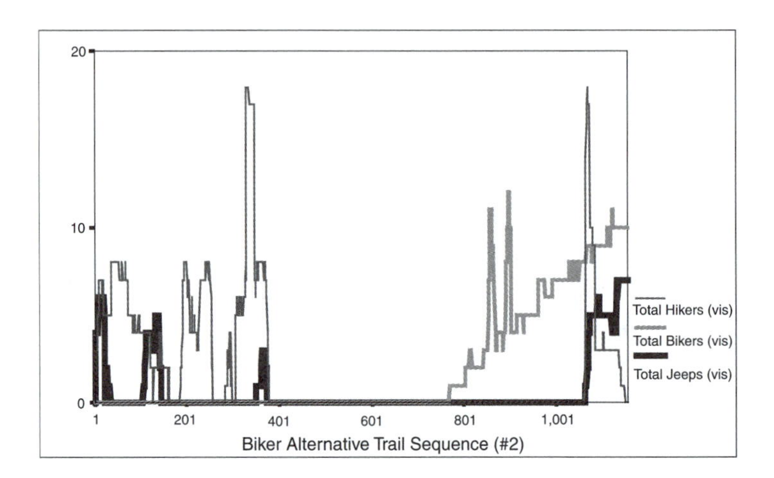

FIGURE 15 Graphed results of biker versus other agent visual encounters using biking trail alternative #2.

number of bike encounters. This is expected since the trail leads outside of the canyon where there are no other trails within view. Since there were no documented negative detractors from jeep recreationists in regards to other jeeps, these types of conflicts can be interpreted to have no effect.

14 SUMMARY OF FINDINGS

There are three main points that summarize the findings of testing alternative trail layouts:

1. The number of encounters significantly decreases over time as a result of using alternative bike and jeep trails;
2. One way traffic through the canyon seems to be a viable alternative for reducing the number of encounters and conflicts; and
3. While conflicts with other agent types declines when testing alternative trail layouts, the experiments reveal that significant encounters do occur within recreationist groups, but according to the results of the survey tend not to be perceived as having a negative impact on the experience.

15 DISCUSSION

The results of the survey indicate that the conflicts most often reported are from bikers having negative encounters with hikers. While jeeps are certainly

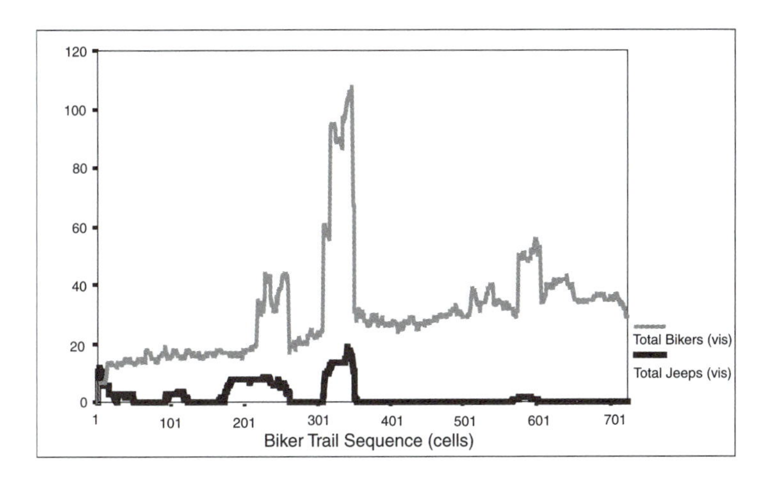

FIGURE 16 Graphed results of biker versus other agent visual encounters using jeep trail alternative #1 for experiment 11.

considered to have a fairly high level of impact on both hikers and bikers, they were not as strong a determinant of a negative recreational experience as anticipated. Bikers and hikers continually clash in the canyons. What is of interest however is how often and where these encounters occur. An examination of the results of the agent simulation runs illustrates that bikers most frequently encounter other bikers. While bikers may have more encounters with other bikers, they do not see this as detracting from their experiences.

The agent simulations seem to be an excellent method for modeling recreationist encounters and ultimately conflicts. While statistical results of the survey used in this study provide an indication of the average number of encounters (viewed as negative detractors), the agent simulations provide a dynamic view of these encounters and identify the spatially explicit locations where they occur. The simulation environment also provides the added benefit of collecting and storing data on encounters over time. Both these data can be evaluated using conventional statistical techniques and compared to explain commonalties and differences.

Of interest in this research, and showing the power of using simulation, is the impact of alternative routes on recreationist encounters. An examination of the biker trail alternatives, with routes as suggested by the respondents to the survey, illustrate the importance of a well-thought-out trail design on recreational encounters. As can be seen in this research both alternative trail designs significantly reduce the number of encounters with other recreationists. In fact, from before the turnaround point to the completion of a biker's journey, they literally have no visual contact with any other recreationist type. If hikers do have an accumulated negative effect on a bikers experience, then

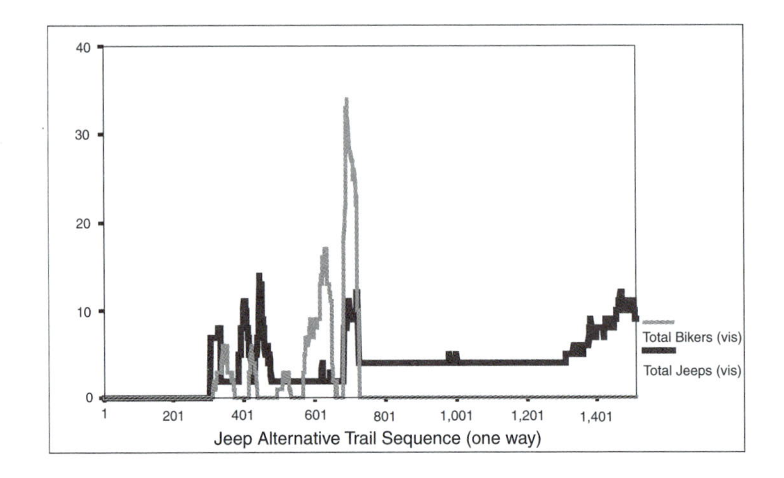

FIGURE 17 Graphed results of jeep versus other agent visual encounters using jeep trail alternative #1 for experiment 11.

it is clear that these alternative routes would alleviate this problem. This situation is identical when assessing encounters using the alternative jeep trail layout. Biker and hiker encounters with jeeps virtually disappear for over 50 percent of the journey. This is a substantial decline in encounters considering the significant number reported by the respondents to the survey. It is clear that the simulation environment can assist in evaluating existing and proposed trails in an attempt to minimize encounters and conflicts which may ultimately lead to a decline in the quality of the recreational experience.

Simulation using personality traits and behavioral rules synthesized from human recreationists provide a way to evaluate and test a variety of recreationist use densities over time. These alternatives can be used to develop new facilities along the trails, and to redirect trail use to maximize user satisfaction while minimizing impact. Being capable of seeing the agents interacting under a variety of constraints can assist the manager in acquiring a better understanding of how human recreationists use and interact on public lands.

Simulation has a number of direct applications to the use of ROS as a management and planning tool. Assessing the number of encounters and where they occur can help to establish limits, or standards, for a particular ROS class, beyond which conditions should not be allowed to change. It is especially advantageous to be able to determine visual encounters, since, in addition to a quantitative assessment, a qualitative evaluation can be made as well. The effects of seeing other people, the activities they are engaged in, or the physical conditions of a setting, including humanmade structures can create inconsistencies and incongruities for the viewer, which, in turn, affects the setting. This can create inadvertent inconsistencies, where, for example, the physical and managerial components of a setting clearly place it in the primitive class, but with increasing numbers of human encounters the social component defines the setting as urban [9]. Simulation can make clear the potential for such inconsistencies before they occur. RBSim is ideally suited as a ROS inventory technique as well. Agent parameters can be set so that agents roam a GIS landscape in search of particular conditions that define ROS classes. The results are mapable ROS data.

This application has made linkages between GIS, multiagent systems, and recreation behavior modeling. While not dealing directly with goal interference theory, it does use it as a foundation for behavior modeling. It is assumed that perceived benefits, and obtaining or maintaining those benefits, directly correlates with goal interference. Encounters, physical or visual, are the main cause of goal interference. The landscape recreationist agent developed in this work was programmed specifically to avoid interference and when threatened passed other agents or avoided stopping at scenic lookouts. This technique allows one to reduce the amount of goal interference, while maximizing benefits. More than that, it allows one to experiment with artificial recreationists to realistically determine thresholds of goal interference and devise management strategies to reduce it. This is one of the advantages of using simulation and the power of such multiagent environments.

16 CONCLUSIONS

This application advances our knowledge and understanding of natural resource assessment using intelligent simulation systems in the following ways:

- Extends the theoretical foundation of recreation and behavior by exploring the concept of benefits-based management for measuring desirable and obtainable benefits of leisure and assessing spatially explicit visual and physical encounters among recreationists in Broken Arrow Canyon.
- Extends the knowledge base of the development, calibration, and use of intentional, multiautonomous agent systems in GIS-represented worlds.
- Demonstrates an entirely new form of intelligent decision support system (IDSS) to assist natural resource managers in assessing and managing human use of natural areas which could be easily extended into a number of other areas such as assessing impacts on wildlife habitats.
- Demonstrates the capabilities of visual operators found in GIS for providing all mobile agents with visual capabilities. This vision system is used for controlling agent movement, goal-seeking, determining locations and distances of potentially conflicting agents, and could be easily modified for identifying significant landscape features.
- Utilizes conventional social science survey techniques with automated field methodologies for validations of agent movement and behavior.

- Demonstrates a user-friendly, parametrized interface for experimenting with alternative trail layouts and a diversity of agent configurations under a variety of conditions.

Much work can be undertaken to improve the predictability and reliability of the modeling framework. To expand our understanding of the dynamic physiological and psychological experience patterns, sampling methods could be used. Dynamic experience patterns can be empirically measured including such factors as visual acuity, focus of attention, mood, psychological benefits, coping strategies, norms of behavior, and physiological changes at strategic locations within a stratified set of landscape settings found throughout the study site. A methodology employing these techniques that provided a way for the visitor to stop, record, and photograph landscapes of importance would provide valuable information and lead to improved understanding of the dynamics of recreation experience. It is important, however, to ensure that wherever and whenever the visitor records such information, the explicit location is captured as well so as to be able to link these changes to physiographic settings.

To improve the modeling of social interactions in a physical environment it is imperative that a more thorough understanding be acquired on how humans translate information from the environment into meaningful actions. Humanlike agent simulations are no different. Once the spatial information is communicated to an artificial agent, it must then be translated from its objective form into the symbolic and cognitive framework from which affective human responses are derived. This area of research needs considerable attention, but will provide meaningful outcomes.

ACKNOWLEDGMENTS

We wish to thank the U.S.D.A. Forest Service, Rocky Mountain Forest and Range Experiment Station, and the Coconino National Forest for their assistance in facilitating this research effort. We also wish to thank Dr. B. L. Driver of the Rocky Mountain Station (now retired) for his helpful review and oversight of this project. This research was supported in part by funds provided by the Rocky Mountain Forest and Range Experiment Station, U.S.D.A. Forest Service.

Note: Instructions for downloading a free copy of RBSim can be obtained from the following website ⟨http://nexus.srnr.arizona.edu/~gimblett/rbsim.html⟩.

REFERENCES

[1] Anderson, J., and M. Evans. "Intelligent Agent Modeling for Natural Resource Management." *Math. & Comp. Model.* **20(8)** (1994): 109–119.

[2] Ball, G. L. "Ecosystem Modeling in GIS." *Envir. Mgmt.* **18(3)** (1994): 345–349.

[3] Behan, J. R. "An Image-Based Assessment of Wildland Recreation Benefits in the Greater Sedona Area, Arizona." Unpublished Master's thesis. School of Forestry, Northern Arizona University, Flagstaff, AZ, 1997.

[4] Bousquet, F., C. Cambier, and P. Morand. "Distributed Artificial Intelligence and Object-Oriented Modeling of a Fishery." *Math. & Comp. Model.* **20(8)** (1994): 97–107.

[5] Briggs, D., J. Westervelt, S. Levi, and S. Harper. "A Desert Tortise Spatially Explicit Population Model." In *Proceedings of the Third International Conference/Workshop on Integrating GIS and Environmental Modeling*, January 21–25, 1996, Santa Barbara, CA: National Center for Geographic Information and Analysis. ⟨http://www.ncgia.ucsb.edu/conf/SANTA_FE_CD-ROM/main.html⟩.

[6] Brown, J. H. "Modeling Ecological Patterns and Processes using Agent-Based Simulations and GIS." In *Proceedings of the Third International Conference/Workshop on Integrating GIS and Environmental Modeling*, January 21–25, 1996, Santa Barbara, CA: National Center for Geographic Information and Analysis. ⟨http://www.ncgia.ucsb.edu/conf/SANTA_FE_CD-ROM/main.html⟩.

[7] Brown, P. J., B. L. Driver, and C. McConnell. "The Opportunity Spectrum Concept and Behavioral Information in Outdoor Recreation Resource Supply Inventories: Background and Application." In *Integrated Inventories of Renewable Natural Resources*, 73–84. General Technical Report RM-55, USDA Forest Service, Ft. Collins, CO, 1978.

[8] Bruns, D., B. L. Driver, M. E. Lee, D. Anderson, and P. J. Brown. "Pilot Tests for Implementing Benefits-Based Management." The Fifth International Symposium on Society and Resource Management, June 7–10, 1994. Ft. Collins, CO.

[9] Clark, R. N., and G. H. Stankey. "The Recreation Opportunity Spectrum: A Framework for Planning, Management, and Research." General Technical Report PNW-98, USDA Forest Service, Portland, OR, 1979.

[10] Conte, R., and N. Gilbert, eds. "Computer Simulation for Social Theory." In *Artificial Societies: The Computer Simulation of Social Life*, 1–15. London: UCL Press, 1995.

[11] Csikszentmihalyi, M., and I. S. Csikszentmihalyi. *Optimal Experience: Psychological Studies of Flow in Consciousness.* New York: Cambridge University Press, 1988.

[12] Csikszentmihalyi, M., and R. Graef. "The Experience of Freedom in Daily Life." *Am. J. Community Psychol.* **8** (1980): 401–414.

[13] Csikszentmihalyi, M., and R. Larson. "Validity and Reliability of the Experience Sampling Method." *J. Nervous & Mental Disease* **175(9)** (1987): 526–536.

[14] Daniels, S. E., and R. S. Krannich. "The Recreation Opportunity Spectrum as a Conflict Management Tool." In *Social Science and Natural Resource Recreation Management*, edited by Joanne Vining, 164–179. Boulder, CO: Westview Press, 1990.

[15] DeAngelis, D. L., D. M. Fleming, L. J. Gross, and W. F. Wolff. "Individual-Based Modeling in Ecology: An Overview." In *Proceedings of the Third International Conference/Workshop on Integrating GIS and Environmental Modeling*, January 21–25, 1996, Santa Barbara, CA: National Center for Geographic Information and Analysis. ⟨http://www.ncgia.ucsb.edu/conf/SANTA_FE_CD-ROM/main.html⟩.

[16] Dibble, C. "Representing Individuals and Societies in GIS." In *Proceedings of the Third International Conference/Workshop on Integrating GIS and Environmental Modeling*, January 21–25, 1996, Santa Barbara, CA: National Center for Geographic Information and Analysis. ⟨http://www.geo.wvu.edu/www/il9/dibble⟩.

[17] Driver, B. L., and S. R. Tocher. "Toward a Behavioral Interpretation of Recreation, with Implications for Planning." In *Elements of Outdoor Recreation Planning*, edited by B. L. Driver, 9–31. Ann Arbor, MI: University Michigan Press, 1970.

[18] Driver, B. L., and P. J. Brown. "The Opportunity Spectrum Concept and Behavioral Information in Outdoor Resource Supply Inventories: A Rationale." In *Integrated Inventories of Renewable Natural Resources*, 24–31. General Technical Report RM-55, USDA Forest Service, Ft. Collins, CO, 1978.

[19] Drogoul, A., and J. Ferber. "Multi-Agent Simulation as a Tool for Studying Emergent Processes in Societies." In *Simulating Societies: The Computer Simulation of Social Phenomena*, edited by N. Gilbert and L. Doran, 127–142. London: UCL Press, 1995.

[20] Ewert, A., D. Chavez, and A. Magil. *Culture, Conflict, and Communication in Wildland-Urban Interface.* Boulder, CO: Westview Press, 1993.

[21] Fege, A. S., C. McCarthy-Ryan, L. Munson, and R. Schreyer. "Managing Visitor Conflicts." Paper presented at Managing America's Enduring Wilderness Resource: A Conference, Minneapolis, MN, September 11–14, 1989.

[22] Ferrand, N. "Multi-Reactive-Agents Paradigm for Spatial Modeling." Contribution to the European Science Foundation GISDATA program, Spatial Models and GIS, ESF-GISDATA. Stockholm: Taylor & Francis. 1996.

[23] Ferrand, N. "Modeling and Supporting Multi-Actor Spatial Planning using Multi-Agents Systems." In *Proceedings of the Third International Conference/Workshop on Integrating GIS and Environmental Modeling*, January 21–25, 1996, Santa Barbara,

CA: National Center for Geographic Information and Analysis. ⟨http://www.ncgia.ucsb.edu/conf/SANTA_FE_CD-ROM/main.html⟩.

[24] Findler, N. V., and R. M. Malyankar. "Emergent Behavior in Societies of Heterogeneous, Interacting Agents; Alliances and Norms." In *Artificial Societies: The Computer Simulation of Social Life*, edited by Gilbert and Conte, 212–236. London: UCL Press, 1995.

[25] Gimblett, H. R., B. Durnota, and R. M. Itami. "Spatially-Explicit Autonomous Agents for Modeling Recreation Use in Complex Wilderness Landscapes." *Complexity Intl. J.* **3** (1996).

[26] Gimblett, H. R., R. M. Itami, and D. Durnota. "Some Practical Issues in Designing and Calibrating Artificial Human Agents in GIS-Based Simulated Worlds." *Complexity Intl. J.* **3** (1996).

[27] Gimblett, H. R. "Simulating Recreation Behavior in Complex Wilderness Landscapes using Spatially-Explicit Autonomous Agents." Unpublished Ph.D. diss., University of Melbourne, Parkville, Victoria, 3052 Australia, 1997.

[28] Gimblett, H. R., and R. M. Itami. "Modeling the Spatial Dynamics and Social Interaction of Human Recreators using GIS and Intelligent Agents." MODSIM 97—International Congress on Modeling and Simulation. Hobart, Tasmania. December 8–11, 1997.

[29] Green, D. G. "Spatial Simulation of Fire in Plant Communities." In *Proceedings of National Symposium on Computer Modeling and Remote Sensing in Bushfire Prevention*, edited by P. Wise, 36–41. Canberra: National Mapping, 1987.

[30] Hammitt, W. E. "Visual Recognition Capacity during Outdoor Recreation Experiences." *Envir. & Behav.* **19(6)** (1987).

[31] Harris, L. K., H. R. Gimblett, and W. W. Shaw. "Multiple Use Management: Using a GIS Model to Understand Conflicts between Recreationists and Sensitive Wildlife." *Soc. & Nat. Res.* **8** (1995): 559–572.

[32] Hormuth, S. E. "The Sampling of Experiences in situ." *J. Personality* **54** (1986): 262–293.

[33] Hull, R. B., and W. P. Stewart. "The Landscape Encountered and Experienced while Hiking." *Envir. & Behav.* **27(3)** (1995): 404–426.

[34] Hull, R. B., W. P. Stewart, and Y. K. Yi. "Experience Patterns: Capturing the Dynamic Nature of a Recreation Experience." *J. Leisure Resh.* **24** (1992): 240–252.

[35] Itami, R. M. "Simulating Spatial Dynamics: Cellular Automata Theory." *Landscape & Urban Plan.* **30** (1994): 27–47.

[36] Ivy, M. I., W. P Stewart, and C. Lue. "Explore the Role of Tolerance in Recreation Conflict." *J. Leisure Resh.* **24(4)** (1992): 348–360.

[37] Jacob, G. R. "Conflict in Outdoor Recreation—The Search for Understanding." *Utah Tour. & Rec. Rev.* **6(4)** (1977).

[38] Jacob, G. R., and R. Schreyer. "Conflict in Outdoor Recreation: A Theoretical Perspective." *J. Leisure Resh.* **12(4)** (1980): 368–380.

[39] Johnson, K. M. "Using Statistical Regression Analysis to Build Three Prototype GIS Wildlife Models." *GIS/LIS '92* (1992): 374–386.

[40] Kohler, T. A., C. R. Van West, E. P Carr, and C. G. Langton. "Agent-Based Modeling of Preshistoric Settlement Systems in the Northern American Southwest." In *Proceedings of the Third International Conference/Workshop on Integrating GIS and Environmental Modeling*, January 21–25, 1996, Santa Barbara, CA: National Center for Geographic Information and Analysis. ⟨http://www.ncgia.ucsb.edu/conf/SANTA_FE_CD-ROM/main.html⟩.

[41] Larson, R., and M. Csikszentmihalyi. "The Experience Sampling Method." In *New Directions for Naturalistic Methods in the Behavioral Sciences*, edited by H. Reis, 41–56. San Francisco: Jossey-Bas, 1983.

[42] Lee, M. E., and B. L. Driver. "Benefits-Based Management: A New Paradigm for Managing Amenity Resources." The Second Canada/US Workshop on Visitor Management in Parks, Forest, and Protected Areas. University of Wisconsin-Madison, Madison, WI. May 13–16, 1992.

[43] Manfredo. M. "The Comparability of Onsite and Offsite Measures of Recreation Needs." *J. Leisure Resh.* **16(3)** (1984): 245–249.

[44] Owens, P. L. "Conflict as a Social Interaction Process in Environment and Behavior Research: The Example of Leisure and Recreation Research." *J. Envir. Psychol.* **5** (1985): 243–259.

[45] Rechel, J. L. "Geographic Information Systems Modeling of Wildlife Movements across Fire and Urban Disturbed Landscapes." *Bull Ecol. Soc. Am.* **73(2)** (1992): 316.

[46] Richards, M.T., and T. C. Daniel. "Measurement of Recreation and Aesthetic Resources in Southwestern Ponderosa Pine Forests." In *Multiresource Management of Southwestern Ponderosa Pine Forests: The Status of Knowledge*, edited by A. Tecle and W. Covington, ch. 7. USDA Forest Service, Southwestern Region, 1991.

[47] Richards, M. T., and H. R. Gimblett. "Recreation Assessment of Forestlands in Ecosystem Management: A Conceptual Model." Research Plan. USDA Forest Service, Rocky Mountain Research Station. July, 1995.

[48] Saarenmaa, H., J. Perttunen, J. Vakeva, and A. Nikula. "Object-Oriented Modeling of the Tasks and Agent in Integrated Forest Health Management." *AI Appl. Nat. Res. Mgmt.* **8(1)** (1994): 43–59.

[49] Santa Fe Institute. "Swarm Multi-Agent Simulation of Complex Systems." 1995. ⟨http://www.santafe.edu/projects/swarm⟩.

[50] Scenic Spectrums Pty Ltd. *Tully Gorge Visitor Impact and Management Study.* Report prepared for Queensland Department of Primary Industries Forest Service. Glen Waverly, Australia, 1995.

[51] Schreyer, R. "Conflict in Outdoor Recreation: The Scope of the Challenge to Resource Planning and Management." In *Social Science and Natural Resource Recreation Management*, edited by Joanne Vining, 12–31. New York: Westview Press, 1990.

[52] Slothower, R. L., P. A. Schwarz, and K. M Johnson. "Some Guidelines for Implementing Spatially Explicit, Individual-Based Ecological Models within Location-Based Raster GIS." In *Proceedings of the Third International Conference/Workshop on Integrating GIS and Environmental Modeling*, January 21–25, 1996, Santa Barbara, CA: National Center for Geographic Information and Analysis. ⟨http://www.ncgia.ucsb.edu/conf/SANTA_FE_CD-ROM/main.html⟩.

[53] Stewart, W. P. "Influence of the Onsite Experience on Recreation Experience Preference Judgments." *J. Leisure Resh.* **24** (1992): 185–198.

[54] Stockwell, D. R. B., and D. G. Green. *Parallel Computing in Ecological Simulation*, edited by A. Jakeman, 540–545. Proceedings of the Simulation Society of Australia, 1989.

[55] Tinsley, H. E., and D. J. Tinsley. "A Theory of the Attributes, Benefits and Causes of Leisure Experience." *Leisure Sci.* **14(3)** (1986): 195–209.

[56] USDA Forest Service. *ROS Users' Guide.* In service document, no publication number. 1982.

[57] Watson, A. E., M. J. Niccolucci, and D. R. Williams. "The Nature of Conflict between Hikers and Recreational Stock Users in the John Muir Wilderness." *J. Leisure Resh.* **6(4)** (1994): 372–385.

[58] Westervelt, J., and L. D. Hopkins. "Facilitating Mobile Objects within the Context of Simulated Landscape Processes." In *Proceedings of the Third International Conference/Workshop on Integrating GIS and Environmental Modeling*, January 21–25, 1996, Santa Barbara, CA: National Center for Geographic Information and Analysis. ⟨http://www.ncgia.ucsb.edu/conf/SANTA_FE_CD-ROM/main.html⟩.

[59] Williams, D., G. Ellis, N. Nickerson, and C. Shafer. "Contributions of Time, Format, and Subject to Variation in the Recreation Experience Preference Measurement." *J. Leisure Resh.* **20 (1)** (1982): 57–68.

An Intelligent Agent-Based Model for Simulating and Evaluating River Trip Scenarios along the Colorado River in Grand Canyon National Park

H. Randy Gimblett
Catherine A. Roberts
Terry C. Daniel
Michael Ratliff
Michael J. Meitner
Susan Cherry
Doug Stallman
Rian Bogle
Robert Allred
Dana Kilbourne
Joanna Bieri

1 INTRODUCTION

In 1979 the National Park Service (NPS) approved a Colorado River Management Plan (CRMP) based on the Grand Canyon Wilderness Recommendation and findings from a comprehensive research program. An amendment to an Interior Appropriations Bill in 1981 prohibited the implementation of this plan and resulted in increased public use levels and continued motorized use in proposed wilderness. In the last 20 years, the demand for whitewater experiences has increased, especially for the self-outfitted public. Today, the NPS is challenged by users and preservationists to provide accessibility while maintaining wilderness integrity.

Whitewater trips along the Colorado River through the Grand Canyon National Park are an excellent example of how increasing human use is impacting a sensitive, dynamic ecosystem and threatening to degrade the quality of experience for human visitors. Although visitation of the Colorado River has remained relatively constant since the 1989 CRMP—at 20,000 to 22,000 visitors and another 3,700 guides, researchers, and park staff traveling through the Grand Canyon each year—figure 1 shows the rapid rise in visitation since 1955 [13]. Visitors travel on over 600 commercial or privately organized river

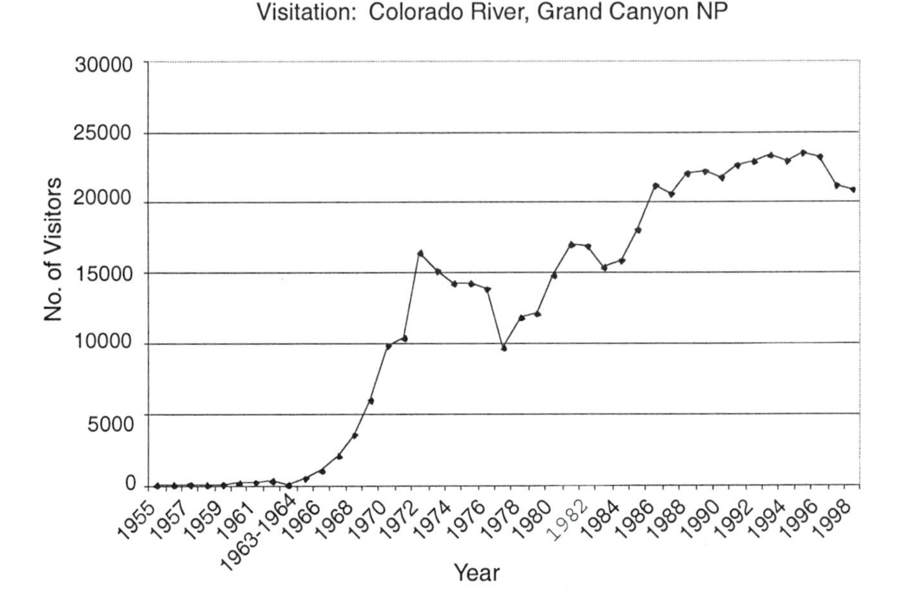

FIGURE 1 Visitation on the Colorado River 1955–1998.

trips on a variety of watercraft powered by oars, paddles, or motors for varying duration.

Most of the recreational use is concentrated in the summer months, resulting in high encounter rates and congestion at riverside attraction sites. Commercially guided operations account for over 80% of the total recreational use, of which 85% is on motorized rafts. The remaining proportion of recreational river trips are undertaken by noncommercial, self-outfitted public. Nearly 60% of the self-outfitted trips occur in the summer months, with an even proportion on use in the spring and fall. Less than 1% of these trips are motorized.

Major drainages and side canyons along the 277-mile river corridor in Grand Canyon National Park provide recreational activities including white water rapids, sightseeing, hiking, and swimming. Well-known attractions and destinations are regular stops for nearly every river trip that passes through the canyon [37]. Crowding and congestion along the river at attraction sites is often extreme and has been shown to affect the character and quality of visitor experience (e.g., Shelby et al. [51, 52, 53, 54]). Park administration is increasingly concerned about the deterioration of the sensitive river and canyon environment. Ongoing monitoring programs record use and environmental impacts at popular destinations and attraction sites, as well as documenting contacts among parties on the river.

At issue is the desire for the NPS to balance the high demand for this recreational treasure against the impact of human use on the natural resource. Moreover, as use increases, the cost to the recreational experience also increases as visitors have higher contact rates and witness a rise in competition for campsites and opportunities for solitude. The Grand Canyon National Park is seeking ways to regulate the rafting use on the Colorado River to minimize the environmental impact while optimizing the recreational value of the experience for humans.

The primary means available to the park for regulating these features of river trips and associated environmental effects of trips is the launch schedule. Scheduling and predicting travel along the 277-mile stretch of river, however, has proven a difficult task. Trips range from 6 to 18 days (from Lees Ferry to Diamond Creek) in the primary Season (10/01–04/30) and up to 30 days in the secondary Season (06/01–08/15). Because of fluctuating water levels imposed on the river from the Glen Canyon dam, nominally identically scheduled trips can vary substantially in their speed of progress through particular reaches of the river. Lower water levels produce slower progress (especially for oar-powered trips), and lost time is usually made up by skipping or reducing the time allocated to attraction sites and/or by spending more time rowing on the river, especially near the end of the trip. On the other hand, higher water results in faster progress, which usually translates to more stops and longer times at both attractions and campsites, resulting in more crowding and greater environmental impacts. An additional feature restricting the motion of some trips are passenger exchanges, which currently occur at Phantom Ranch and/or Whitmore Wash on predesignated dates and times.

Campsites are not assigned and fixed itineraries are not required, so negotiations between parties for the limited campsites frequently occur on river. Campsites in corridors approaching attraction sites routinely fill with trips setting up for their next day's adventure, and conflicts among parties are common. The progress of individual trips is affected by these interactions with other parties, and there is a general assumption (based on early research) that "encounters" degrade the "wilderness character" of the river trip, and that they have adverse effects on the quality of experience for individual recreationist. Little is known, however, about how to predict or control the numbers of encounters (except generally to limit the number of people/parties on the river), or whether all encounters are alike regardless of the types of parties involved, the locations on the river, and the contexts in which they occur [31].

Pressures from both privates and commercial outfitters remain high because of increased demand for river access. Commercial trips are typically scheduled two years in advance and there is over a 12–15 year waiting list for private river trips. The effects of increasing trips or altering schedules are difficult to predict or evaluate due to the complexity of the variables involved, and the ambiguity about what factors affect the quality of the river experience and/or the levels of adverse impacts on the sensitive river environment.

Environmental impacts at popular recreational and camping sites are already of great concern to the park [37].

In connection with the Grand Canyon National Park's first Wilderness Management Plan [29], the agency is revisiting public and agency concerns about wilderness management. The obvious question is how established motorized use fits into the wilderness management concept. Related and significant issues such as group size, length of stay, and resource preservation are also being addressed within the wilderness management framework. One of the greatest challenges being met by the NPS is the fact that wilderness policy and management practices have been disregarded in the last 20 years.

2 GRAND CANYON RIVER TRIP SIMULATOR

The renewed public interest in wilderness river management, along with more rigorous attention to agency wilderness policy, has guided the NPS toward the development of a comprehensive approach to visitor use planning for the proposed wilderness, and specifically the Colorado River. Included in this comprehensive approach is the need by recreation planners and natural resource users for more sophisticated tools to help them understand the human-environment interactions along the Colorado River, and to effectively respond to their mandate to manage and protect this unique environment and the highly valued human experiences it supports [38].

Very few tools exist that provide the power and flexibility to handle such a complex management problem. The Wilderness Use Simulation Model [50] (WUSM) is an important example of early efforts to develop a tool to model complex human-environment interactions in an outdoor recreation context. The WUSM was developed to model hikers' use of trail segments, cross-country travel routes and camping areas, with an emphasis on estimating the numbers of encounters and potential conflicts among parties. Underhill and associates [8, 9, 58, 59] adapted the original WUSM for application to rafting parties on the Colorado River. The effects of launch schedules and variable water flows were major input components of the river trip model [8]. The primary outputs of the model were the rates of raft progress down the river and a tabulation of the associated (projected) numbers of encounters among modeled raft parties on the river and at attractions and camping stops. Even though the model was implemented on a mainframe computer, considerable simplification of the actual river trip situation was required. River trip itineraries were fixed, with only launch and take-out dates allowed to vary between trips. Outputs were restricted to aggregate summaries of the frequencies of encounters among raft parties (classified by general types). This initial Grand Canyon river trip model is no longer in use and the Park is still in need of better means for evaluating alternative trip schedules [38].

While most computer simulation efforts of the past have more typically focused on biophysical environmental processes [28], "human dimensions" of environmental systems have also been successfully developed. Many of these successful simulation modeling efforts employ a number of artificial intelligence techniques combined with geographic information system (GIS) functions to address human-environment interactions (e.g. Green [30], Slothower [55], Gimblett et al. [23, 24], and Briggs et al. [11]). Exploratory studies (e.g., Berry et al. [3], Gimblett et al. [22], Saarenmaa et al. [49], Flynn [21], and Gimblett [25]). Tobler [57] and Itami [34] have suggested the use of cellular automata (CA) as a method for simulating dynamic environmental processes over large-scale landscapes, and applications of this approach have been successfully demonstrated (e.g., Green et al. [30] and Manneville et al. [45]). Individual-based models (IBM) have recently been applied to develop spatially explicit models of ecological phenomena. IBMs are "organisms-based models capable of modeling variation among individuals and interactions between individuals" [55]. IBMs offer potential for studying complex behavior and human/landscape interactions within a spatial framework. One form of individual-based modeling approaches, "agent-oriented programming," facilitates representation of dynamic interactions among multiple agents that coexist in an environment. Included in this approach is the study of complex adaptive systems, where tools and techniques are being developed to study emergent behavior: for example, Swarm [32, 43], Echo [41, 20], Gensim [1], and RBSim [25, 26, 27, 5, 6] (see Gimblett et al. [27]).

The combination of spatially explicit IBMs, reactive agents, artificial intelligence (AI), and GIS offer a powerful alternative to previous modeling techniques for exploring emergent, complex, evolutionary processes. The ability to model the differences among groups, local interactions, and variability over time and space, as well as the complex, decision-making processes of individuals, make IBMs an ideal technique for simulating recreation behavior and interactions in contexts like that of the Colorado river rafting trips.

This chapter describes a recently developed simulation system that integrates statistical, AI, and geospatial computer modeling techniques to analyzing the complex human-environment interactions in the context of river rafting trips. The Grand Canyon river trip simulator (GCRTS) was developed using GIS and intelligent agent modeling techniques to simulate rafting trips down the Colorado River through Grand Canyon National Park. The following sections will describe the simulation system and its application for river management.

3 RIVER TRIP EXPERTISE

The rule base in the GCRTS was developed from the experience and knowledge of a number of experts [48]. The Grand Canyon river guides, Grand Canyon private boaters association, and the Grand Canyon river outfitters associations provided insight into the complex interactions occurring on the Colorado River. The information provided by these experts was essential in the development of the rule base for the simulation. Evidence gathered from interviews with these groups was critical in formulating the general decision structure for a trip leader. Several trip types were identified (private, commercial, and research; oar, motor, and dory powered; various trip lengths) and interviews documented the hourly decisions that the trip leader makes for each particular trip type.

Our initial intuition was that the model would require many different types of trips, each with specific decision/action profiles. The expert interviews, however, revealed a higher-level structure in which substantial numbers of trips follow the same general planning strategy.

The daily goal of any trip is determined by campsite selection, trip length, exchanges, and setting up for attraction sites. The trip length is modified or adjusted to account for any fixed points like exchanges that have to be met on specific dates/times. The daily goal is also modified to account for any key attraction sites—in case the trip wants to "set up" for a key attraction site for the following morning. The daily goal is also modified if the straight calculation dumps you in a part of the river with a paucity of campsites, then the trip needs to modify its plans so that it is in a campsite rich area by late afternoon. So, while campsites play a role for sure, these other aspects (trip length, exchanges, setting up for attraction sites) play a key role.

As a trip traverses the river during each day, certain sites are available to stop, hike, lunch, swim, or relax. The decision to visit a particular site is primarily based upon three ordered criteria: (1) how special or attractive the site is; (2) the number of people already visiting that site; and (3) whether the trip has time to spare given its goal for the day. Thus, for example, a trip will almost certainly stop at key attraction sites regardless of the number of people there or their progress toward the planned campsite. Factors of weather and flood level play a lesser role in the global planning and decision-making process. The experience of the boaters/trip leaders often helps them determine a priori how they need to adjust their current progress and/or trip plan in order to avoid crowding at the most attractive sites. Trips can adjust the length of stay at camp and attraction sites as well as adjusting their rate of travel over the course of each day. These and other insights gained from the interviewed experts provided a well-defined set of rules and parameters that formed the initial basis for the model and the structure for actual trip data collection.

4 THE SIMULATION SYSTEM

The GCRTS is composed of two interacting but standalone modules. The River Trip database component uses an ACCESS database to store, analyze, and report on parameters of actual and/or simulated river trips. This component contains a set of algorithms for querying individual or multiple trips to obtain summary statistics and to make comparisons among actual and/or simulated trips. This provides the quantitative data for the validation of the simulation model, as well as providing important insights into the characteristics of actual and simulated trips.

The second major module is a simulation system, programmed using object-oriented classes, properties, and object structure. GCRTS is comprised of sets of interconnected agent classes designed to simulate trip behavior and represent the associated river environment. These agent classes are programmed to interface with the River Trip database, reading and writing information into a variety for formats.

5 TRIP CLASS

The *Trip Class* stores and passes information to the simulation environment that describes attributes of river trips such as length of trip, trip id, name of outfitter, number of boats, type of boats, boatman, number of passengers, number of crew, etc. The trip class also includes important information about exchanges—whether or not the trip will be exchanging passengers at Phantom Ranch and/or Whitmore Wash and, if so, the dates and number of passengers engaged in the exchange. This class was designed to provide maximum flexibility in creating agents to represent actual commercial and private trips that leave Lees Ferry and travel along the Colorado River in the Grand Canyon National Park.

6 RIVER CLASS

The *River Class* stores and controls information about the river and physical (camp and attraction) sites along the river. Information is read from a GIS database consists of geographic features 90-meter raster cells. This data structure allows the decisionmaker to accurately represent geographic features of the river environment and to easily update features as changes occur (e.g., eroding beaches, varying water levels, addition or deletion of attractions or campsites, river flow rates). This class stores attributes such as river mile, beach locations (river right or river left), beach carrying capacities, attractiveness values for camps and attraction sites (based on historic data), and number of people currently occupying a beach. In addition, this class keeps track of numbers of trips on the river.

7 ENCOUNTER USER CLASS

The *Encounter User Class* keeps track of the number of encounters each trip has with other trips on and off the river. This class monitors numbers and types of encounters that are occurring on a cell-by-cell basis, and writes data out to a database. Encounters are time marked and stored for each trip, for each cell and for each camp/attraction site. Types of encounters that are measured include on-river encounters, off-river (beach) encounters, or passing encounters where one trip on river passes by another trip that is off-river at a site.

8 VISUALIZATION CLASS

The *Visualization Class* provides a way to view and control elements of simulations on the river. GIS data, shaded relief maps, and dynamic pan and zoom utilities display the progress of trips as they move down the river. In addition to watching the movement and interaction of trips on the river, the user can query beaches/attraction sites at any time in the simulation to obtain information about each.

9 SIMULATION CLASS

The *Simulation Class* is the heart of the simulator. It provides the interface for running the simulation, controlling trip movement, reading trip, and launch schedules and it controls an enormous amount of file management tasks. This class controls the decision making of the individual simulated trips to ensure that the decisions reflect, to the best of our ability, the natural decisions that would be made if a real person was confronted with the same scenarios. Trip simulation outcomes are stored in the River Trip database for subsequent analysis.

The software "engine" that was developed uses the launch schedule in simulating raft parties (trips) down river. Launch schedules are derived two years in advance [?]. Based on the 1989 CRMP, the launch schedule year-to-year has been fairly consistent. A template launch schedule that captures the essence of what launch schedules look like under the 1989 CRMP has been created and implemented in the simulation system. This launch schedule follows all the NPS regulations based on limits of acceptable change which include such guidelines as no more than 6 motor rigs launching on a given day, no more than 150 commercial passengers on a day (except on Wednesdays when there is a double private launch and the upper limit is 134 commercial passengers), etc.

The 277-mile stretch of the Colorado River that is used for Grand Canyon river trips is divided into 90-meter "cells" and each trip is dealt with on a cell-by-cell basis. The simulator only treats the miles from Lees Ferry (mile 0) to Diamond Creek (mile 225.7) since the park only regulates up to Diamond. The underlying structure of the engine is the use of a priority time structure. This structure permits attention to only one trip object at a time and prioritizes the trips to be handled next. This one-by-one analysis is dependent on the time at which each trip will next enter a new cell in the river. The trip that is at the top of the priority stack determines, using artificial intelligence algorithms to be described later, what action to take (to stop or not and, if so, for how long). It then returns control to the engine. As each trip passes or stops at an identified site, the information is logged for subsequent analysis and display.

10 REPRESENTATION OF RIVER TRIPS

The basic unit of analysis in the GCRTS is the trip. Each trip has a specified duration, determined by the launch date from Lees Ferry, and the take-out date and location. While the take-out location for a trip may occur at Diamond Creek or farther down river, the current system of accounting for trips needs to keep track of the day when the trip is at Diamond Creek. So, the simulation pays closest attention to the launch date and the Diamond date. Trips may vary widely in composition, ranging from a single large motorized raft holding 30 or more passengers to flotillas of individual kayaks. Party sizes can vary from a few to 40 or more. Commercially outfitted trips tend to be relatively uniform within the outfitter company, but differ between companies. Whether composed of many or a single watercraft, each trip is modeled as a single point moving down the river. Park rules limit how far craft associated with a given trip can be dispersed on the river, and the full party must camp together.

In the model, the primary data about any trip, whether actual or simulated, is which of the 5000+ 90-meter river cells the trip (or trip "centroid") is occupying at any given time. This "who-where-when" data forms the basis for all of the trip-to-trip encounter, attraction site visitation, and campsite occupancy information in the model.

An individual trip may be represented by the relationship between river location and time. Either actual calendar time or trip lapsed time may be used, depending upon the analysis desired. Figure 2 shows "where-when" lapsed-time data for a selection of (nominally) seven-day commercial trips from the 1998 trip survey data. The horizontal axis indicates the trip day while the vertical axis represents the river mile (marked off in 50 mile lines). The dots indicate the campsite locations. The lines connecting the dots indicate the daily miles traveled, and the slope of that line gives us the rate of travel of the trip (miles/day). The lines are always increasing because due to the strong river current, the trips only travel down river—they do not reverse direction and travel back up river. If a portion of a graph for a trip appears horizontal, then that trip did not make any miles between two camps. This is defined as a "layover," when the trip remains at the same campsite for two subsequent nights. Layovers are more common for trips of longer duration. The linear pattern is typical for all trip types, with the exception of some of the longer (up to 30 days) private trips. Analysis of figure 2 reveals a consistency in that this trip type (seven-day motorized trips) tends to travel at similar rates down river. While the consistency between trips generally appears quite high, it is important to note that some of the differences between trips of nominally the same type can be over 25 miles (for a given time) or a full day (for a given location). Encounters are defined in some cases in terms of five-minute intervals of contact. Some campsites are less than a mile apart and often a few hundred yards on the river can make the difference between seeing another party or not. Clearly then, given the variability in the actual data, model projections of encounters between trips or visitation times to specific attractions or campsites must be probabilistic. Still, the consistency of the actual trip data was encouraging to the modeling effort. For example, it is apparent in figure 2 that trips of this type tend to travel in approximately equal daily distances. This information is useful in the modeling effort because trips tend to plan on a target campsite for each upcoming evening, and this data helps us determine the appropriate daily goals for the simulated trips of similar type.

A straightforward quantitative representation of trip data like that in figure 3 is a simple linear regression of (in this case lapsed) time on location. Such regression lines for different classes of trips can be compared to provide a statistical picture of similarities and differences. For example, figure 3

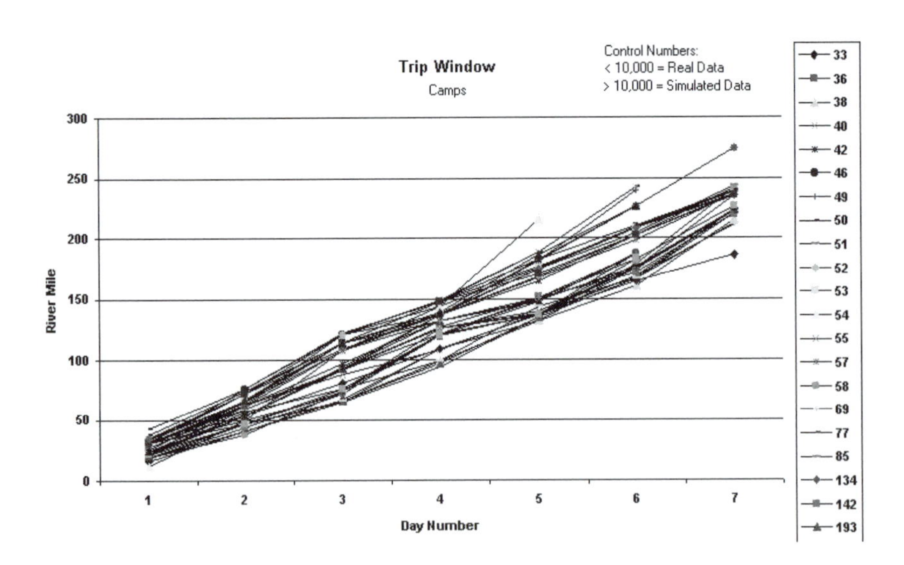

FIGURE 2 Lapsed-time data for a selection of (nominally) seven-day commercial trips.

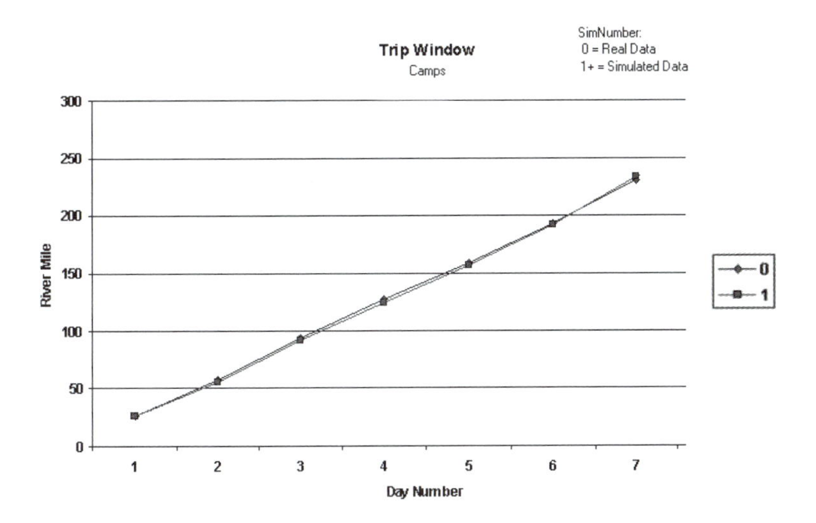

FIGURE 3 Simulating typical trips using data reduction algorithms.

compares the regression line for actual seven-day commercial trips with the line based on 41 simulated seven-day commercial trips. While not providing a perfect match, and evidencing a consistent underestimation of actual trip progress, the simulation does show substantial predictive ability even at this early stage of development.

This simple regression approach is used in the model to represent different classes of actual and simulated trips, and to represent individual simulated trips based on multiple implementations (runs) of the model. Because of the obvious probabilistic nature of individual trip progress, any single implementation of the model can only represent one possible outcome for any simulated trip. A second run of the model, with the exact same launch schedule and other conditions will produce a different where-when outcome for a given trip. Thus, the GCRTS requires multiple simulation runs to project the outcome for any specified trip-launch schedule scenario. The best statistical representation of a simulated trip is the where-when regression line based on the multiple iterations for that trip.

The regression line provides an excellent statistical representation of trip classes and of individual trips based on multiple simulation runs. This provides the basis for general statistical comparisons between actual and simulated trips, and for comparing simulated trips under different sets of management policies and assumptions. For other purposes the regression line has all the disadvantages of any mean—it may not represent any particular trip in any detail, or even a possible trip. For example, while it may make statistical sense for an "average trip" to camp at some "average location," actual trips must camp at very specific locations where there is an actual beach to camp on. This feature of the regression representation is particularly problematic for the visualization component of the GCRTS. For this purpose a different representation of individual trips is required.

11 REPRESENTING THE "TYPICAL" TRIP

Specificity is a universal problem in environmental visualization. Most environmental models, whether of trees in the forests or recreators on a river, provide statistical summaries of conditions. For example, a forest growth and development model may project that there will be 9.43 trees per acre (± 1.68) of a particular species and size over a modeled area. This is useful statistical information, but it is not sufficient for creating a visualization of the projected forest—some specific number of trees must be rendered, and each must be placed at a specific location in the modeled landscape. Frequently random distribution algorithms are used to solve this problem, but real trees (or other environmental features, or human recreators) are rarely randomly distributed. Many tree species tend to grow in clumps of similar sizes/ages, and others may tend to grow only on certain aspects or in drainages or on ridge tops. Similarly, it is not appropriate to visualize river trips as distributed randomly along the river, or to have individual trips camping at "average" locations.

The solution in the GCRTS is to represent any individual trip by the most "typical" where-when outcome for that trip. The selection of the typical trip is based on the where-when regression line over the multiple model runs for a given trip/launch-schedule scenario. For example, for a given launch schedule/management scenario a particular seven-day commercial trip may be represented by many different individual (where-when) outcomes, depending upon the number of model iterations chosen. The best statistical representation of that particular trip is the where-when regression line. The best, most "typical" outcome to visualize is the particular model iteration that produces the closest fit to the (average) regression line—the specific outcome that has the minimum squared deviations from the where-when regression line (minimum residuals). For the visualization component of the GCRTS, then, an individual trip is represented by the most typical (least-squared residuals) where-when outcome over the multiple model runs for that trip. This most typical trip will be shown to camp only at actual campsites, and to have specific encounters with specific other trips at specific locations on the river. Current work is underway to examine the optimal number of simulation runs needed to create the "most typical" simulation.

11.1 REPRESENTING MULTIPLE-TRIP OUTCOMES

Often the motivation for running the GCRTS is to investigate the overall effects of a change in launch schedules or management policies. In these cases no particular individual trip is the focus of concern; rather, interest is directed at the aggregate effects over the many trips potentially affected by the scenario being simulated. For statistical comparison purposes, average numbers of encounters, average numbers of visits to a given site, and average progress (where-when) of different classes of trips is perfectly appropriate and highly informative. Again, however, average outcomes are not appropriate for visualization.

Similar to the solution used for individual trips, typical multiple trip outcomes are selected based on the trip-by-trip fit to their respective where-when regression lines. In this case, squared residuals are aggregated over all individual trips for a given simulation run. The specific modeled multiple-trip outcome (the individual model run) that produces the minimum squared aggregated residuals (i.e., the outcome closest to the mean over all model runs) is selected for visualization. In this way each of the individual trips in the visualization, and their interactions are "realistic." Every visualized trip camps at an actual campsite. Interactions and encounters between trips occur where and when the model algorithms and decision rules prescribed (not some average time and place). For example, two trips will not be shown to camp at the same site unless those two trip-agents in the model specifically choose to share that site on that night.

12 ARTIFICIAL INTELLIGENCE ALGORITHMS FOR AGENT DECISION MAKING

When considering appropriate approaches for treating complex human behavior in the environment, a robust simulation must be adaptive to changes in both human behavior and the natural environment. It is clear, for instance, that the trip leaders on the Colorado River operate on experience culled, in part, by the nature of the current management policies. While an initial model can be established based upon this known experience set, intelligent agents within the simulation must have the ability to adapt their expertise to the changing conditions of policy, and thus replicate the appropriate human decision processes under these new scenarios. While some methods of classical mathematical and stochastic modeling can encapsulate and reproduce, to a high degree of accuracy, the current environment conditions, these methods are not well suited to the predictive simulations that are required here. An essential component to the dynamic of this simulation model is the human decision-making processes, and how that influences the resource impact and use. In order to model this rule structure and enable a simulated trip to behave rationally given reasonable changes in the management conditions, we employed a computational artificial intelligence technique known as fuzzy logic [7, 60]. This technique, coupled with the stochastic rules developed from the database, allowed us to encapsulate the decision process.

Fuzzy logic can methodically encapsulate the rule structure of an "expert system" and provide a computationally simplified method for making the same type of gradations in decision making that a human user might. In traditional decision-making process, the lines between "yes" and "no" are fixed and inflexible. Previously, conditions for a decision process were modeled and discretized with finite domains of input. With fuzzy logic, the boundaries between discretized conditions to be blurred and overlapped.

For example, previously a beach would have been assessed as either crowded or not crowded. The computer program would have then characterized the crowding level as either "good" or "bad." In fuzzy logic, a beach can be characterized with as many levels as desired (e.g., as very crowded, fairly crowded, somewhat crowded, or not crowded). These looser characterizations of the variable "level of crowding" permit the simulation to consider a broader range of conditions and, consequently, a broader range of outcomes. Even the characterization of whether the level of crowding is "good" or "bad" can be given gradations (e.g. "very bad," "sort of bad," "ok," "not very bad"). The weight for each characterization does not have to be equally distributed among the various levels of crowding. For example, if an attraction site is highly attractive, then even a high level of crowding is considered acceptable. Yet, at a less attractive site, even a small amount of crowding can weigh significantly on the decision of the boatman not to stop. To summarize, fuzzy logic vastly broadens the level of how much subtly can be captured and can more closely model the sophisticated decision-making approaches used by humans.

This very closely replicates the sense of the linguistic variables that are common in the every day human decision making and precisely the type of decision process involved in the river trip dynamic. Boats determine where they stop and for how long based upon a number of conditions. The rules for evaluating these conditions, however, remain relatively stable regardless of how over the long term these conditions vary. This provides a strong basis for a fuzzy logic system in modeling the overall behavior of the trips that traverse the river corridor. Extrapolating the rule structure for trips on the river will allow the application of a computational technique that is well understood to perform well over a broad range of perturbance and simulation conditions.

Implementing the model of the decision process for boatman on the river consists of three primary components revealed from the data analysis and interview processes: site attractiveness, crowding, and schedule. Three groups of nondisjoint sets describing the three parameters were constructed. As a trip approaches a site in the simulation, the historical frequency of visitation is pulled from the database to serve as the site attractiveness factor. The crowding of the beach is evaluated, along with the degree to which a trip is ahead or behind schedule to reach its daily goal. These three parameters serve as inputs into the fuzzy system. The degree to which each condition parameter belongs to the appropriate fuzzy set is calculated and then all decision rules are evaluated with the appropriate weighting values. The centroid of the sum of the output values is taken and utilized as the probability for which a trip will land on that site. To summarize, the fuzzy logic considers three main conditions, weighs them appropriately, and then provides the probability for a certain decision to be made (such as whether to stop for an attraction).

A probability factor was chosen to simplify the way in which output from the fuzzy logic calculation are treated and then converted, in the end, to a binary "yes" or "no" decision. A nonstochastic approach would designate some fixed cutoff point in the range of the fuzzy engine output to serve as the positive or negative result. We recognized that this could restrict the robustness of the simulation to handle new situations, and therefore chose to maintain a stochastic or probabilistic approach. Moreover, the potential always exists for a factor outside of our three main factors to occur that might result in a trip making a decision that otherwise might seem fickle or uninformed. Even if each factor suggests that a trip would choose to stop at a site (it is not crowded, it is historically popular, and the trip has plenty of time left in the day), there are occasions in reality when a trip might still pass that site by. Perhaps, for example, the passengers are really interested in a long hike that isn't provided by this site or, perhaps another group is at the site that this trip really wants to avoid. Rather than try to take into account each and every possible alternative scenario that *could* happen, instead the fuzzy logic engine in this situation kicks back a 95% probability that the trip will stop—and then a die is rolled. Overwhelmingly, the trip will choose to stop there, but occasionally (5%) of the time, the trip won't stop there. This helps the simulation take into account the seemingly countless unusual exceptions or unusual events that were presented to us during our expert interviews.

It is important to point out that the fuzzy logic algorithm is only utilized when a decision needs to be made that is, on the surface, unclear. There are a large number of factors that will predetermine the action of a trip and will temporarily suspend the fuzzy logic algorithm in the simulation. For example, a trip will not go through the fuzzy logic algorithm to decide if it should stop at Phantom Ranch if, indeed, that trip is already required to stop there to engage in a passenger exchange. A trip will not consider stopping to camp at a beach whose capacity is lower than the number of people who need to camp. We recognize that there are a number of other possible factors that could override the decision to visit a perfectly nice site or, conversely, visit a generally unsatisfactory one. Stochastic modeling permits us to entertain the possibilities of these overriding factors, and in some way encapsulate the imperfect rationality of the human decision process as well as account for the many factors that our research may have missed.

Neither a completely deterministic model nor a completely stochastic one could fully encapsulate the variety of factors that drive the river dynamic. A careful mixing of the two schools of mathematical modeling is used and they complemented one another in the development of the simulation model.

13 GCRTS FEATURES

The simulation model is comprised of three major components: trip planner and simulation interface, simulation analysis, and visualization interface. Figure 4 provides a view of the main menu encountered by the user. Under the "run simulation" option, the user must first set up the launch schedule and set the "rules" under which the trips must operate. The user can build a simulation using the trip planner capitalizing on the launch schedule as the basis for the simulation. A launch schedule representing the typical launch schedule under the 1989 CRMP can be used for the simulation or it can be modified. The launch schedule can be modified by adding of deleting a variety of trips and enforcing certain restrictions on the simulation such as limiting the number of visitors, restricting takeouts, specifying launch times and access to camp and attraction sites, as well as the number of simulation runs to be performed. It is also possible to build an entirely new launch schedule from scratch.

FIGURE 4 Grand Canyon river trip simulator (GCRTS).

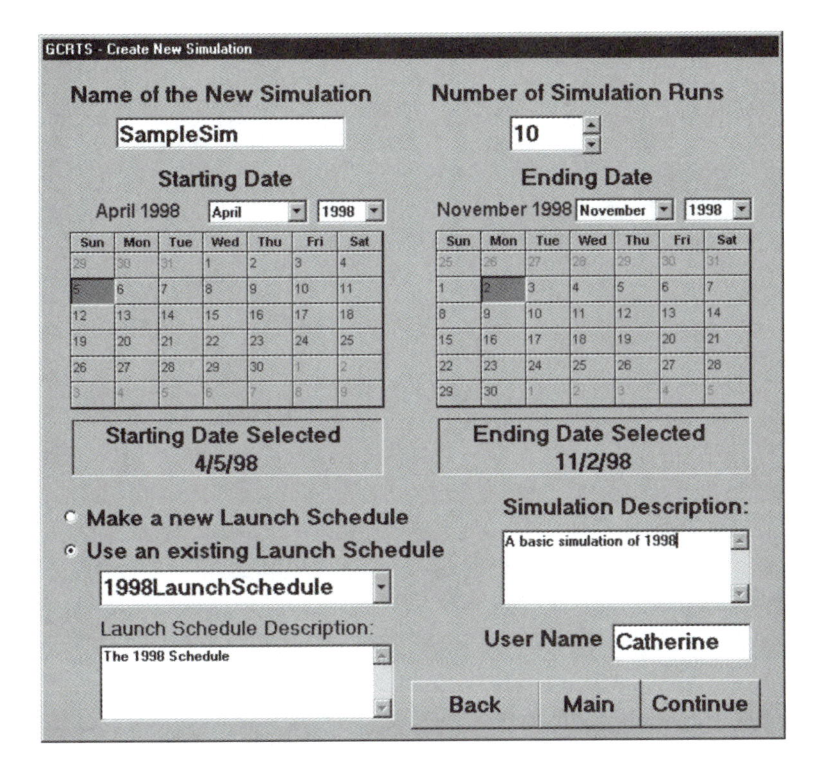

FIGURE 5 Simulation builder.

The number of simulation runs that need to be undertaken to get an accurate representation of typical trips on the river is still an open question among simulation researchers. This issue will be discussed below, along with a statistical method developed in this research for determining the best representative trip.

Once all the parameters for the model are specified the simulation is performed. There is an extensive amount of bookkeeping undertaken on simulation results. The simulation not only keeps track of all spatial and temporal movement of trips for the specified time period, but all encounters and levels of visitor use in the canyon. All this data is stored in a database, along with actual data collected on river use and subsequently analyzed using statistical analysis. To give the reader a sense of the amount of computing time required to run these simulations, a 500 MHz desktop personal computer can run 10 full seasons of simulations and then choose the most representative simulation in approximately 3 hours.

Figure 5 provides a view of the launch schedule editor. The click-and-choose menu provides an easy method for the user to develop the specifica-

GCRTS - Launch Schedule Editor

Sunday	Monday	Tuesday	Wednesday	Thursday	Friday	Saturday
	1 HATC 7W>L OUTD 5P8L TOUR 6W3L WEST 6W3L WEST 6W3L	2 ARIZ 6W2L DIAM 4P5D EXPD 5P8D HATC 7W>L WEST 6W3L	3 ARIZ 8L WEST 6W3L WRAD 7W>L	4 CANX 15D GRCE 8L HATC 7W>L OARS 6P8W4L TOUR 6W3L	5 AZRA 6P8D AZRA 6P9D GRCE 8L HATC 7W>L WRAD 4P5W>L	6 ARIZ 7W2L COLO 8L DIAM 7W>D GRCE 8L
7 ARIZ 13D CANY 3P5L CANY 7L HATC 9L OARS 5P9D	8 7P12D ARIZ 6W2L AZRA 6P8D DIAM 8D HATC 7W>L	9 ARIZ 6W2L WEST 6W3L WEST 6W3L WRAD 7W>L	10 WEST 6W3L WEST 6W3L WRAD 8W>L	11 DIAM 8D GRCE 8L HATC 7W>L OARS 6P8W4L TOUR 6W>D	12 AZRA 8D>L DIAM 8D GRCE 8L TOUR 12W>D WRAD 4P5W>L	13 ARIZ 7W2L COLO 8L GRCE 8L MOKI 8W>D
14 ARIZ 7W2L CANY 3P5L CANY 7L COLO 7D>L HATC 8W>L	15 HATC 7W>L MOKI 6P9L OUTD 5P8L TOUR 6W3L WEST 6W3L	16 ARIZ 6W2L AZRA 6P8D DIAM 7W>D EXPD 5P8D WEST 6W3L	17 18D CANX 7P9D DIAM 4P5D WEST 6W3L WRAD 6P7W>D	18 GRCE 8L HATC 7W>L HATC 7W>L TOUR 6W3L	19 18D AZRA 6P8D GRCE 8L HATC 7W>L WRAD 4P5W>L	20 ARIZ 7W2L COLO 4P6L DIAM 7W>D GRCE 8L
21 CANY 3P5L CANY 7L HATC 9L SLEI 8L	22 ARIZ 6W2L AZRA 6P8D HATC 7W>L OUTD 5P8L TOUR 6W3L	23 14D ARIZ 6W2L WEST 6W3L WEST 6W3L WRAD 6P7W>D	24 16L CANX 15D MOKI 6P9L WEST 6W3L WEST 6W3L	25 CANX 15D GRCE 8L OARS 6P8W4L TOUR 6W>D WRAD 7W>L	26 DIAM 4P5D GRCE 8L GRCE 14L HATC 7W>L WRAD 4P5W>L	27 AZRA 8D>L COLO 8L GRCE 8L HATC 7W>L
28 7P12D ARIZ 7W2L ARIZ 6W2L CANY 7L HATC 7W>L	29 8D AZRA 6P9D DIAM 10W>D OUTD 5P8L TOUR 12W>D	30 ARIZ 6W2L EXPD 5P8D SLEI 5P8W>D WEST 6W3L WEST 6W3L				

June, 1998

Month

< June >

Name

1998LaunchSchedule

View Info

Add a Launch

Delete a Launch

Group Add /Delete

Back Main Continue

Disabled	Type	Outfitter	SCHED	TripLength	Motorized	TakeOut	LaunchDay	LaunchTime	Passengers	Crew	DiamondDate	PhantomExDay	PhantomPxIn	Phant
False	P		7P12D	18	False	DIAM	6/8/98		13	0	6/25/98	7	2	
False	C	ARIZ	6W2L	7	True	LAKE	6/8/98		28	4	6/13/98	0	0	
False	C	AZRA	6P8D	13	False	DIAM	6/8/98		20	8	6/20/98	6	20	
False	C	DIAM	8D	8	True	DIAM	6/8/98		26	4	6/15/98	0	0	
False	C	HATC	7W>L	7	True	LAKE	6/8/98		26	4	6/14/98	0	0	
False	C	OUTD	5P8L	12	False	LAKE	6/8/98		21	8	6/18/98	5	21	
False	C	TOUR	6W3L	8	True	LAKE	6/8/98		29	4	6/14/98	0	0	

FIGURE 6 Launch schedule editor.

tions for a simulation run. Dates and time periods for the simulations are specified and options provided to view a launch schedule that a resource manager may wish to modify. Figure 6 illustrates the functionality built into the launch schedule editor. Once the initial parameters are selected in figure 5, the specified launch schedule is retrieved and each of the trips and specified launch days and times are revealed. This menu provides the user with the capability to modify trip, or add or delete trips. The bottom section of this menu provides all detailed information known about the trip including date and time of launch, number of passengers, type of watercraft, etc. All of these parameters can be modified to test a variety of launch scenarios. It should be pointed out that for the simulation to run successfully, a number of data fields must be provided (launch date, group size, exchange information, Diamond date, etc.). Without this information, the simulation cannot direct these trips down the Colorado River in a reasonable and correct manner.

Figure 7 provides a view of the simulation analysis interface. This interface is obtained from selection the "Analyze Sim Output" button on the main menu shown in figure 4. The Grand Canyon River Trip Simulation Analysis menu provides a mechanism for the user to query the actual or simulated data and compare both to answer questions about river use. Numerous queries have been developed to provide the users with answers to typical questions they

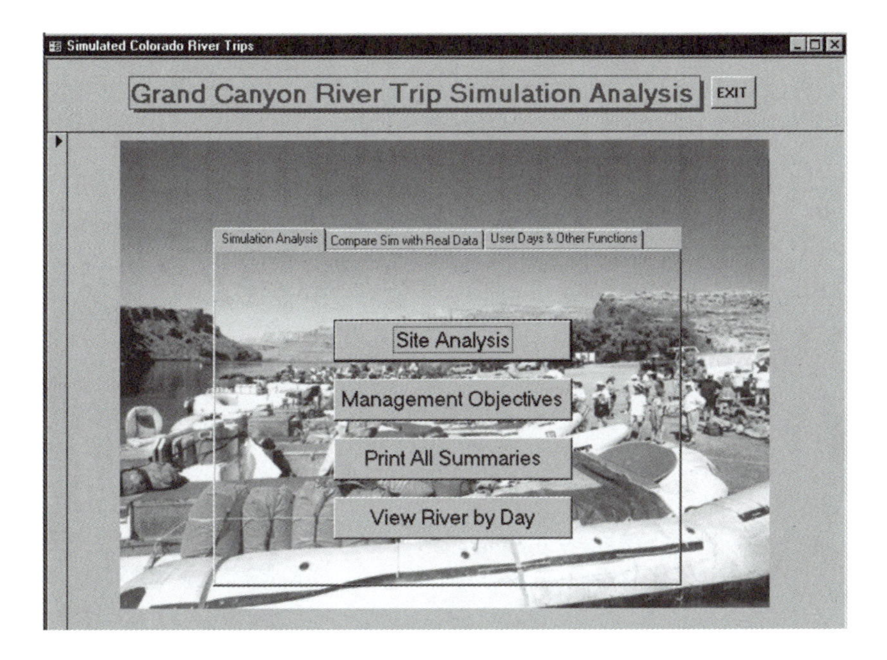

FIGURE 7 Interface to the river database and customized queries.

have about river use. In addition, there are numerous statistical operations and data visualization techniques utilized to provide the user with a view of those queries.

For example, one question that a user might have is to determine to what degree the launch schedule meets current management objectives or standards established by the Grand Canyon National Park for river contacts/day. The official objective states that there is *"80% probability that a party will make contact with 7 or less parties, with up to 90 minutes in sight of less than 125 people"* [14]. By selecting this option, either actual or simulated trips will be statistically assessed to determine how well and to what extent this objective is being met. This procedure, coupled with the simulation tool provides possible ways to not only examine current management objectives, but perhaps a method to derive new more realistic standards. This procedure also invites a level of judgment into the simulation...it is essential that standards are available with which to judge the quality of an alternative launch scenario.

In addition to the management objective queries, many forms of data analysis can be performed. Figure 8 provides a view of the distribution of 12-day commercial trips from a selected period in the 1998 season. The horizontal axis defines the day that the trip is on the river and the vertical axis the river mile. This is one of the first known views of the distribution of use on the river. This figure also clearly shows that commercial trips are not as

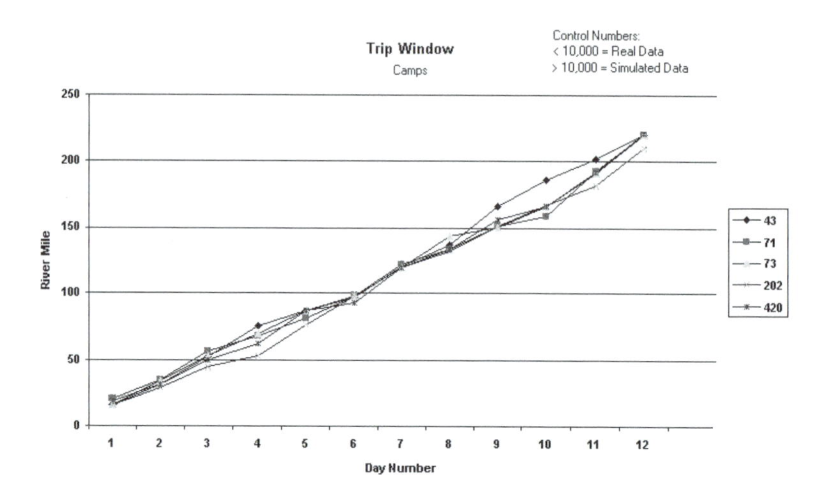

FIGURE 8 Example of 12-day commercial trips.

widely dispersed and different as was initially thought. The dots in the image represent the camp locations for each night on the river. Clearly this figure shows consistency and regular patterns of commercial use on the river.

River managers using this tool need to know at any particular point during the simulation run, where certain trips are camped, how many are camped at certain locations, and the number of contacts they have had. Figure 9 provides information from the simulation on the type of trip, the trip length, when they launched, name of outfitter, and information about where they camped, what activity sites they stopped at, and where they had contacts. This information is gathered on all trips on the river and can be viewed together, individually and both as a statistical summary as well as graphed out over the length of the trip. This provides the manager both with information about the patterns of use for each of the commercial outfitters and a means to help the simulation programmers calibrate the simulated trips against actual data collected on private and commercial use patterns.

Encounters and contacts on the river are of great concern to the National Park Service. Crowding and encounters can have adverse impacts both on the environment and also on visitor experience. Historically, standards have been established for numbers of trips and for maximum party sizes so as to minimize social and environmental impacts. These standards have been in existence for some time, but have less than adequate scientific bases. What is needed is a better understanding of the numbers of encounters and where they occur on the river. A river user can use the GCRTS to experiment with alternative schedules/policies in search of better means to meet current standards, or to arrive at more realistic standards.

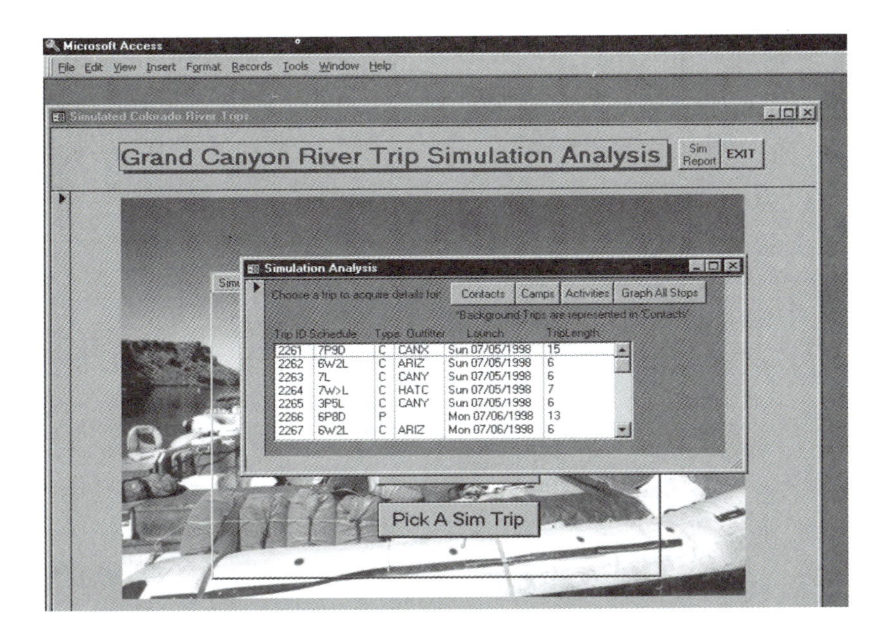

FIGURE 9 Obtaining detailed information (contacts, camps, and activities) for individual trips.

As a simulation is executing, the encounters are recorded in classes [e.g., both trips on-river, river to shore, and off-river (beach) encounters along with the spatial locations, duration and the specific trips involved]. Figure 10 provides a statistical view of encounters typical data.

Visualization of both relational and spatial data facilitates users' outcomes of simulations. Most importantly, visualization of river trips and encounters among trips can reveal more clearly where bottlenecks occur and can suggest how management actions affect the distribution of use and reduce encounters. The GCRTS provides a mechanism through visualization to view trips as they dynamically traverse the river, stop at camp or attraction sites, encounter each other, and take out at specified locations. Figure 11 provides a view of the visualization interface to the simulation environment. Trips can be seen launching, laying over and in particular when they encounter other watercraft on the river. The visualization interface allows the user to stop the simulation and zoom in on specific locations on the river or specify a moving window for following a particular trip. At any point, individual trips, campsites, or attraction sites can be queried for relevant characteristics and states.

SimTripID	RiverContact	RiverMile	PeopleContact	Duration	Type	Outfitter	SCHED	LD	OtherTripID
2129	Mon, 06/08/199	3.1	30	2	C	HATC	7W>L	Mon, 06/08/199	2127
2129	Mon, 06/08/199	8.	13	15	P		7P12D	Mon, 06/08/199	2130
2129	Mon, 06/08/199	16.4	29	1	C	OUTD	5P8L	Mon, 06/08/199	2124
2129	Mon, 06/08/199	16.4	30	1	C	DIAM	8D	Mon, 06/08/199	2126
2129	Mon, 06/08/199	17.	13	1	P		7P12D	Mon, 06/08/199	2130
2129	Tue, 06/09/199	19.9	13	1	P		7P12D	Mon, 06/08/199	2130
2129	Tue, 06/09/199	29.3	32	1	C	ARIZ	6W2L	Tue, 06/09/199	2133
2129	Tue, 06/09/199	29.3	34	1	C	WRAD	7W>L	Tue, 06/09/199	2134
2129	Tue, 06/09/199	29.3	21	1	C	WEST	6W3L	Tue, 06/09/199	2132
2129	Tue, 06/09/199	30.4	21	1	C	WEST	6W3L	Tue, 06/09/199	2132
2129	Wed, 06/10/199	31.6	21	1	C	WEST	6W3L	Tue, 06/09/199	2132
2129	Wed, 06/10/199	37.7	40	1	C	WEST	6W3L	Tue, 06/09/199	2131
2129	Thu, 06/11/199	47.2	40	1	C	WEST	6W3L	Wed, 06/10/199	2135
2129	Thu, 06/11/199	47.2	40	1	C	WEST	6W3L	Wed, 06/10/199	2136
2129	Sat, 06/13/1998	107.3	34	9	C	WRAD	8W>L	Wed, 06/10/199	2137
2129	Sat, 06/13/1998	107.8	9	1	C	TOUR	6W>D	Thu, 06/11/199	2140
2129	Sun, 06/14/199	107.8	34	1	C	TOUR	6W>D	Thu, 06/11/199	2138
2129	Sun, 06/14/199	108.2	23	1	C	HATC	7W>L	Thu, 06/11/199	2141
2129	Sun, 06/14/199	119.8	30	1	C	DIAM	8D	Thu, 06/11/199	2143
2129	Mon, 06/15/199	120.	23	1	C	HATC	7W>L	Thu, 06/11/199	2141
2129	Mon, 06/15/199	131.1	30	1	C	DIAM	8D	Thu, 06/11/199	2143
2129	Mon, 06/15/199	135.1	30	2	C	DIAM	8D	Thu, 06/11/199	2143
2129	Tue, 06/16/199	137.	32	1	C	GRCE	8L	Thu, 06/11/199	2142
2129	Tue, 06/16/199	141.5	32	5	C	GRCE	8L	Thu, 06/11/199	2142
2129	Wed, 06/17/199	166.6	33	1	C	GRCE	8L	Fri, 06/12/1998	2147
2129	Thu, 06/18/199	175.5	34	4	C	AZRA	8D>L	Fri, 06/12/1998	2145
2129	Thu, 06/18/199	189.7	32	1	C	DIAM	8D	Fri, 06/12/1998	2148

FIGURE 10 Detailed information on river encounters.

14 USING GCRTS TO DERIVE ALTERNATIVE RIVER MANAGEMENT SCENARIOS

To regulate rafting traffic on the Colorado River there are many issues to consider including congestion, access, and environmental impact. Moreover, there are competing political interests with an interest to investigate the potential outcomes of their individual agendas. GCRTS provides a useful tool for users to gain insight into how their ideas might play out on the river corridor [47]. Consequently, alternative launch scenarios of many sorts are currently in development. The hope is that some solutions to the tensions between congestion, access, and environmental impact can be addressed.

A broad set of alternative launch schedules can be created—for example, they could reflect interests to extend or reduce the boating season, to restrict the use of motorized watercraft, or to increase or decrease the allocation to one or more of the user groups. To illustrate the potential uses of GCRTS, launch schedules have been developed by the authors that represent a 100% use level, a 50% use level, and an oars-only use [4]. The simulation run for 100% use level used a launch schedule that represents a typical schedule and is considered a fair representation of current use. The 50% use launch schedule has only half the number of trips and passengers. Creating this level launch schedule involved some judgment calls, but it does represent an even cut of all trip types, of passenger numbers, and of user-days. The oars-only use maintains current use levels but disallows motorized boats. Again, creating this oars-only scenario involved many judgment calls and most likely does not fairly represent a launch schedule that would realistically arise under an oars-

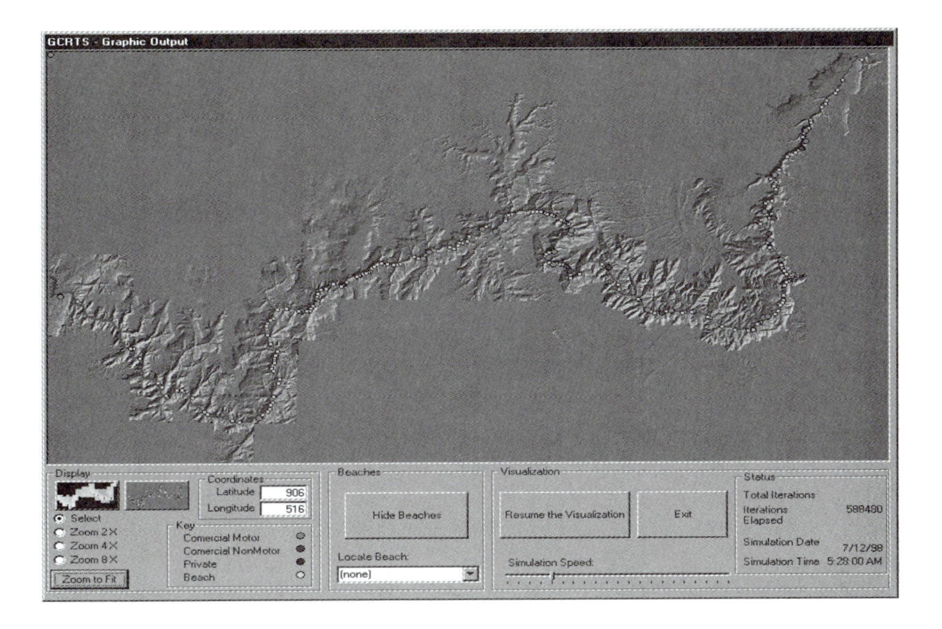

FIGURE 11 Visualization interface to interactively view progression of trips on colorado river.

only restriction. Nonetheless, these three scenarios were created for illustrative purposes. These simulations were analyzed and compared against each other.

To illustrate the type of analysis that is possible once an alternative launch schedule has been run through the simulation, the GCRT analysis tools can provide, for example, a list of top camp sites or attraction sites and their visitation frequencies. The top campsites from the historical 1998 trip reports and the ones resulting from an alternative launch schedule can also be compared. Comparison allows us to understand the ability of the simulation to replicate the current situation on the river and the capability of the simulation to predict the outcome of change. To date, the simulated trips are choosing the same top activity sites but at a lower frequency. Although one might expect the frequency to reduce by 50% across the board, this is not the case from our computer model. The top attraction sites are preserved. Comparing the different use levels gives some insight into the effect of reducing the number of launches. The historical popularity of the top activity sites is maintained at the 50% use level, but the amount of use at particular sites is reduced (fig. 12).

The oars-only and 100% use levels seem to indicate that the trips, on average, travel down river faster in the oars-only setting. Figure 13 examines trips of 16-day duration. It shows the average travel pattern, graphing the day number along the horizontal axis and the river mile (on average) along the

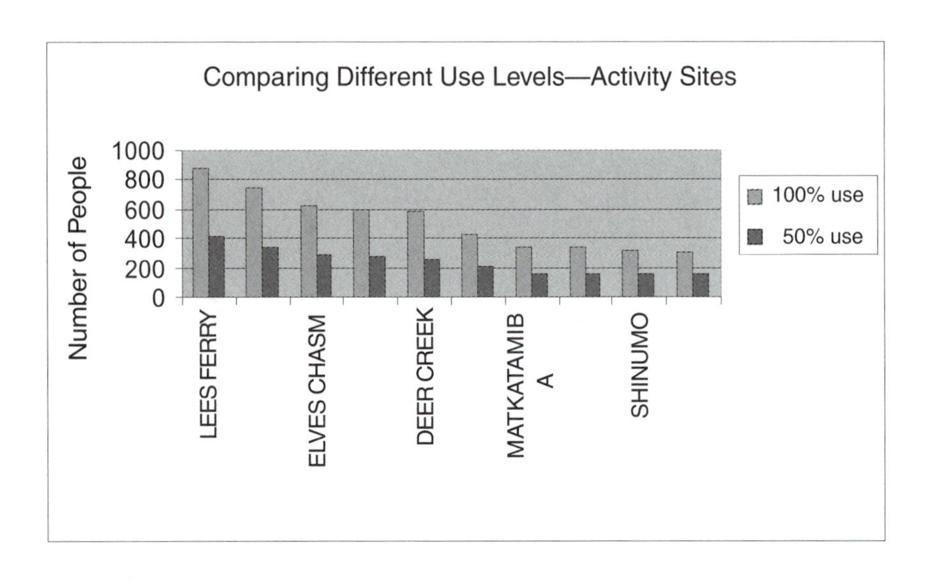

FIGURE 12 Different use levels at activity sites.

vertical axis. The data points represent the average camp location for each night of the trip.

In both simulations, trips that are 16 days in duration are nonmotorized (e.g., oar). The surrounding settings for these two simulations, however, were different. Under the 100% use level, these oar trips negotiate for campsites with a mixture of motorized and nonmotorized trips. In the oars-only scenario, trips are negotiating only with other oar-powered trips. The simulation suggests that the oars-only setting pushes trips farther downriver sooner than is the case in the mixed setting. This could be explained as due, in part, to the potential for more campsite competition among boats traveling with a similar, slower speed. Typically, when trips encounter each other, they exchange information and negotiate campsites for the upcoming evening. When the boats travel at similar speeds, there are fewer opportunities for communication. In the oars-only scenario, there are no faster motorboats passing oar trips and engaging in communication and planning about the upcoming campsite selections. Consequently, boats could find themselves coming upon more unexpected occupied sites in the late afternoon. Trips that are going similar speeds don't have as much flexibility when camp time arrives—and as they discover occupied campsites, they will move further down river to find an available site.

An important difference to note when comparing simulated data and real data is in the difference in the number of trips in the query. When comparing real data with simulated data, it is important to recognize that the real data

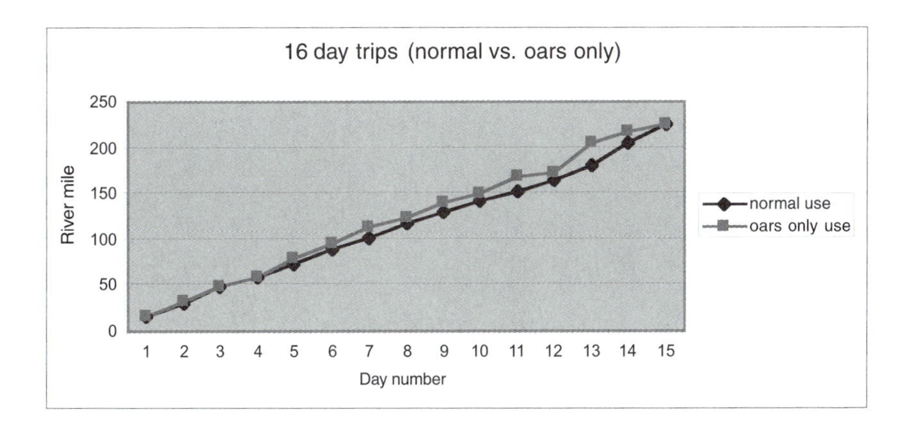

FIGURE 13 Examines trips of 16-day duration.

represent only about 50% of the trips, whereas the simulated data have information for every trip on the launch schedule. One way to keep this in mind is to look at only the trend, or average, of the river trips. Another approach is to look at frequency rates that are scaled to the number of trips represented overall. These approaches allow the user to see the differences between the simulated trips and the real data more accurately. A third alternative is to compare a 100% use simulation with the 50% use simulation.

At the writing of this chapter the 1999 trip reports have not been analyzed, this is an ongoing project of refinement and analysis to create the best possible representation of the behavior of river trips on the Colorado River. The real test of GCRTS will be in the use of the simulator to examine the outcomes of various alternative launch schedules. The insight that can be provided by GCRTS is expected to be a valuable contribution to a complex situation, that of managing rafting traffic on the Colorado River in an optimal way for both the river rafters and the natural resource itself.

15 SUMMARY AND CONCLUSIONS

This chapter has outlined some initial attempts at integrating multi-intelligent agents and rule-based decision-making algorithms within a GIS-based environment to model complex human-environment interactions. Using agents to represent individuals or parties, incorporating GIS to represent the environment, and utilizing intelligent agent technology in modeling of river management issues provides a number of important advantages. Agents used to represent trips can be programmed with strategies, goals, intentions, and negotiation strategies that learn about and can be adaptive to their surroundings and others they encounter. Using GIS to represent environmental setting provides the user with an effective way to view agent interactions and assess the number of encounters and where they occur. In addition, a GIS approach incorporating visualization makes simulation outcomes easier for policymakers, planners, managers, and the public to understand. Using intelligent agent technology allows resource managers to develop and test "what if" scenarios and to explore management options for better resolving recreation use and interaction problems. The model aids managers in establishing more appropriate limits of use to meet mandated standards for social, environmental, and economic impacts

This chapter has described the current state of development of the computer simulation system for modeling complex human-environment interactions in the context of river rafting trips. The goal of the system was to achieve the quantitative precision and specificity of a multivariate statistical model at the same time maintaining the flexibility of AI-based rule systems. An important advantage of representing complex models in a computer simulation is that the results of the model can be displayed, or visualized in a way that can provide additional insights into the effects and interactions that are represented. Information about how to construct such a system could contribute to the solution of a wide range of scientific and practical problems raised in human-environment interactions.

ACKNOWLEDGMENTS

This project was a cooperative effort between the University of Arizona, Northern Arizona University, and Grand Canyon National Park. The authors would like to thank the USDI National Park Service, Grand Canyon National Park for their vital support and assistance. Special thanks are due to Linda Jalbert, Lauri Domler, and Jennifer Burns for their personal perserverance and contributions to the effort. Part of this research was supported by cooperative agreements between the University of Arizona, Northern Arizona University, and National Park Service (cooperative agreement numbers 1443-CA8601-97-006 and CA8210-99-002 respectively), McIntire Stennis project number ARZT-139238-M-12-142; NAU-NASA Space Grant; and NAU Organized Research Grant.

Information on the River or project model can be obtained at ⟨http://odin.math.nav.edu/msl/⟩ or ⟨http://tdserv.psych.arizona.edu/grand-canyon/⟩ or ⟨http://www.nps.gov/grca/colorado/newsltr/aug98/pageone.htm/⟩.

REFERENCES

[1] Anderson, J., and M. Evans. "A Generic Simulation System for Intelligent Agent Designs." *Appl. Art. Intel.* **9(5)** (1995): 527–562.

[2] Ball, G. L. "Ecosystem Modeling in GIS." *Envir. Mgmt.* **18(3)** (1994): 345–349.

[3] Berry, J. S., G. Belovsky, A. Joern, W. P. Kemp, and J. Onsager. "Object-Oriented Simulation Model of Rangeland Grasshopper Population Dynamics." In *Proceedings of Fourth Annual Conference on AI, Simulation, and Planning in High Autonomy Systems*, edited by B. Zieglier and S. Rozenblit, 12–108. Los Alamitos, CA: IEEE Computer Society Press, 1993.

[4] Bieri, J., and C. Roberts. "Using the Grand Canyon River Trip Simulator to Test New Launch Schedules on the Colorado River." *Assoc. for Women in Sci. Newslett.* (August 2000).

[5] Bishop, I. D., and H. R. Gimblett. "Modeling Tourist Behaviour: Geographic Information Systems and Autonomous Agents." Paper presented at the 1st International Scientific Congress on Tourism and Culture for Sustainable Development, May 19–21, Athens, Greece, 1998.

[6] Bishop, I, D., and H. R. Gimblett. "Management of Recreational Areas: Geographic Information Systems, Autonomous Agents and Virtual Reality." *Envir. & Plan. B.* **27** (2000): 423–435.

[7] Bojadziev, G. *Fuzzy Sets, Fuzzy Logic, Applications.* New Jersey: World Scientific, 1995.

[8] Borkan, R. E. "Simulating the Effects of Glen Canyon Dam Releases on Grand Canyon River Trips." Unpublished Masters Thesis, School of Renewable Natural Resources, University of Arizona, Tucson, AZ, 1986.

[9] Borkan, R. E., and A. H. Underhill. "Simulating the Effects of Glen Canyon Dam Releases on Grand Canyon River Trips." *Envir. Mgmt.* **13(3)** (1989): 347–354.

[10] Box, P. "Spatial Units as Agents: Making the Landscape an Equal Player in Agent-Based Simulations." This volume.

[11] Briggs, D., J. Westervelt, S. Levi, and S. Harper. "A Desert Tortise Spatially Explicit Population Model." In *Proceedings of the Third International Conference/Workshop on Integrating GIS and Environmental Modeling*, January 21–25, 1996, Santa Barbara, CA: National Center for Geographic Information and Analysis. ⟨http://www.ncgia.ucsb.edu/conf/SANTA_FE_CD-ROM/main.html⟩.

[12] Brown, J. H. "Modeling Ecological Patterns and Processes using Agent-Based Simulations and GIS." In *Proceedings of the Third International Conference/Workshop on Integrating GIS and Environmental Modeling*, January 21–25, 1996, Santa Barbara, CA: National Center for Geographic Information and Analysis. ⟨http://www.ncgia.ucsb.edu/conf/SANTA_FE_CD-ROM/main.html⟩.

[13] Cherry, S. "Modeling of River Trips on the Colorado River in Grand Canyon National Park." Unpublished paper, School of Renewable Natural Resources, University of Arizona, Tucson, AZ, 1997.

[14] Colorado River Management Plan. Grand Canyon National Park. Limits of Acceptable Change, Page B-4-B-6, 1989.

[15] Conte, R., and N. Gilbert, eds. "Computer Simulation for Social Theory." In *Artificial Societies: The Computer Simulation of Social Life*, 1–15. London: UCL Press, 1995.

[16] DeAngelis, D. L., D. M. Fleming, L. J. Gross, and W. F. Wolff. "Individual-Based Modeling in Ecology: An Overview." In *Proceedings of the Third International Conference/Workshop on Integrating GIS and Environmental Modeling*, January 21–25, 1996, Santa Barbara, CA: National Center for Geographic Information and Analysis. ⟨http://www.ncgia.ucsb.edu/conf/SANTA_FE_CD-ROM/main.html⟩.

[17] Drogoul, A., and J. Ferber. "Multi-Agent Simulation as a Tool for Studying Emergent Processes in Societies." In *Simulating Societies: The Computer Simulation of Social Phenomena*, edited by Gilbert and Doran, 127–142. London: UCL Press, 1995.

[18] Ferrand, N."Multi-Reactive-Agents Paradigm for Spatial Modeling." Contribution to the European Science Foundation GISDATA program, Spatial Models and GIS, ESF-GIS DATA. Stockholm: Taylor and Francis, 1996.

[19] Findler, N. V., and R. M. Malyankar. "Emergent Behavior in Societies of Heterogeneous, Interacting Agents; Alliances and Norms." In *Artificial Societies: The Computer Simulation of Social Life*, edited by N. Gilbert and R. Conte, 212–236. London: UCL Press, 1995.

[20] Forrest, S., and T. Jones. "Modeling Complex Adaptive Systems with Echo." In *Complex Systems. Mechanism of Adaptation*, edited by Russel J. Stonier and Xing Huo Yu, 3–20. The Netherlands: IOS Press, 1994.

[21] Flynn. M. M. "A Method for Assessing Near-View Scenic Beauty Models: A Comparison of Neural Networks and Multiple Linear Regression." Unpublished Master's Thesis, School of Renewable Natural Resources, The University of Arizona, Tucson, AZ, 1997.

[22] Gimblett, H. R., G. L. Ball, and A. W. Guisse. "Autonomous Rule Generation and Assessment for Complex Spatial Modeling." *Landscape & Urban Plan. J.* **30** (1994): 13–26.

[23] Gimblett, H. R., B. Durnota, R. M. Itami. "Spatially-Explicit Autonomous Agents for Modeling Recreation Use in Complex Wilderness Landscapes." *Complex. Intl. J.* **3** (1996).

[24] Gimblett, H. R., R. M. Itami, and B. Durnota. "Some Practical Issues in Designing and Calibrating Artificial Human Agents in GIS-Based Simulated Worlds." *Complex. Intl. J.* **3** (1996).

[25] Gimblett, H. R., and R. M. Itami. "Modeling the Spatial Dynamics and Social Interaction of Human Recreationists using GIS and Intelligent

Agents." Paper presented at MODSIM 97—International Congress on Modeling and Simulation, Hobart, Tasmania, December 8–11, 1997.

[26] Gimblett, H. R. "Simulating Recreation Behavior in Complex Wilderness Landscapes using Spatially-Explicit Autonomous Agents." Unpublished Ph.D. diss., University of Melbourne, Parkville, Victoria, 3052 Australia, 1998.

[27] Gimblett, H. R., R. M. Itami, and M. Richards. "Simulating Wildland Recreation Use and Conflicting Spatial Interactions using Rule-Driven Intelligent Agents." This volume.

[28] Goodchild, M., B. O. Parks, and L. T. Steyaert, eds. *Environmental Modeling with GIS.* New York: Oxford University Press, 1993.

[29] Grand Canyon National Park's first Wilderness Management Plan. Grand Canyon National Park, Grand Canyon, AZ, 1998.

[30] Green, D. G. "Spatial Simulation of Fire in Plant Communities." In *Proceedings of National Symposium on Computer Modeling and Remote Sensing in Bushfire Prevention*, edited by P. Wise, 36–41. Canberra: National Mapping, 1987.

[31] Hall, T., and B. Shelby. "Evaluating Social Conditions on the Colorado River in the Grand Canyon: Effects on Experiences and Changes over Time." In *Proceedings of 1999 Congress on Recreation and Resource Capacity*, Conference held at Snowmass Village in Aspen, CO, on November 29–December 2, 1999.

[32] Hiebeler, D. "The Swarm Simulation System and Individual-Based Modeling." In *Decision Support 2001. 17th Annual Geographic Information Seminar and the Resource Technology '94 Symposium*, edited by J. Michael Power, Murray Strome, and T. C. Daniel, vol. 1, 474–494. Toronto, Canada: American Society for Photogrammetry and Remote Sensing, 1994.

[33] Itami, R. M. "Simulating Spatial Dynamics: Cellular Automata Theory." *Landscape & Urban Plan. J.* **30** (1994): 27–47.

[34] Itami, R. M. "Cellular Worlds: Models for Dynamic Conceptions of Landscapes." *Landscape Architecture* **78(5)** (1988): 52–57.

[35] Jalbert, L. M. "Monitoring Visitor Distribution and Use Patterns along the Colorado River Corridor: River Contact Survey and Attraction Site Monitoring." Resource File in Grand Canyon Science Center, Grand Canyon National Park, 1990.

[36] Jalbert, L. M. "Monitoring Visitor Distribution and Use Patterns Along the Colorado River Corridor: River Contact Survey and Attraction Site Monitoring." Resource File in Grand Canyon Science Center, Grand Canyon National Park, 1991.

[37] Jalbert, L. M. "The Influence of Discharge on Recreational Values including Crowding and Congestions and Safety in Grand Canyon National Park." Division of Resources Managment in Cooperation with Glen Canyon Environmental Studies, Grand Canyon National Park, 1992.

[38] Jalbert. L. M. "Colorado River Management; Resource Monitoring Program: Recreational Impacts to Camp and Attraction Site Quality, River Rehabilitation and Visitor Experience, Program Summary." Resource File in Grand Canyon Science Center, Grand Canyon National Park, 1993.

[39] Jalbert. L. M. "Grand Canyon River Management: Balancing Use, Preserving Wilderness Values, and Establishing the Need for Research." *Proceedings of 1999 Congress on Recreation and Resource Capacity*, Conference held at Snowmass Village in Aspen, CO, on November 29–December 2, 1999.

[40] Johnson, K. M. "Using Statistical Regression Analysis to Build Three Prototype GIS Wildlife Models." *GIS/LIS* (1992): 374–386.

[41] Jones, T., and S. Forrest. "An Introduction to SFI Echo." Working Paper 93-12-074, Santa Fe Institute, Santa Fe, NM, 1993.

[42] Kohler, T. A., C. R. Van West, E. P. Carr, and C. G. Langton. "Agent-Based Modeling of Preshistoric Settlement Systems in the Northern American Southwest." In *Proceedings, Third International Conference/Workshop on Integrating GIS and Environmental Modeling*, Santa Fe, NM, January 21–26, 1996. Santa Barbara, CA: National Center for Geographic Information and Analysis.

[43] Langton, C., N. Minar, and R. Burkhart. 1995. "The Swarm Simulation System: A Tool for Studying Complex Systems." Santa Fe Institute, Santa Fe, New Mexico, USA. ⟨http://www.santafe.edu/projects/swarm/~swarmdoc/swarmdoc.html.⟩.

[44] Langton, C., I. Lee, and Swarm Team. "Spatially Explicit Modeling with the Swarm Multi-Agent Software Package." Paper presented at GIS '98 Toronto, Ontario, Canada, 1998.

[45] Manneville, P., N. Boccara, G. Y. Yichniac, and R. Bidaux, eds. *Cellular Automata and Modeling of Complex Physical Systems.* New York: Springer, 1989.

[46] Rechel, J. L. "Geographic Information Systems Modeling of Wildlife Movements across Fire and Urban Disturbed Landscapes." *Bull. Ecol. Soc. Am.* **73(2)** (1992): 316.

[47] Roberts, C., and R. Gimblett. "Computer Simulations for Rafting Traffic on the Colorado River." In *Proceedings of 5th Biennial Conference of Research on Colorado Plateau.* U.S. Department Interior, U.S. Geological Survey, 2000.

[48] Roberts, C. "Expert Interviews with Grand Canyon River Guides." Unpublished paper, Northern Arizona University, Flagstaff, AZ, 1998.

[49] Saarenmaa, H., J. Perttunen, J. Vakeva, and A. Nikula. "Object-Oriented Modeling of the Tasks and Agent in Integrated Forest Health Management." *AI Appl. Natl. Res. Mgmt.* **8(1)** (1994): 43–59.

[50] Shechter, M., and R. L. Lucus. *Simulation of Recreational Use for Park and Wilderness Management.* Washington, DC: Johns Hopkins University Press for Resources for the Future, 1978.

[51] Shelby, Bo, and J. M. Nielsen. "Design and Method of the Sociological Research in the Grand Canyon." Technical Report #1, Colorado River Research Program Report Series, U.S. Department of Interior, National Park Service, 1976.

[52] Shelby, Bo, and J. M. Nielsen. "Motors and Oars in the Grand Canyon." Technical Report #2, Colorado River Research Program Report Series, U.S. Department of Interior, National Park Service. 1976.

[53] Shelby, Bo, and J. M. Nielsen. "Use Levels and Crowding in the Grand Canyon." Technical Report #3, Colorado River Research Program Report Series, U.S. Department of Interior, National Park Service, 1976.

[54] Shelby, Bo, and J. M. Nielsen. "Private and Commercial Trips on the Grand Canyon." Technical Report #4, Colorado River Research Program Report Series, U.S. Department of Interior, National Park Service, 1976.

[55] Slothower, R. L., P. A. Schwarz, and K. M. Johnson. "Some Guidelines for Implementing Spatially Explicit, Individual-Based Ecological Models within Location-Based Raster GIS." In *Proceedings of the Third International Conference/Workshop on Integrating GIS and Environmental Modeling*, January 21–25, 1996, Santa Barbara, CA: National Center for Geographic Information and Analysis. ⟨http://www.ncgia.ucsb.edu/conf/SANTA_FE_CD-ROM/main.html⟩.

[56] Sylvester, S. M., and L. J. Gross. "Integrating Spatial Data into an Agent-Based Modeling System: Ideas and Lessons from the Development of the Across-Trophic-Level System Simulation (ATLSS)." This volume.

[57] Tobler, W. R. "Cellular Geography." In *Philosophy in Geography*, edited by S. Gale and G. Olsson, 379–386. Dordrecht, The Netherlands: D. Reidel, 1979.

[58] Underhill, A. H., and A. B. Xaba. "The Wilderness Simulation Model as a Management Tool for the Colorado River in Grand Canyon National Park." Technical Report 11, CPSU/UA, Tucson, AZ, 1983.

[59] Underhill, A. H., A. B. Xaba, and R. E. Borkan. "The Wilderness Simulation Model Applied to Colorado River Boating in Grand Canyon National Park, USA." *Envir. Mgmt.* **10(3)** (1986): 367–374.

[60] Welstad, S. *Neural Net and Fuzzy Logic Applications in C/C++*. New York: John Wiley & Sons, 1994.

[61] Westervelt, J., and L. D. Hopkins. "Facilitating Mobile Objects within the Context of Simulated Landscape Processes." In *Proceedings of the Third International Conference/Workshop on Integrating GIS and Environmental Modeling*, January 21–25, 1996, Santa Barbara, CA: National Center for Geographic Information and Analysis. ⟨http://www.ncgia.ucsb.edu/conf/SANTA_FE_CD-ROM/main.html⟩.

Agent-Based Simulations of Household Decision Making and Land Use Change near Altamira, Brazil

Kevin Lim
Peter J. Deadman
Emilio Moran
Eduardo Brondizio
Stephen McCracken

1 INTRODUCTION

Individuals who influence decisions regarding the use of land, operate within a complex environment comprised of interacting elements that include both natural systems and human institutions. Individually, the elements of the natural and human systems that influence land-use decisions may be very complex. Within natural systems, dynamic processes, such as the hydrological cycle, and the distribution of biophysical resources, such as soil fertility, influence land-use decision making. Elements of an individual's institutional environment can also influence the options and incentives that are available to an individual, and thus the land-use decisions that they make. Understanding the nature of these complex processes and interactions is a nontrivial task. However, agent-based simulation offers researchers a tool to better understand the nature of these complex systems.

The recent development of computer simulation technologies by social scientists has provided a tool for not only predicting social phenomena, but also for better understanding the nature of these human systems. Replicative validity is not the goal of many social simulation efforts. Instead, researchers have focused on developing relatively simple simulations as tools for understanding the properties of social systems and the way in which interactions between actors at the local level results in the emergence of behaviors or phenomena at the global level [8]. In this role, simulation becomes a tool for evaluating assumptions and exercising theories of action [1].

Many of the techniques applied to social simulation can be traced back to earlier developments in the physical or natural sciences. For example, computer simulation has a relatively long history in the natural sciences in applications related to fisheries, forest environments, and watersheds. But recent advances in computer hardware and software technologies have made these technologies accessible to social scientists. Recently, we have seen simulation efforts that have included models of not only the natural system in question, but also the human system with which it interacts. In fields such as anthropology [6, 13, 14] and resource management [3, 5, 27], human systems simulations are being developed which directly address the actions of human individuals or groups as they interact with a natural system.

This approach to simulation is pursued in this chapter. A model of individual human decision making at the household level is linked through a geographical space to a model of ecosystem behavior. The goal of this modeling exercise is to explore the potential of a spatially referenced agent-based model, for understanding how behavior at the local level interacts with natural processes to produce observable phenomena at a higher level. We explore this goal with an application that focuses on the land-use decisions made by individual households within a region of the Amazon rainforest near Altamira, Brazil. The simulation described in this chapter is the product of a pilot effort between the University of Waterloo and Indiana University designed to explore proposed theories of land-use change in this region. Although still preliminary in its scope, this simulation illustrates the potential of such a spatially referenced agent-based approach for better understanding the complex human and natural processes that interact within this region.

The next section of this chapter discusses the history of land use in the Altamira region and outlines the importance of understanding land-use processes at the farm level. Subsequent sections describe the structure of the land-use change in the Amazon (LUCITA) simulation system, and the initial findings that have emerged from an analysis of the model's behavior. A final discussion addresses the strengths and weaknesses of this simulation in the context of land-use change and social simulation research.

2 LAND-USE CHANGE IN THE BRAZILIAN AMAZON

The Brazilian Amazon has been experiencing marked changes in the past 30 years [10, 23, 28]. From an area that in 1975 had less than 1 percent of its forest cover removed, the Basin is already 15 percent deforested. Deforestation has proceeded from east to west, along roads and along an "arc of deforestation" along the southern periphery. In these areas, rates of deforestation have been in excess of one percent per year in the past two decades with its peak in

1995 [12]. This massive change in land cover is a result of national decisions to integrate the region into national economic development, by means of a two pronged approach that combined massive road building with colonization and resettlement projects [22, 26]. Attractive subsidies and tax incentives, land title, and access to extension services made moving to the Amazon economically profitable for both large and small landholders.

Before this most recent set of events, the history of land use in the Brazilian Amazon had been characterized by economic development along river banks, which limited occupation to a small portion of the Basin. In the colonial period this took the form of searching for spices, slaves, and some valuable wood species. Under Jesuit tutelage, some of the missions successfully developed cocoa plantations, cattle ranches, and other surplus production, but they collapsed in the eighteenth century following the expulsion of the Jesuits from Brazil. In the national period, the Amazon experienced massive population dislocations to exploit natural rubber (1880–1920), but in which great wealth was achieved by a few at the cost of the many. A shorter-lived rubber boom took place during World War II when the Malaysian rubber plantations' supplies to the Allies were cut off and Brazilian natural rubber was desperately needed. Following these booms, the Amazonian towns stagnated economically, lost population, and persisted by barter and subsistence production.

All this began to change after World War II as nationalist leaders began to see the vast Amazon frontier, accounting for 58 percent of the Brazilian territory, as an important component of achieving world power status. The March to the West began to be seen as a valuable geopolitical objective. With the assumption of power by the military in 1964, implementation of these objectives began to take place very quickly. The Transamazon Highway, running east-west across the Basin, was a particularly important component of this geopolitical plan, and it was backed up with a coordinated plan of incentives to attract both small and large interests to the region.

The showcase for the colonization part of the project was the Altamira Integrated Development Project that began in 1971. From a town of about 1,000 people in 1970, the town grew to over 10,000 in one year [22], and it has continued to grow steadily since then to over 85,000 in the 1990s. Of all the colonization projects along the Transamazon Highway, Altamira was the only one blessed with above-average quality soils, less stagnant water due to a rolling terrain, and therefore less malaria. Communities of 48 homes were built every 10 km to facilitate community life, while still maintaining reasonable distances to the properties. The project was laid out systematically into 100-hectare properties in a rasterized fashion. The layout has come to be called "fishbone pattern" because land was allocated along the main trunk of the highway, as well as along side roads spaced symmetrically every 5 km. Small landholders from throughout Brazil came to the area to claim their properties with over 6,000 families coming in the first decade to the Altamira region. Brazil being a very large country with very different climates and cultural traditions, the immigrants brought with them varied approaches to land use that require that attention be paid to household behavior, rather than assume that they all behave in ethnically equivalent terms.

Farmers from northeast Brazil, accounted for about 30 percent of initial settlers. They came from a land characterized by cyclical droughts, and irrigated agriculture in very small plots along river banks. These proved to be among the most and least educated of the settlers, with a combination of previous landowners of small, irrigated plots and sharecroppers on large properties and plantations. They differed among themselves as much or more than they differed as a group from those of other regions of Brazil. Farmers from the Amazon region accounted for another 30 percent, and they were mostly descendants of rubber tappers, swidden cultivators living along river banks who moved to the roadside properties to get a legal title to land. They were familiar with the local forest species and, with effective ways to recognize good soils, preferred cultivation of manioc and cowpeas as foolproof crops for the area. Another 25 percent came from southern Brazil, and while the government hoped that they would be prime examples of modern agriculture, many of them turned out to be coffee plantation sharecroppers who had left the northeast seeking their fortune in southern Brazil in the previous generation. The remaining group, from the Central-West region, was mostly familiar with cattle ranching at small scale and was seeking to expand their holdings.

From this mix of immigrants the Altamira project started and, over the past 30 years, there has been significant turnover in ownership, with less than 30 percent of the original households remaining on the land. Recent analysis of our data suggests that original households who selected the best soils in the area (i.e., the alfisols) have been remarkably successful in holding on to their land, and that most of those properties have not entered the real estate market to benefit later-arriving settlers [19]. This agent-based simulation benefits from household-level data collected by one of the co-authors in the first three years of settlement [20, 21, 22], and subsequent and more extensive household survey research in 1997–1999 [4, 16, 17, 19]. It also benefits from very intensive studies of land use and land cover analysis, with a focus on the dynamics of secondary succession in the first half of the 1990s [15, 24, 25].

These previous analyses, however, have not undertaken the challenges posed by agent-based modeling within a spatially explicit framework as is proposed here. While the LUCITA simulation system described here is based on these previous analyses, its focus on simulating land-use decisions at the individual household level is inspired by a model outlined in McCracken et al. [16] that focuses on frontier occupation and environmental change as process. This model proposes that land-use changes in the Altamira region should be understood, not only as a result of large-scale, temporally defined effects such as changing credit policies, but also as a product of local household-level effects, such as the age and gender characteristics of farm families. This model maps out a trajectory for families, which relates the type of agricultural practices pursued to a number of factors including the available labor pool within each household. The model describes five stages in the life of a household. In the early stages of household development, limited family labor supplies lead to a reliance on annual crops and associated high rates of deforestation. In the later stages of household development, larger labor and capital resources allow for the development of pastoral lands and/or perennial crops.

Actual trajectories of household agricultural strategies are not as clear as those suggested by the conceptual model [16]. This raises the question of how families make land-use decisions, given the characteristics of their natural environment (such as soils, topography, and water availability), their economic environment (such as distance to markets, credit policies, and commodity prices), and their own households. It is intended that the development of the LUCITA simulation system will eventually provide researchers with an additional tool for exploring these questions. The structure of the initial version of LUCITA, and some observations of its initial behavioral characteristics, are described in the following sections.

3 LUCITA MODEL DESIGN

3.1 OVERVIEW

The LUCITA model was developed using the Swarm simulation system [18], a set of software libraries written in the Objective-C object-oriented programming language to help facilitate the modeling and simulation of complex adaptive systems. LUCITA is comprised of two submodels that interact with one another through a spatially referenced raster landscape. These two submodels are utilized to capture both the ecological and human dynamics and processes characteristic of the target system. Not only do complex feedback loops exist within each submodel, but also indirectly between the two submodels through the landscape. It is the representation of these intrafeedback and interfeedback loops of the target system that makes the LUCITA model unique from other spatially referenced agent-based models (ABMs).

The basis of the ecological submodel is derived from the KPROG2 model, originally developed by Fearnside [7] to estimate human carrying capacities in regions of the Transamazon Highway. Multiple regression equations for changes in soil characters and the estimation of crop yields were adopted from the KPROG2 model for use in LUCITA. The dynamics associated with changes in soil fertility due to varying agricultural practices and the process of secondary succession could not have been modeled without the multiple regression equations provided by KPROG2. Thus, the ecological submodel is capable of modeling the impacts of deforestation on soil properties, the relationship between soil fertility and successful crop yields, and the effect of soil properties on the rates of natural reforestation.

The human system submodel can be best described by the architecture of an autonomous household agent. Each household agent is representative of a colonist family and is defined by the composition of the family, available family and male labor pools, and available liquid capital. Decision making, with respect to what agricultural land strategy should be adopted for any given patch of land, is governed by a nonevolving classifier system. Land-use strategies, or rules, are represented by binary strings based on principles of genetic algorithms and compete with one another for selection. Those rules that are successful, defined as rules satisfying some type of a threshold level, are rewarded. Ideally, successful rules are reinforced through simulated time and poor rules excluded from future agent decision making. This agent architecture provides a framework to test the conceptual model of household transition [16], where it is hypothesized that as colonist households age in the frontier, decision-making shifts from deforestation intensive strategies, which require minimal labor and capital requirements, and have lower economic returns, to those with high economic returns, that demand less deforested land, but require greater labor and capital inputs.

3.2 THE SWARM SIMULATION SYSTEM

The Swarm simulation system, generally referred to simply as Swarm, was originally developed by a team of researchers at the Santa Fe Institute to assist the study of complex adaptive systems [18]. The motivation for the research and development of Swarm was the recognition of the importance of computer models as a research tool, the fact that most researchers are not software engineers and that too much time was being wasted on writing poor software code rather than focusing on research, and the need for a standardized suite of tools to facilitate the development of reproducible computer models. Swarm is a set of software libraries written in Objective-C, an object-oriented programming language, and makes no formal assumptions of the type of model being developed. This implies that Swarm can be used in a wide array of scientific disciplines such as chemistry, economics, and anthropology.

The basic unit of a Swarm simulation is the agent, where an agent is defined as any type of actor within a system that is capable of generating events that are able to impact itself, other agents, or the surrounding environment [18]. Interactions of an agent with itself, other agents, and its surrounding environment are made via discrete events. A Swarm simulation is comprised of a schedule of discrete events defining a series of processes taking place with a collection of agents. Drawing from the object-oriented programming paradigm, a swarm agent is modeled as an object. Any object has both a state and behavior. Object variables are used to describe the state of an object, where the behavior of an object is defined by the class from which it was instantiated.

3.3 SPATIAL DATA LANDSCAPE IN GEOGRAPHIC INFORMATION SYSTEMS

The raster landscape is representative of the intensive study area documented in the KPROG2 model [7]. The study area is situated in the vicinity of Agrovila (village) Grande Esperança, in the municipality of Prainha, in the state of Pará. The area is approximately 50 km west of Altamira. The primary reason for selecting this study area was because of the availability of soils data, such as pH. Although the study area differs from the study area documented by Moran [22], from which the source of household behavior data is obtained, it can be argued that the conceptual model of household transition applies to all regions of the Amazon Basin, irrespective of a specific geographic location. For example, given the availability of soils data in the Moran study area, there is no reason why LUCITA could not spatially reference that location.

An area comprised of 236 properties, each 100 hectares in area, is represented by a raster landscape. Properties that are adjacent to the Transamazon Highway have a lot dimension of 500 m by 2000 m and those located off on feeder roads with a lot dimension of 2500 m by 400 m. Each raster cell has a grid resolution of 100 m, representative of an area of 1 hectare. For the purpose of generating a raster landscape, each property lot is assumed to be rectangular in shape. Several property lots in the soils data maps were not rectangular in shape and, therefore, a geometric transformation of the property lots were required so that both the property layout in the data maps matched the property lot generated within the LUCITA model. Given the assumption of rectangular-shaped property lots in the landscape and that the data maps were not to scale, meaning that from a visual analysis, all property lots were not 100 ha in area, digitizing the data maps and subsequently converting them to a raster format for initial soil parameter input into LUCITA was not an option. Instead, a text file, representative of the Swarm landscape, was generated and imported into the ARC/INFO GRID module, where a process of manually adjusting cell values to match the soils data maps was performed. The GRID data layers of pH, carbon, nitrogen, phosphorus, and aluminum, were converted back into a text file for import into the LUCITA model. The untransformed landscape layout, the transformed landscape generated by Swarm as displayed by ARC/INFO GRID, and an example of a soils data map is depicted in figure 1.

For each landscape grid cell, a one-to-one reference exists between a grid cell and an environment object instantiated from an environment class. This implies that for any given household property lot, there exists one hundred environment objects since one property lot is composed of one hundred grid cells. The purposes of an environment object is to provide the spatial grid coordinates of a particular patch of land with respect to the artificial landscape, to differentiate one patch from another through the use of unique internal keys, to store the current land cover for a particular grid cell, to keep a tally of the number of years a grid cell has been used continuously and for what

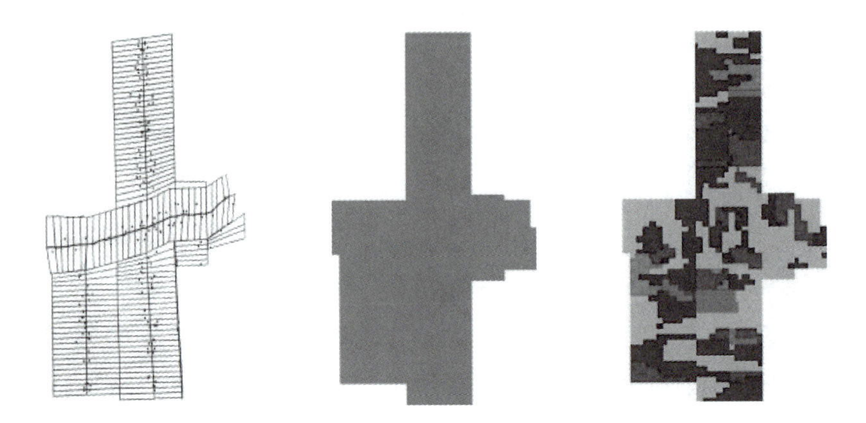

FIGURE 1 Illustration of property lot layout in data maps (left), where dots are representative of sampled areas, the raster layout in ARC/INFO GRID (center), and a sample LUCITA data map (right), in this case representative of pH categories.

land use, and to act as an interface to spatially link a soil object to a grid cell for KPROG2 purposes described in the following section.

During any simulation, the landscape is responsible for tabulating land cover frequencies and for managing the transition of land from one type of land cover to another, while enforcing rules such as the number of maximum years a patch of land may be used in continuous agriculture and the minimum number of years a patch of land must remain in fallow prior to reuse. For instance, manioc, which has a growing season just over a year, renders a patch of land unavailable for use for two years. Accordingly, the landscape will monitor and identify when this patch has satisfied the growing requirements of the manioc land-use strategy and subsequently release the patch of land for future use at the appropriate time event. Similarly, stages of secondary succession, categorized by age, are also defined internally by the landscape. Conceptually, the landscape can be thought of as a land manager, which simply monitors what proportion of land is being used for a particular land use and what state those patches of land should be set to for the next time event.

3.4 KPROG2 MULTIPLE REGRESSION EQUATIONS

As mentioned in the previous section, for any given grid cell an environment object exists and it is through a reference in the environment object that a soil object is linked to a grid cell. The KPROG2 multiple regression equations and the parameters required by these equations to model crop yields, soil changes caused by burning, soil changes under land covers that are not burned (i.e., secondary succession), soil changes under pasture, and soil changes as a result of the application of fertilizers and lime, are contained with the definition of the soil class. The specific soil parameters that are required by the multiple regression equations required for calculating crop yields and soil processes include levels of pH, nitrogen, carbon, phosphorus, and the concentration of aluminum ions.

In slash-and-burn agriculture, a farmer will deforest a patch of land and subsequently burn in an attempt to alter soil conditions to improve crop yields. Three types of land covers are considered by a farmer for clearing and burning and they include virgin forest, secondary forest, and weedland covers. Each of these three land covers has a set of multiple regression equations that describes how the soil parameters are to change if it were to be cleared and burned. Under circumstances where no burning is required, such as land in pasture, land in secondary succession, or land in continuous agricultural use, changes in soil parameters are governed by other sets of multiple regression equations. Maintenance of cacao and black pepper plantations often require the application of fertilizers to raise phosphorus concentrations and lime to reduce the acidity of soil, both critical criteria for good perennial crop yields. Changes in soil parameters resulting from the application of fertilizers and lime is no different than other soil processes that are modeled using a distinct set of multiple regression equations. For instance, the equation used to model changes in nitrogen after a virgin forest burn is described in equation (1) below. A complete review of all the multiple regression equations adopted from the KPROG2 model is documented by Fearnside [7].

Changes in nitrogen are represented by

$$Y = 5.80 \times 10^{-2} - 0.654A + 4.89 \times 10^{-2}B + 2.63 \times 10^{-2}C$$

where

$$\begin{aligned} Y &= \text{nitrogen change (\% dry weight)} \\ A &= \text{initial nitrogen (\% dry weight)} \\ B &= \text{initial carbon (\% dry weight)} \\ C &= \text{initial pH} \end{aligned} \tag{1}$$

3.5 AGENT ARCHITECTURE AND DECISION MAKING

For any given property lot of 100 grid cells on the raster landscape, an instantiation of a household object class exists and is referenced to those cells. Each agent has an internal representation of its environment and itself. An agent has an internal representation of the environment in that each agent is aware of the boundaries of its artificial world within which it exists, the components of that artificial world that it is capable of impacting, and the types of land covers characterizing its immediate surroundings. Further, an agent, using its internal representation of itself, is capable of describing its family composition, both total family members and the number of males in the family, the available capital resources, and the land-use strategies it is capable of implementing.

The behavior of an agent can be described by a set of actions that an agent is capable of executing repeatedly throughout a simulation. In general, these actions tend to deal with the clearing of land, the burning of deforested land, the growing of agricultural crops, and the harvest of crops sown in a given year. In most instances, labor and capital resources are required by these actions. The ability of an agent to identify which patches of land from its property to deforest and burn is determined by a set of clearing preferences defined by the end user at the start of a simulation. Obviously, variations in clearing preferences can affect the rate of deforestation on a given property lot or across the landscape for that matter. For instance, if an agent's first clearing preference is virgin forests prior to any land cover, it can be expected that all virgin forest land will be deforested prior to any other land covers being considered. Following the clearing of a patch of land, an agent must make a decision regarding which crop should be planted based on previous experiences. This decision-making process is governed by a classifier system and is described below. Following a full growing season, crops are harvested and crop yields calculated.

Land-use strategies or rules in LUCITA are encoded as 270-bit (1s and 0s) genetic algorithm strings and are stored in a rule base. Booker et al. [2], Goldberg [9], and Holland [11] provide an overview of genetic algorithms. Each agent has a distinct rule base comprised of eight rules reflecting the land-use strategies for the agricultural crop production of rice, beans, manioc, maize, cacao, and black pepper. The monthly family labor, monthly male labor, and capital requirements of any land-use rule and the action to be triggered given that the requirements of that particular rule are satisfied is encoded in the structure of the 270-bit string. Each monthly family and male labor requirement is translated from base 10 to base 2 and is represented by a series of 10 bits. Therefore, the twelve months of family and male labor requirements for a given land-use rule is encoded in the first 240 bits (i.e., 24 months multiplied by 10 bits/month) of the 270-bit string. The capital requirements for a given land-use rule are encoded no differently than monthly labor requirements; however, a series of 20 bits is required for encoding instead of the previous 10 bits. Small base 10 numbers, characterized by monthly labor requirements, when converted into base 2, requires very few bits. In contrast, large base 10 numbers, characterized by capital requirements, when translated into base 2 strings, requires many more bits for representation. The difference in the number of bits required to encode monthly labor and capital requirements is attributed to the type of values associated with each type of variable. For both monthly labor and capital requirements, an estimate of the greatest possible value for these variables were made, and subsequently translated into base 2 to identify the minimum number of bits that would be required for encoding. The final 10 bits of the 270-bit string are used to represent an effector, or the action to be taken if the conditions of a string are satisfied, and in this

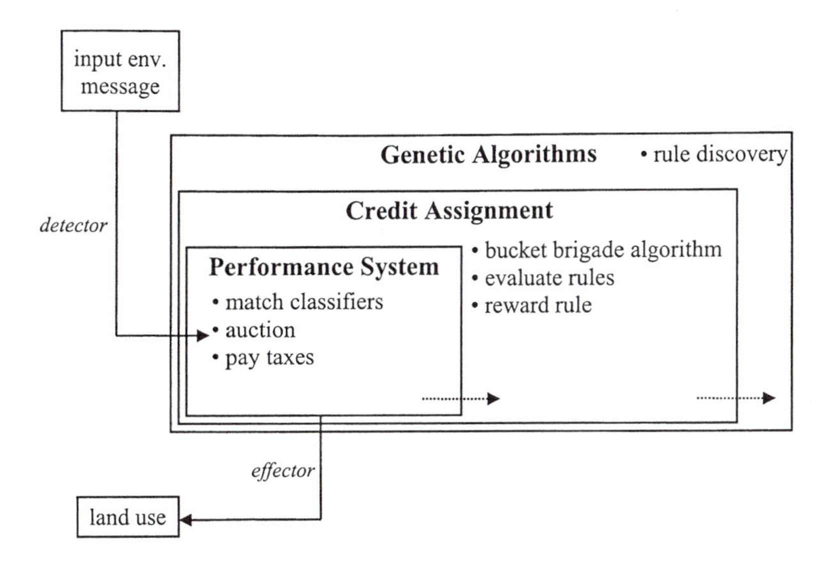

FIGURE 2 General organization of a classifier system used in LUCITA.

case simply represents that type of land-use strategy to be implemented. A strength value is associated with each rule and reflects the fitness of that rule or the effectiveness of that rule. An effective land-use rule, meaning those that generate satisfactory crop yields, should be reinforced through simulated time and should be observed through an increase in strength. A base 2 representation of rules was selected for ease of implementation with the classifier system.

Providing that an agent has enough labor and capital resources to deforest a patch of land for agricultural production, the decision-making process of which of the eight land-use strategies to implement is determined by a classifier system. The classifier system used in LUCITA is designed and implemented following the concepts described by Booker et al. [2]. A classifier system is typically comprised of three components: a performance system, a credit assignment system, and a rule discovery system (fig. 2).

The performance system is responsible for matching rules in the rule base to an encoded message composed of 260 bits, which describes the monthly family labor, monthly male labor, and capital available to an agent at a particular instance in time. The structure of an encoded 260-bit message is no different than the above-described 270-bit rule structure, except for the omission of an effector tag. Those rules that are matched are entered into an auction. A rule is considered to be a match if and only if every monthly labor and capital available to an agent is greater than or equal to the respective monthly labor and capital requirements as defined by that rule. This implies that under circumstances where very labor and capital intensive land-use rules are matched, those rules that are less labor and capital intensive are also matched and entered into the auction. For instance, if an agent can afford to grow cacao, which has high labor and capital requirements, an agent is most likely also capable of growing rice or beans. During the auction process, a matched rule makes a bid based upon its strength. An effective bid is subsequently calculated by simply adding a random value between 0 and 1 to the original bid made by a rule. An effective bid is calculated to avoid the situation where two or more rules have the same fitness and in turn make identical bids. The rule with the greatest effective bid is then selected and used by the agent.

Effective rules in the rule base must be somehow reinforced through time and those that are poor, eventually excluded from the decision-making process. A household agent should have a history of land-use experiences and ideally be able to learn from those experiences and identify those types of land-use strategies that are more effective than others. This process of weeding out effective rules through simulated time can be accomplished by taxing and rewarding those existing rules in a rule base, meaning decrementing or incrementing the strength of a rule. It is the credit assignment system that is responsible for the actual rewarding process. Because only one matching rule can be selected for one patch of land, the bucket-brigade algorithm is not required. In the LUCITA classifier system, at any time event, all rules existing in the rule base must pay a life tax rate (i.e., 1%). Further, rules that match an encoded agent message must pay a bid tax rate to compete in the auction process (i.e., 5%). The winning rule pays its tax in the form of its bid, sometimes 10% of its strength. In general, the effectiveness of the winning rule is evaluated against some criteria, such as the ability of that rule to generate an expected crop yield per hectare. Under circumstances where that winning rule satisfies the defined criteria, it is rewarded and thereby reinforced. However, under circumstances where that rule is deemed ineffective, that particular rule would have paid a significantly larger tax than all other rules, thereby reducing its strength or fitness and affecting its future of being reselected. The dynamics associated with the competition of rules and the reinforcement of effective rules through simulated time will be illustrated in a case scenario of the one-household version of LUCITA.

The functionality of the rule discovery system is not implemented in LUCITA since it cannot be applied to the land-use rules given the nature of the information encoded in the structure of the rules. The purpose of the rule discovery system of a classifier system is to try to evolve new rules by applying genetic operators, such as mutation or crossover, to the most fit rules in a rule base.

In more traditional applications of classifier systems, rule strings are often an encoding of a series of conditions that either evaluates to a true or false state, where a series of conditions triggers some type of response or action. The reason why applying genetic operators to the LUCITA rules would be inappropriate is because that the conditions of any given rule string does not evaluate to some state of true or false, but rather translates to some specific type of numeric labor or capital value. The defined labor and capital conditions defined by a rule are static and must be satisfied prior to the implementation of any land-use strategy. For instance, it would be incorrect to crossover two rule strings, each representing completely different land-use strategies, since this would imply that the labor and capital requirements for each of these rules are variable which is clearly not the case. Each land-use rule has one and only one set of family labor, male labor, and capital requirements that must be satisfied prior to implementation. For this reason, the classifier system utilized in LUCITA is referred to as a nonevolving classifier system.

3.6 SCHEDULING OF EVENTS

For any given simulated year in LUCITA, a series of events is scheduled to simulate the actions of a frontier colonist who practices slash-and-burn agriculture and the associated impacts of those practices on an artificial landscape. In the versions of LUCITA described in this chapter, dynamic scheduling of events is not considered although possible using Swarm. At this present stage of development two versions of LUCITA exist—the one-household version and the landscape version. The one-household version of LUCITA focuses on exploring simulations at a local scale (one property), so as to provide a basic understanding of how an agent makes decisions, how decision making is affected by variability in environmental conditions, what relationships or feedback loops exist between both submodels, etc. In contrast, the landscape version of LUCITA focuses at a regional scale (236 properties), where only the regional land-use trends are of interest. The rationale of this approach is that if the one-household version of LUCITA is explored to a point that local interactions can be explained and understood, then at the regional scale, there is no need to consider local interactions but rather emphasis can be placed on observing the emergence of regional land-use trends. Processes or actions relevant to the KPROG2 submodel and the human system submodel are scheduled as events. The two versions of LUCITA only differ in the number of agents scheduling events and the number of properties affected by agent actions. A flow chart diagram illustrating the scheduling of events is provided in figure 3.

At the start of each year, an event is scheduled to tabulate the frequency of each land cover occurring on the landscape and archived in a data file. At the conclusion of a simulation, this data file can be used to describe the trajectory of land uses both at a local and regional scale, depending on which version of LUCITA was simulated. Following this tabulation, an event is scheduled to identify which patches of land need to be shifted to an alternate land cover based on the transition of land covers internally programmed. The scale of a

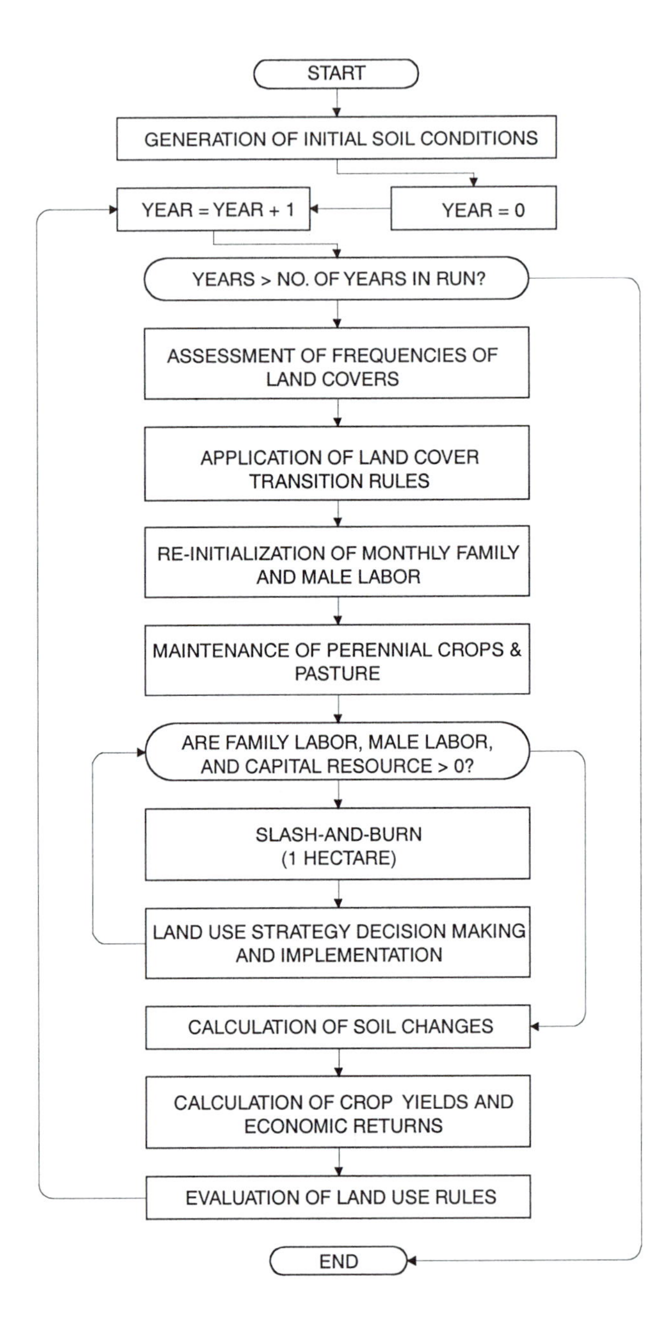

FIGURE 3 Summary flow chart of schedule of events for any given agent and its property lot.

patch of land is always a grid cell, or 1 hectare. The criteria used to determine the transition of one land cover to another is based on the previous land cover of a patch of land and the number of years that patch of land has been in continuous use. For example, a patch of land that has exceeded the maximum number of years of continuous cultivation are processed to a stage of fallow and a patch of land that is some stage of secondary succession is shifted to a further advanced stage of secondary succession.

It is important to note that family and male labor pools are reset at the start of each new year. It is assumed in LUCITA that any given patch of land used for the production of perennial crops or for pasture with or without cattle is maintained prior to any considerations made by an agent to implement a new land-use strategy. Hence, an event is scheduled for each agent to check within its property for patches of land that need to be maintained and to commit the necessary labor and capital resources required by those land uses. Any new land-use strategies are not considered until after this maintenance event is processed.

Providing that labor and capital resources are still available, an event is scheduled for the clearing and burning of one patch of land. An agent will identify, based on its available labor and capital, which of the available three land covers (i.e., virgin forest, secondary forest, and weed) it is capable of burning. The criteria used in selecting a patch of land to be cleared and burned are the clearing preferences defined by an agent, the identified land covers that an agent is capable of clearing, and the patches of land free for use in a given property lot. Land is cleared and burned prior to any consideration by an agent of what type of land use is to be implemented. Following the clearing and burning of a patch of land, the decision-making process of an agent is invoked and with the help of the classifier system, a land-use strategy is recommended and implemented. Under circumstances where there is insufficient labor and capital to implement a land-use strategy after clearing and burning a patch of land, that land is simply abandoned by an agent and enters some early stage of secondary succession. An event of clearing and burning a patch of land, followed by an event for agent decision-making purposes is repeated until either one of the criteria of family labor, male labor, or capital is exhausted.

After the process of land allocation is complete, an event is scheduled to calculate the soil changes for each patch of land in a given property. Not only do soil changes need to be calculated for patches of land that have been cleared and burned for new agricultural land uses, but also for those patches of land undergoing some stage of secondary succession. For any given patch of land under any land cover, a set of KPROG2 multiple regression equations exists to determine the appropriate changes in soil parameters. Using these changes in soil parameters, an event is initiated to calculate crop yields for each and every patch of land in agricultural use. The crop yield for each land-use strategy is evaluated against the expected crop yield for the number of patches used for production to determine the effectiveness of that particular land-use strategy. For any given land-use strategy, providing that the expected crop yield is satisfied, a reward is sent to the classifier system to reward that land-use strategy in the rule base.

4 LUCITA SIMULATION RESULTS

4.1 EXPLORATORY RESEARCH

The purpose of this section of the chapter is to provide simulation results of two case scenarios for each of the one-household version and the landscape version of LUCITA for a time period of 40 years. The two case scenarios differ in the definition of the land-use clearing preferences. The household parameters were constant for all simulations runs, irrespective of which version of LUCITA was simulated. As described in an above section, the purpose of both versions of LUCITA and the output generated from each version are different; however, from a perspective of how events are scheduled in both versions, they are fundamentally identical. The purpose of the one-household version of LUCITA is to explore the dynamics between an agent and its surrounding environment at a local (property) scale. This is accomplished by generating extensive output data files tracing changes in the environment, changes in labor and capital resources, and the dynamics of decision making using the LUCITA classifier system. The purpose of the landscape version of LUCITA is to observe the emergence of regional land-use trends. For this reason, little output data is generated except for annual land cover frequencies. The landscape version of LUCITA can be considered a container of nested one-household versions.

4.2 INITIAL AVAILABLE LABOR AND CAPITAL

The initial available monthly family labor for any given month is calculated by taking the product of the number of family members and the days in a month. Similarly, the initial monthly available male labor for any given month is calculated by taking the product of the number of males in a family and the days in a month. For instance, for a month with 31 days, if a family was composed of four individuals, from which two were of the male gender, the available family labor would be 124 man-days equivalent per hectare and the available male labor would be 62 man-days equivalent per hectare. Issues regarding fertility and mortality within a household are not considered in the two versions of LUCITA. For this simulation, a family size is calibrated to six individuals, from which three are of the male gender. All individuals are assumed to be of a mature age and capable of contributing labor. Initial available monthly family and male labor is summarized in table 1. With respect to initial capital, each household agent is initialized with Cr$0 (Brazilian cruzeiro currency). The initial family composition and initial capital of an agent were arbitrarily selected and were not based on any data sources.

TABLE 1 Number of man-days equivalent of monthly family labor per hectare. Number of man-days equivalent of monthly male labor per hectare.

Monthly Available Family Labor												
Month	**Jan**	**Feb**	**March**	**April**	**May**	**June**	**July**	**Aug**	**Sept**	**Oct**	**Nov**	**Dec**
Family Labor	186	168	186	180	186	180	186	186	180	186	180	186
Monthly Available Male Labor												
Month	**Jan**	**Feb**	**March**	**April**	**May**	**June**	**July**	**Aug**	**Sept**	**Oct**	**Nov**	**Dec**
Male Labor	93	84	93	90	93	90	93	93	90	93	90	93

TABLE 2 The pH in initial soil quality generation for 236 property lots in study area. From *Human Carrying Capacity of the Brazilian Rainforest* by P. M. Fearnside © 1986 Columbia University Press. Reprinted by permission of the publisher.

Class	pH range	Frequency (%)	Mean pH
1	< 4.0	33.0	3.7
2	4.0 – 4.4	30.2	4.1
3	4.5 – 4.9	15.3	4.7
4	4.0 – 5.4	12.5	5.2
5	5.5 – 5.9	5.3	5.6
6	6.0 – 6.4	3.6	6.3
7	≥ 6.5	0.1	7.1

4.3 INITIAL SOIL CONDITIONS

For the purpose of this simulation, an assumption is made that the soil conditions across a property lot are homogeneous for the one-household version of LUCITA. The mean of the most frequently observed class for each initial soil quality data sampled by Fearnside [7] is assigned to each respective soil parameter. For example, using the initial pH soil quality data for the study area, summarized in table 2, the mean of class 1 is assigned as an initial pH value since that particular class is the most frequently observed.

In the case of the landscape version of LUCITA, the spatial data maps translated from Fearnside [7] were used to calibrate each soil parameter. Therefore, the landscape soil parameters for each of the 236 property lots are spatially variable unlike the homogeneous one-household property lots.

4.4 LAND COVERS AND CLEARING PREFERENCES

Nineteen possible land covers exist in LUCITA and are described in table 3. Prior to a simulation run, an agent's clearing preferences must be defined. The land covers that are considered for clearing include virgin forest, weeds, and any patch of land in some stage of secondary succession (i.e., land IDs from 1 through 9 only). Categories of secondary succession are based on the criterion of age only and are defined following the same classification used in

TABLE 3 The nineteen possible land covers in LUCITA.

Land-Use Code ID	Land Cover
0	House
1	Virgin Forest
2	Weeds & Bare Land (less than 1 year of age)
3	Secondary Succession 1 (greater than 1 but less than 2 years of age)
4	Secondary Succession 2 (2 to 3 years of age)
5	Secondary Succession 3 (4 to 6 years of age)
6	Secondary Succession 4 (7 to 11 years of age)
7	Secondary Succession 5 (12 to 16 years of age)
8	Secondary Succession 6 (17 to 20 years of age)
9	Secondary Succession 7 (over 20 years of age)
10	Rice
11	Beans
12	Maize
13	Manioc
14	Fallow
15	Cacao
16	Black Pepper
17	Pasture without Animals
18	Pasture with Animals

the original KPROG2 model. For any simulation, any patch of land can only be used continuously for a maximum cultivation period of two years. When the maximum number of years of continuous cultivation are exceeded, a patch of land must enter a fallow stage for a minimum period of three years prior to reuse. Clearing preferences in combination with whether or not labor and capital requirements are available to slash-and-burn are used to identify which type of patch of land in a given property should be selected for use.

The two case scenarios have different clearing preference definitions, defined in table 4. The first case scenario assumes that an agent's first preference is for mature secondary forests followed by less mature stages of secondary succession and only deforest virgin forest land when no other secondary land covers are available. This scenario is representative of circumstances where frontier colonists recognize the importance of secondary succession in regenerating soil fertility. The second scenario assumes that an agent has some type of an incentive to deforest virgin forests. When virgin forests have been depleted, it is assumed that agents will select to clear the patches of the land that require the least labor and, therefore, preferences range in order from weeds and bare land to secondary succession 7. This scenario is representative of a situation where agents may be influenced by social or economic factors to clear virgin forest lands and do not have an understanding of the impor-

TABLE 4 Definition of land clearing preferences for two case scenarios.

Scenario 1		Scenario 2	
Order	**Land Cover**	**Order**	**Land Cover**
1	Secondary Succession 7 (over 20 years of age)	1	Virgin Forest
2	Secondary Succession 6 (17 to 20 years of age)	2	Weeds & Bare Land (less than 1 year of age)
3	Secondary Succession 5 (12 to 16 years of age)	3	Secondary Succession 1 (greater than 1 but less than 2 years of age)
4	Secondary Succession 4 (7 to 11 years of age)	4	Secondary Succession 2 (2 to 3 years of age)
5	Secondary Succession 3 (4 to 6 years of age)	5	Secondary Succession 3 (4 to 6 years of age)
6	Secondary Succession 2 (2 to 3 years of age)	6	Secondary Succession 4 (7 to 11 years of age)
7	Secondary Succession 1 (greater than 1 but less than 2 years of age)	8	Secondary Succession 6 (17 to 20 years of age)
8	Weeds & Bare Land (less than 1 year of age)	8	Secondary Succession 6 (17 to 20 years of age)
9	Virgin Forest	9	Secondary Succession 7 (over 20 years of age)

tance of allowing cultivated land to regenerate through secondary succession to maintain productive soil fertility.

4.5 PATCH SELECTION

It is important to note that the type of patch of land identified for clearing is not randomly selected from the property lot. A bubble sort is applied to all patches of land in a property lot to order them in a south-to-north ordering for properties located on feeder roads, and in an east-to-west ordering for properties adjacent the Transamazon Highway (fig. 4). This implies that the direction of clearing is predefined; however, the magnitude of land cleared remains unaffected and remains dependent on labor and capital resources of an agent. Household behavior regarding the spatial location of the selection of land for use is beyond the scope of current versions of LUCITA due to limited data and knowledge. The ordering of property lots also has implications with respect to spatial patterns of deforestation, but it is important to note that in these early stages of LUCITA, the spatial pattern of deforestation is not as important as the trajectory of household land uses through simulated time.

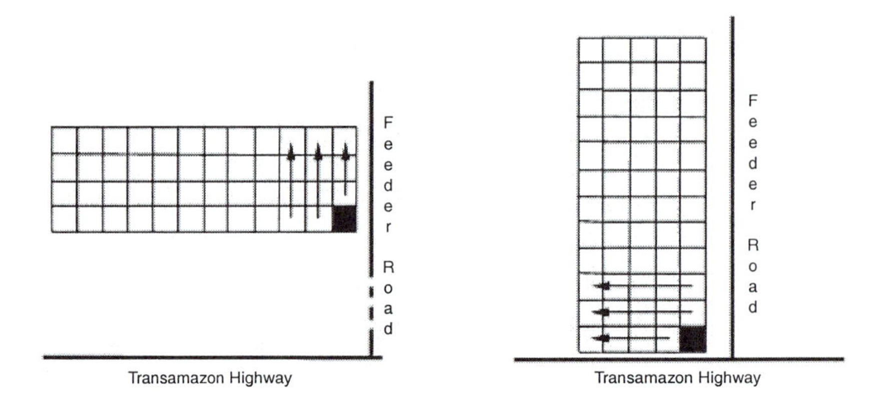

FIGURE 4 The direction that free patches of land are searched by an agent on property lots off on feeder roads (left) and on property lots adjacent the Transamazon Highway (right) for slash-and-burn agriculture. Black grid represents location of house.

4.6 CLASSIFIER SYSTEM PARAMETERS AND REWARD CRITERIA

Any rule contained in an agent's rule base must pay a life tax of 0.5% of its existing rule strength. Rules that are matched based on the available labor and capital possessed by an agent must pay a bid tax rate of 1% of its existing rule strength to be considered in the auction process. The winning rule from the auction process must, in turn, pay 5% of its existing strength when selected for implementation. Classifier system parameters are constant for all simulation runs.

Ideally, effective rules should be reinforced through time. Annual cash-crop strategy rules are rewarded when the expected yield for all patches of land in a given annual cash-crop rule is satisfied. For instance, if the expected yield for rice is 1500 kg/ha and 5 hectares of land are used for crop production of rice by an agent, the annual cash crop rice rule is only rewarded if 7500 kg for the 5 hectares of land is produced. The actual reward value is calculated by adding the total bid amount made for all the hectares of land use for that successful land use, plus half of that total. No data is available for perennial crops and, given the soil condition of the study area, even with liming and fertilizing, it is often difficult to produce satisfactory yields. For this reason, a perennial crop rule is assumed to be successful or effective if it produces a yield greater than 0. The reward value is calculated in the same fashion as that of annual cash crops. Pasture land uses are not evaluated as to their effectiveness and are therefore never rewarded. The rational for this approach is that pasture land uses in most instances are always implemented when little labor or capital remains after implementing other land uses. Therefore, pasture land uses will

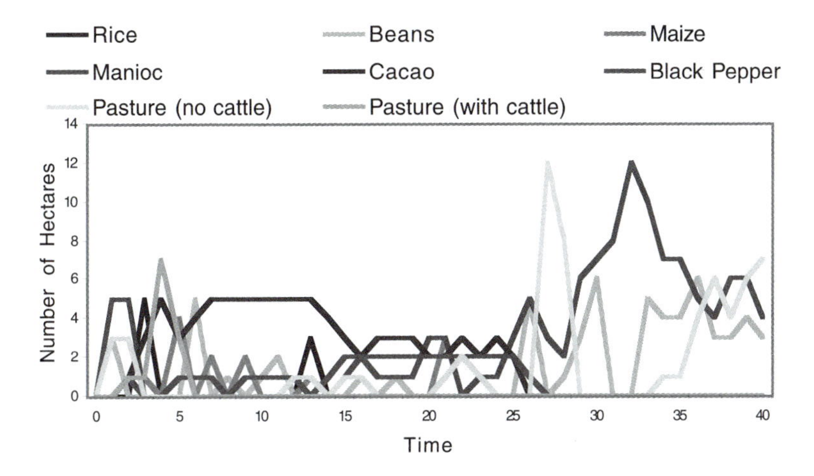

FIGURE 5 Land-use frequencies for scenario 1 using the one-household version of LUCITA.

most likely be implemented more frequently than others resulting in pasture land uses emerging as dominating land uses. Instead, pasture land uses are allowed to compete for selection; however, when selected such uses are not reinforced since pasture land use is often implemented to deplete all labor or capital resources of an agent before moving on to the next time step.

4.7 SCENARIO 1: ONE-HOUSEHOLD VERSION OF LUCITA RESULTS

The results from simulating scenario 1 are presented in figure 5. Virgin forestland, stages of weeds, and all stages of secondary succession are not depicted in the graph for scaling reasons. Only 29% of the original forest were cleared for agricultural and pasture use. During the first seven years, all land-use rules can be seen to compete. This time period, where an agent experiments with rules, is somewhat shorter than expected. Maize and bean crops can be seen to be the first crops produced, followed by black pepper, rice, cacao, and maize. It is interesting to note that sufficient labor and capital are available as early as year 1 to implement perennial crop rules.

Both pasture land uses are similarly implemented early in this time period, as expected, as very little labor and capital are required by these land uses. It is important to emphasize that pasture land uses compete in the auction process, but do not earn rewards since no criteria is available to evaluate if they are effective or not. This implies that pasture land uses are only implemented when very little labor remains following the implementation or maintenance of other crop rules.

From year 7 through to year 24, cacao establishes itself as the dominating land-use rule because of a combination of the fact that it possesses the highest strength and because it is assumed that any patch of land in perennial use or pasture is maintained for the maximum cultivation period prior to any consideration of the implementation of new land-use rules. Other crops are implemented less frequently as illustrated under the cacao trend line.

By year 26, cacao noticeably ceases to be the dominating land use. This event can be explained by the decrease in soil fertility in the patches of land used. Based on the clearing preferences, land uses cycle on only 29 patches of land from the one-hundred-patch property lot as seen in snapshots of the simulation presented in figure 9. The most advanced stage of secondary succession that is ever reached with these land patches prior to being recleared is the secondary succession 3. By analyzing the crop-yield data file, cacao crop yields are shown to be negative, implying that a decrease in soil fertility has occurred, resulting is those land patches no longer being productive for the production of cacao. An examination of the multiple soil data files of the patches of land in use reveal that pH levels decreased to a very low value.

Following this decline in cacao production, most of the land is implemented in the pasture with no cattle land use. This surge of the pasture with no cattle land use can be explained by observing the previous year's land use. Manioc, beans, cacao, and black pepper are produced in year 26, and because of manioc's two-year growing season and the assumption that perennial crops are maintained for the maximum cultivation period, most of the labor for the subsequent year is already committed to maintenance of these crops. The little remaining labor is sufficient only for the implementation of the pasture with no cattle land use resulting in the surge of that land use.

Come year 28, maize and bean land uses emerge as the dominating land-use trends, complemented by the pasture with no cattle land use. These two annual cash crops emerge as the dominating land uses because of their ability to produce on soils with poor fertility. By examining the crop production data files, it is revealed that while other land uses began to generate negative crop productions near the same time of the decline of cacao production, maize and beans continued to produce crops.

In summary, it appears that how clearing preferences are defined may affect soil fertility, in turn affecting the production of crops. In this scenario, a stage is reached when soil fertility drops to a point where many land uses are no longer able to produce crops. The maize and bean land uses simply emerge as the most effective land uses at the end of the simulation because of their ability to produce under low soil fertility. Based on the definition of the clearing preferences for this scenario, an agent is incapable of deforesting all virgin forests on its property lot and only does so when no other alternative is available. Therefore clearing preference in this scenario plays a large role in the emergence of effective land uses and the magnitude of virgin forest deforestation.

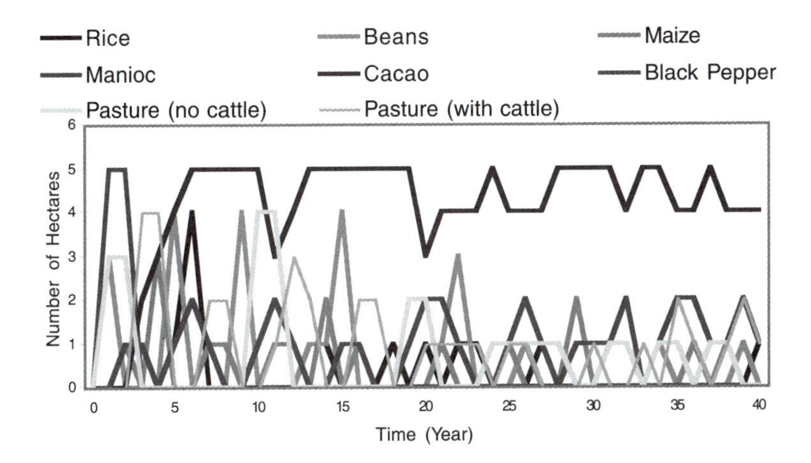

FIGURE 6 Land-use frequencies for scenario 2 using the one-household version of LUCITA.

4.8 SCENARIO 2: ONE-HOUSEHOLD VERSION OF LUCITA RESULTS

Simulation results for scenario 2 are presented in figures 6 and 10. Like scenario 1, virgin forests, weeds, and stages of secondary succession land uses are not displayed for scaling reasons. Virgin forests are completely depleted by year 17 and mature secondary forests fully regenerated by year 20. Land-use trends are very similar to scenario 1. A common trend in both scenarios is the dominance of the cacao land-use rule early in the simulation. Other rules continue to compete for implementation as shown under the cacao trend line, but cacao is the most frequent land use implemented and maintained on the property lot.

The first seven years of the simulation are similar to scenario 1 where results of the experimentation of land-use rules only differ from scenario 1 in the order of the experimentation chosen by an agent. Black pepper is only produced during years 2 to 4 and is never produced again in the simulation. This event can be explained in terms of the dominance of the cacao land use. Both cacao and black pepper have high labor and capital requirements. Given that cacao has the highest strength and wins most bids, when no sufficient labor or capital remains for implementation of the cacao land use, then, in turn, any consideration of implementation of black pepper is excluded because of the similar labor and capital requirements. All land uses, excluding black pepper, continue to compete throughout the simulation.

Cacao continues to be the dominating land use throughout scenario 2, in contrast to scenario 1, because of the definition of clearing preferences. A clearing preference for virgin forest is emphasized throughout the simulation and, as a result, patches of land are rarely recleared for use since new land is continuously deforested. This implies that soil fertility remains consistently high and, in turn, is capable of maintaining the soil conditions for cacao production unlike in scenario 1 where a decrease in soil fertility occurred. With the maintenance of cacao production, meaning that the cacao land use was being reinforced through time, little opportunity was available for other land uses to dominate.

In summary, scenario 2 further reinforces the importance of clearing preferences in affecting the emergence of dominating land uses and the magnitude of deforestation. Further, at the end of the simulation, mature secondary forests regenerate to cover 77% of the original primary forest extent. In comparison to scenario 1, 71% of the original virgin forest was left undisturbed. Comparing the amount of forest cover in both scenarios, irrespective of the type of forest cover, suggests an agent, based on its description of family composition and capital, can only sustainably manage one quarter of its property lot.

4.9 SCENARIO 1 AND SCENARIO 2: LANDSCAPE VERSION OF LUCITA RESULTS

The landscape version of LUCITA generates regional land-use trends based on a spatially explicit soil landscape. Land-use results from simulating scenario 1 and scenario 2 using the landscape version of LUCITA are presented in figure 7 and figure 8, respectively. Snapshots of both landscape scenarios are depicted in figures 11 and 12.

Regional land-use trends are very similar to those at the local property scale for scenario 1. Virgin forests, weeds, and secondary stages of succession are once again omitted from figure 7 for scaling reasons. From the total virgin forest extent, 66% remains undisturbed and is very close to the 71% at the local scale. The most advanced stage of secondary succession reached is stage 4. Cacao emerges as the dominant land-use trend for approximately the same duration as that of the local scale. Manioc and pasture with no cattle land uses emerge as the dominating land-use trends near the end of the simulation. Instead of the bean land-use trend also emerging near the end of the simulation, like in the case of the local scale simulation, maize emerges as the third dominant land-use trend. Granted that the clearing preferences were identical for simulations using each version, the slightly different land-use trends in the landscape results must be affected by the spatial variability of the soil data characterizing the landscape. Higher or lower soil fertility in particular regions of the landscape may lead to increases or decreases in crop production, which in turn may affect the evaluation of rule strengths. The general similar trends can be explained by the generation of initial soil quality for the local scale property lot, where the most frequent soil value of each parameter observed on the landscape were assigned to the property lot's soil parameters.

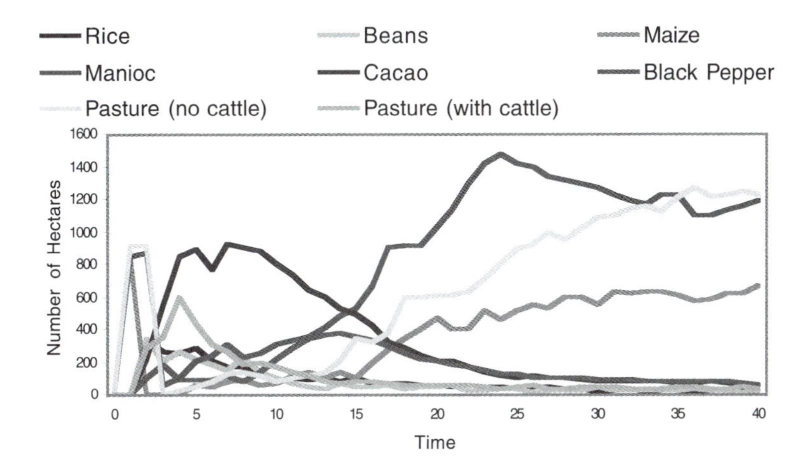

FIGURE 7 Land-use frequencies for scenario 1 using the landscape version of LUCITA.

In scenario 2 for the landscape, depicted in figure 8, general land-use trends are nearly identical to those at the local scale. Once more, only crop and pasture land uses are depicted in the graph. Virgin forest depletion occurs at year 33, much later than year 17 at the local scale. Mature secondary forests begin regenerating at year 21, one year later than the case at the local scale. Figure 8 shows the dominance of the cacao land use early in the simulation to the end of the simulation. Other rules continue to compete as shown under the cacao trend line and match the trends at the local scale results. Following depletion of primary forest cover, mature secondary forests regenerate to cover 75% of the original virgin forest extent, almost identical to the value obtained at the local scale. Despite the spatial soil variability of the landscape, the land clearing preferences ensures that cacao is continuously implemented only on fertile soil and, so long as soil fertility is adequate for cacao production, it will continue to dominate.

5 DISCUSSION

An examination of the results presented in this chapter reveals both the strengths and limitations of LUCITA, and the need for additional data collection within the Altamira study area. This section examines the replicative validity of the simulations, as compared to the observed situation in Altamira, and explains some of the behavioral anomalies of the simulations. These observations point to the need for additional data collection on the natural, institutional, and demographic and behavioral characteristics of the region.

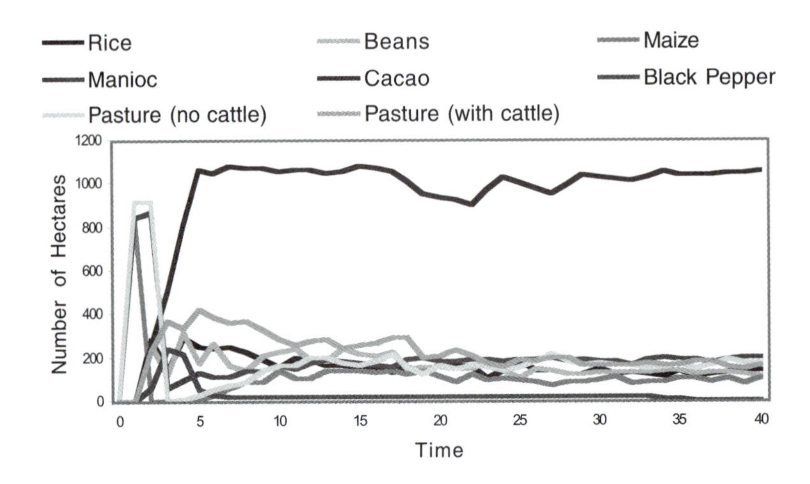

FIGURE 8 Land-use frequencies for scenario 2 using the landscape version of LUCITA.

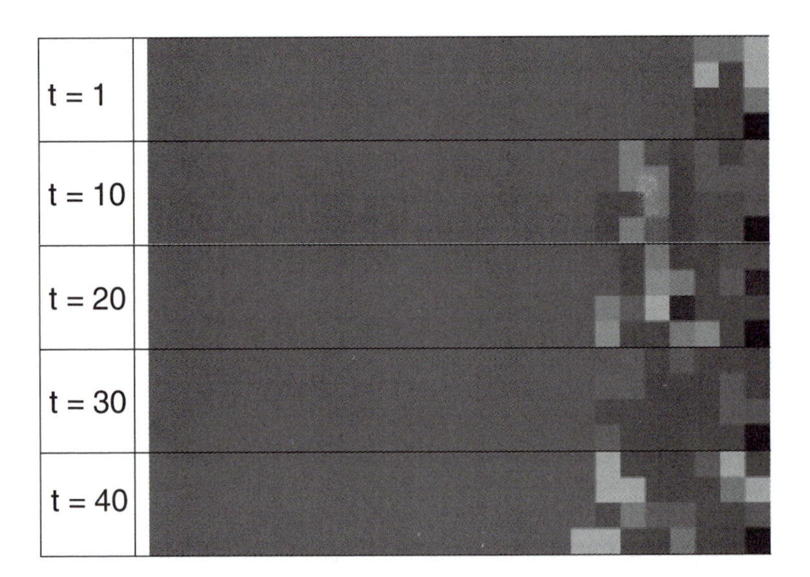

FIGURE 9 Snapshots of scenario 1 simulation using one-household version of LUCITA.

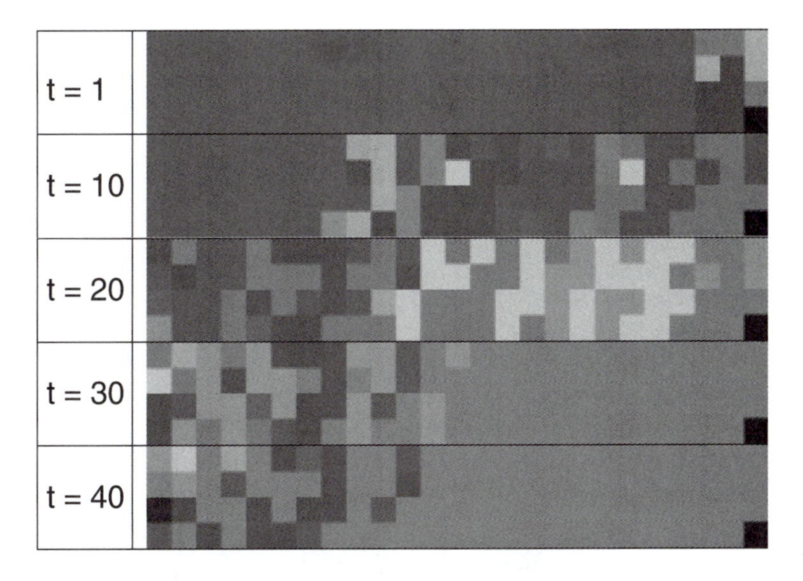

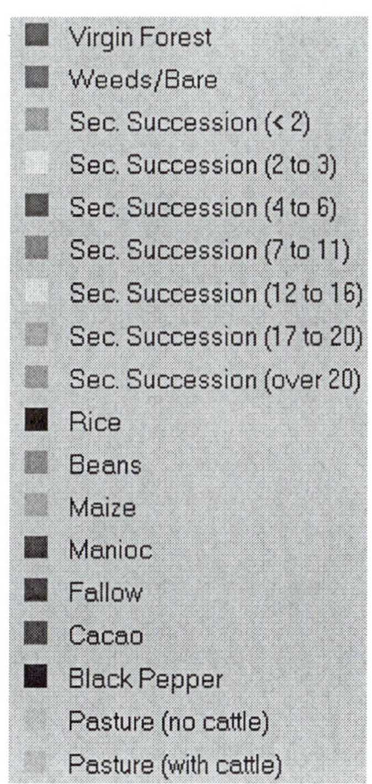

FIGURE 10 Snapshots of scenario 2 simulation using one-household version of LUCITA.

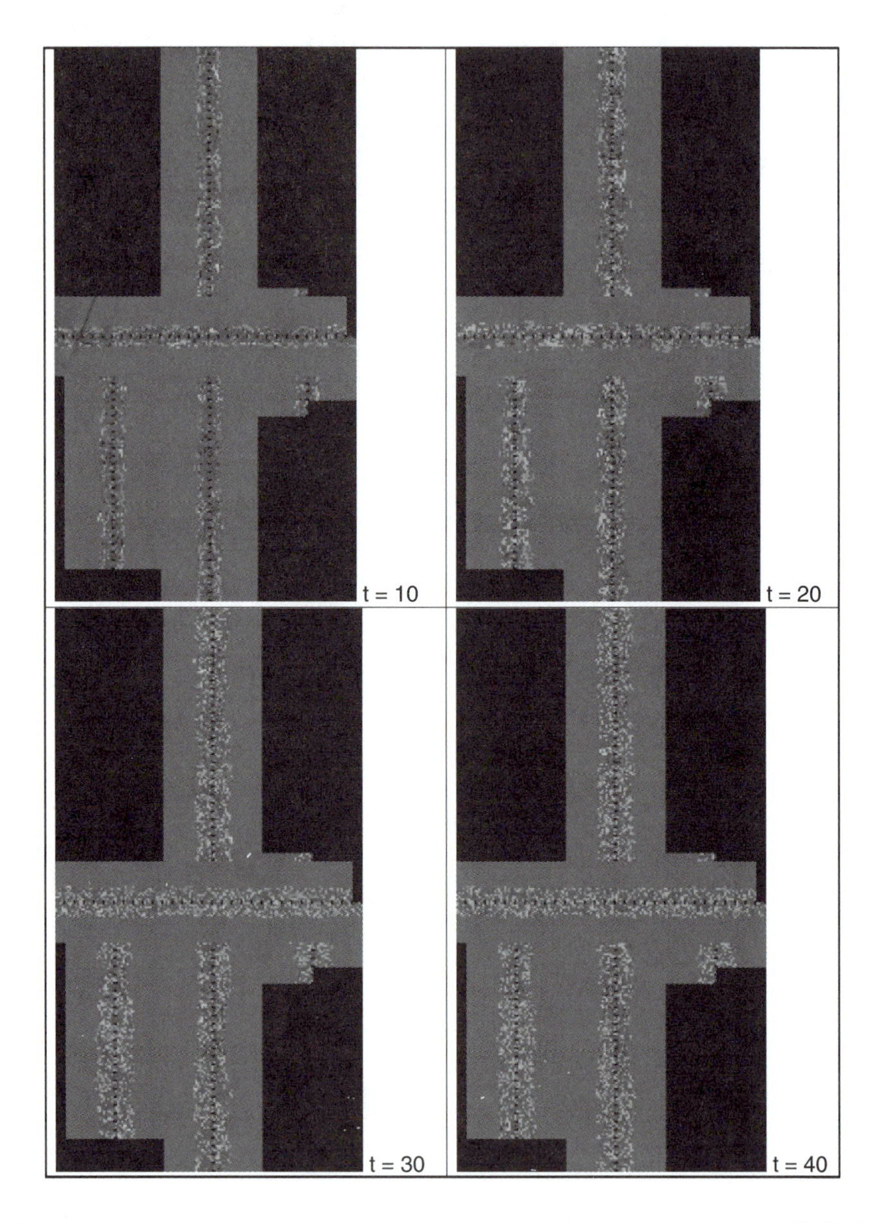

FIGURE 11 Snapshots of scenario 1 simulation using landscape version of LUCITA. Legend in Figure 10.

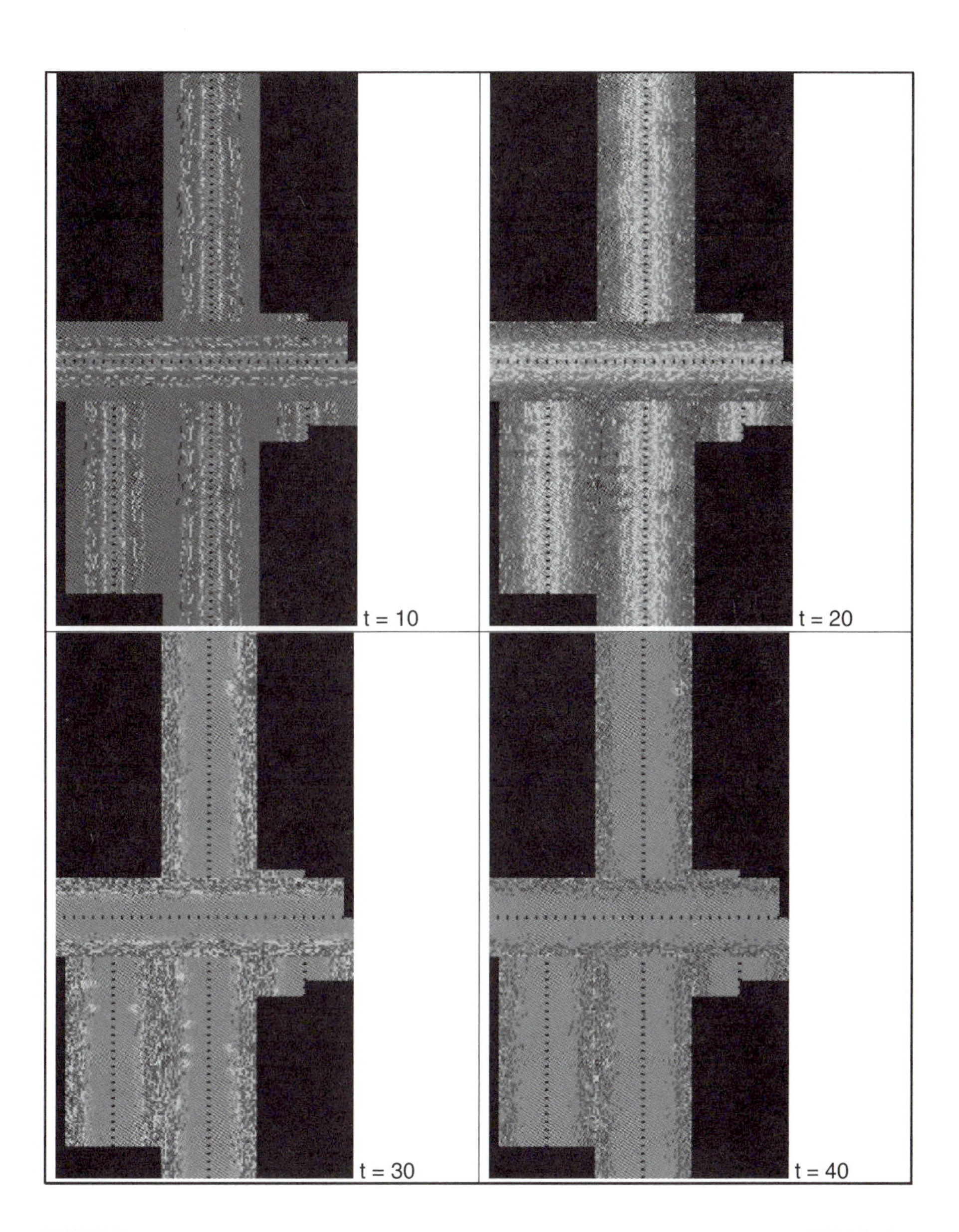

FIGURE 12 Snapshots of scenario 2 simulation using landscape version of LUCITA. Legend in Figure 10.

The land-use trajectories generated from scenario 1 and scenario 2 using both versions of LUCITA do not resemble the trajectory proposed in the conceptual model of household transition outlined in McCracken et al. [16] for several reasons, the most important being the definition of the reward criteria used to evaluate the effectiveness of rules following implementation by an agent. In general applications of classifier systems, at any given time step, one rule is selected, evaluated, and reinforced prior to the next time step where a new rule is selected. In the case of LUCITA, at any time step, several rules can be selected and implemented depending on an agent's available labor and capital resources. Emphasis is placed on the word "selected," meaning that we are not referring to the matching of rules to an environmental message encoded by an agent representing its available labor and capital, but rather those rules that actually win a distinct auction process in the performance system of a classifier system. Those rules that are selected are not rewarded, nor evaluated until all decision making has been completed at the end of a time step for the simple reason that, during a growing year, a farmer does not know if the production of a particular crop is successful until the end of the growing season. Rewarding a rule before all decision making has been completed will, in effect, bias decision making toward effective rules selected early on, since their rule strength will be reinforced before other rules have an opportunity to be tested within that same time step.

The adoption of an approach where a set of rules are not evaluated and rewarded until all decision making has been completed introduces several problems, which has emerged from the simulation of both scenarios presented in this chapter, ultimately affecting land-use trajectories. Selected rules must pay a bid value depending on the defined bid ratio; however, because rule strengths cannot be modified until all decision making has been completed, these bid amounts must be archived in some type of a data structure. Similarly, reward payments must be archived to prevent changes in rule strengths. A problem that has emerged in the LUCITA simulations is that, if a land use is implemented on more than one patch of land and is evaluated as an ineffective rule, its rule strength is often depreciated to a strength below previously identified less effective rules. This is because, for each implementation of a land-use rule, a bid is paid based on the defined bid ratio. The more times a land use is implemented during a given time step, the larger the paid bid amount and, if the land-use rule is evaluated to be ineffective, there is no recovery of the paid bid through rewards.

Based on the definition of the reward criteria for land-use rules in the existing version of LUCITA, a land-use rule cannot be evaluated as partially effective. A rule either pays its total bid amount or is rewarded the paid bid amount plus a percentage. There is no consideration of the spatial variability of crop yields in the existing version of LUCITA and, hence partial recovery of a proportion of total paid bids by a land-use rule is not possible. For instance, an extremely successful crop yield on one hectare of land may compensate for poor yields on another two hectares and yet still achieve the total expected crop yield for three hectares of land, ultimately resulting in the reinforcement of the rule. Consider that same situation, but in this case the total expected yield for three hectares of land is not achieved; despite the expected yield for one hectare of land being satisfied, the rule itself is not reinforced. Therefore, as mentioned above, rules that are implemented on several patches of land can lose a large proportion of its rule strength if evaluated as ineffective based on the definition of the reward criteria and the lack of consideration of the spatial variability of crop yields. Given the general poor initial soil qualities and the rapid decrease in soil fertility following slash-and-burn, many hectares are unable to achieve the expected crop yields. In turn, this inability to achieve expected crop yields is reflected by a continuous net decrease of rule strengths toward zero as opposed to a dichotomy of rules emerging, comprised of effective and noneffective rules.

Simulated land-use trajectories are also affected by how the labor and capital requirements of a land use are defined, and in turn determining which ones can be implemented by an agent. A general idea of the conceptual transition of households is that, when a household arrives to the frontier, that household on average has very little liquid capital and, hence, can only pursue a subset of the total available land-use rules, which often excludes perennial crop production. What has emerged from the simulations is that agents are often capable of beginning to implement perennial land-use rules as early as the second production year since enough liquid capital has been produced in the first year of crop production. The general idea of the transition of households is modeled correctly since, during the first time step, we observe no competition of perennial crop rules and mostly only annual cash crop rules and pasture land-use rules. This suggests that a discrepancy exists in how net income from the sale of crops is calculated. Because of a lack of data describing the costs or expenses of a household family on an annual basis, it was assumed that from the total annual income generated from crop sales, 75% of the income was lost to expenses, such as medicine, transportation, and crop seed costs. The capital requirements for each land-use rule defined in the genetic algorithm strings only factor in building materials and chemical costs for maintenance. The ability of an agent to implement perennial crop rules as early as the second year is attributed to the above assumption of net income given the unavailability of data. In the LUCITA simulations presented in this chapter, capital was not a constraint for decision making as the conceptual model suggests since agents were acquiring capital at an exponential rate despite the deduction of 75% of income, but rather labor, specifically male labor, was the limiting factor in determining how many and what land uses could be implemented in a given time step. With more detailed information regarding colonist expenses, net income should be able to be better calculated, resulting in less-biased land-use trajectories.

6 CONCLUSIONS

These weaknesses of LUCITA have important effects on the land-use trends that were simulated for both scenarios. However, the replicative validity of the existing versions of LUCITA was not a priority at this stage. Instead, the utility of using agent-based modeling techniques integrated with geographic information systems (GIS) spatial data is explored in this chapter. In this regard, LUCITA has the potential for a high degree of structural validity. It approached the study of land-use change from the bottom up, addressing the behavior of the individual households that make land-use decisions and allowing overall landscape patterns to emerge as a result of the many actions of these individuals. The goal was never to test or explore the conceptual model of household transition, but rather to develop a model that considered both the ecological domain and the human domain and how they interact with each other. Further, the development of this preliminary simulation system highlighted the need for additional data collection efforts within the Altamira region, designed specifically to support simulation development. The modeling and simulation efforts described here relied on data that was originally collected for other purposes. Additional data collection efforts specific to simulation development would focus on collecting biophysical information such as soils, topography, and drainage patterns. Additional data on economic factors such as the history of credit policies, crop prices, and economic conditions, which have not yet been implemented, as well as frontier family demographic and behavioral characteristics, such as the cultural and behavioral factors that influence land-use decision making, would also be required.

Natural resource management modeling efforts in many instances consider the ecological domain and the human domain in isolation, or at best simulate the actions of one domain while holding the characteristics of the other constant. However, given the complex interactions that exist within each domain and between each domain, integrated modeling approaches are needed and must be developed. Such an approach will assist researchers in better understanding the complex nature of the interactions between human and natural systems. There are very few attempts similar to LUCITA, where two submodels interacting through a spatially explicit landscape with adaptive agents exist in the literature. It is hoped that this chapter presents a preliminary methodology that other researchers can use as a starting point and learn from some of the challenges that we have presented from our results and discussion of the simulations.

REFERENCES

[1] Axelrod, R. *The Complexity of Cooperation: Agent-Based Models of Competition and Collaboration.* Princeton, NJ: Princeton University Press, 1997.

[2] Booker, L. B., D. E. Goldberg, and J. H. Holland. "Classifier Systems and Genetic Algorithms." *Art. Intel.* **40** (1989): 235–282.
[3] Bousquet, F., C. Cambier, C. Mullon, P. Morand, and J. Quensiere. "Simulating Fishermen's Society." In *Simulating Societies: The Computer Simulation of Social Phenomena*, edited by N. Gilbert and J. Doran. London: UCL Press, 1994.
[4] Brondizio, E. S., S. McCracken, E. F. Moran, A. D. Siqueira, D. Nelson, and C. Rodriguez-Pedraza. "The Colonist Footprint: Towards a Conceptual Framework of Deforestation Trajectories among Small Farmers in Frontier Amazonia." In *Patterns and Processes of Land Use and Forest Change in the Amazon*, edited by C. Wood et al. Gainesville, FL: University of Florida Press, in press.
[5] Deadman, P., and H. R. Gimblett. "The Role of Goal-Oriented Autonomous Agents in Modeling People-Environment Interactions in Forest Recreation." *Math. & Comp. Model.* **20(8)** (1994): 121–133.
[6] Doran, J. and N. Gilbert, eds. "Simulating Societies: An Introduction." In *Simulating Societies: The Computer Simulation of Social Phenomena*, edited by N. Gilbert and J. Doran. London: UCL Press, 1994.
[7] Fearnside, P. M. *Human Carrying Capacity of the Brazilian Rainforest.* New York: Columbia University Press, 1986.
[8] Gilbert, N. and K. G. Troitzsch. *Simulation for the Social Scientist.* Buckingham: Open University Press, 1999.
[9] Goldberg, D. E. *Genetic Algorithms in Search, Optimization, and Machine Learning.* Reading, MA: Addison-Wesley, 1989.
[10] Hall, A., ed. *Amazonia: The Challenge of Sustainable Development.* London: ILAS/Macmillan, 2000.
[11] Holland, J. H. *Hidden Order: How Adaptation Builds Complexity.* Reading, MA: Addison-Wesley, 1995.
[12] INPE/IBAMA. "Amazonian Deforestation 1995–1997." São José dos Campos, SP, Brasil: INPE/IBAMA, 1998.
[13] Kohler, T. A. and E. Carr. "Swarm-Based Modeling of Prehistoric Settlement Systems in Southwestern North America." In *Proceedings of Colloquium II, UISPP, XIIIth Congress, number 5 in Sydney University Archaeological Methods*, edited by I. Johnson and M. North. Forli, Italy, 1996.
[14] Kohler, T. A., C. R. Van West, E. P. Carr, and C. G. Langton. "Agent-Based Modeling of Prehistoric Settlement Systems in the Northern American Southwest. In *Proceedings of the Third International Conference/Workshop on Integrating GIS and Environmental Modeling*, January 21–25, 1996, Santa Barbara, CA: National Center for Geographic Information and Analysis. ⟨http://www.ncgia.ucsb.edu/conf/SANTA_FE_CD-ROM/main.html⟩.
[15] Mausel, P., Y. Wu, Y. Li, E. F. Moran, and E. S. Brondizio. "Spectral Identification of Successional Stages Following Deforestation in the Amazon." *Geocarto Intl.* **4** (1993): 61–71.

[16] McCracken, S. D., E. S. Brondizio, D. Nelson, E. F. Moran, A. D. Siqueira, and C. Rodriguez-Pedraza. "Remote Sensing and GIS at Farm Property Level: Demography and Deforestation in the Brazilian Amazon." *Photogram. Eng. & Remote Sensing* **65(11)** (1999): 1311–1320.

[17] McCracken, S., A. D. Siqueira, E. F. Moran, and E. S. Brondizio. "Land-Use Patterns on an Agricultural Frontier in Brazil: Insights and Examples from a Demographic Perspective." In *Patterns and Processes of Land Use and Forest Change in the Amazon*, edited by C. Wood et al. Gainesville, FL: University of Florida Press, in press.

[18] Minar, N., R. Burkhard, C. Langton, and M. Askenazi. "The Swarm Simulation System: A Toolkit for Building Multi-Agent Simulations." Overview paper. Santa Fe Institute, Santa Fe, New Mexico, 1996.

[19] Moran, E. F., E. S. Brondizio, and S. McCracken. "Trajectories of Land Use: Soils, Succession, and Crop Choice." In *Patterns and Processes of Land Use and Forest Change in the Amazon*, edited by C. Wood et al. Gainesville, FL: University of Florida Press, in press.

[20] Moran, E. F. *Agricultural Development in the Transamazon Highway.* Bloomington: Indiana University Latin American Studies Center, 1976.

[21] Moran, E. F. *Human Adaptability: An Introduction to Ecological Anthropology.* North Scituate: Duxbury Press, 1979. Reissued in 1982 by Westview Press.

[22] Moran, E. F. *Developing the Amazon.* Bloomington: Indiana University Press, 1981.

[23] Moran, E. F., and E. S. Brondizio. "Land-Use Change after Deforestation in Amazonia." In *People and Pixels: Linking Remote Sensing and Social Science*, edited by D. Liverman, E. F. Moran, R. R. Rindfuss and P. C. Stern, 94–120. Washington, DC: National Academy Press, 1998.

[24] Moran, E. F., A. Packer, E. S. Brondizio, and J. Tucker. "Restoration of Vegetation Cover in the Eastern Amazon." *Ecol. Econ.* **18** (1996): 41–54.

[25] Moran, E. F., E. Brondizio, P. Mausel, and Y. Wu. "Integrating Amazonian Vegetation, Land-Use, and Satellite Data." *BioScience* **44(5)** (1994): 329–338.

[26] Smith, N. J. H. *Rainforest Corridors: The Transamazon Colonization Scheme.* Los Angeles: University of California Press, 1982.

[27] Tillman, D., T. A. Larsen, C. Pahl-Wostl, and W. Gujer. "Modeling the Actors in Water Supply Systems." *Water Sci. & Tech.* **39(4)** (1999): 203–211.

[28] Wood, C., and D. Skole. "Linking Satellite, Census, and Survey Data to Study Deforestation in the Brazilian Amazon." Volume edited by D. Liverman, E. Moran, R. Rindfuss, and P. Stern, 70–93. Washington DC: National Academy Press, 1998.

Index

Note: page numbers with the letter "t" refer to a table, the letter "f" to a figure. Names of authors and researchers appear in small caps.